D1330988

ENZYME INHIBITORS AS DRUGS

Previously published volumes
1970 Aldridge: *Mechanisms of Toxicity*
1971 Rabin and Freedman: *Effects of Drugs on Cellular Control Mechanisms*
1972 Rang: *Drug Receptors*
1973 Callingham: *Drugs and Transport Processes*
1974 Parsons: *Peptide Hormones*
1975 Grahame-Smith: *Drug Interactions*
1976 Roberts: *Drug Action at the Molecular Level*
1977 Hughes: *Centrally Acting Peptides*
1978 Turk and Parker: *Drugs and Immune Responsiveness*

BIOLOGICAL COUNCIL
The Co-ordinating Committee for Symposia on Drug Action

ENZYME INHIBITORS AS DRUGS

Edited by

MERTON SANDLER, M.D., F.R.C.P., F.R.C.Path.

Institute of Obstetrics and Gynaecology,
Queen Charlotte's Maternity Hospital,
University of London

First published 1980 by
THE MACMILLAN PRESS LTD
London and Basingstoke
Associated companies in Delhi Dublin
Hong Kong Johannesburg Lagos Melbourne
New York Singapore and Tokyo

Typeset by Reproduction Drawings Ltd., Sutton, Surrey

Printed in Great Britain
by Unwin Brothers Limited
The Gresham Press Old Woking Surrey

ISBN 0 333 28984 6

Biological Council Co-ordinating Committee for Symposia on Drug Action

Report of a symposium held on 9 and 10 April 1979 in London at The Middlesex Hospital Medical School

Sponsored by:

Biochemical Society
British Pharmacological Society
British Society for Antimicrobial Chemotherapy
British Society for Immunology
Chemical Society
Nutrition Society
Pharmaceutical Society of Great Britain
Physiological Society
Royal Society of Medicine
Society for Applied Bacteriology
Society for Drug Research
Society for Endocrinology
Society of Chemical Industry, Fine Chemicals Group

The organisers are grateful to the following for the generous financial support which made the meeting possible:

Bayer AG
Bayer UK
Berk Pharmaceuticals Limited
Boehringer Ingelheim Limited
The Boots Company Limited
Fisons Limited
Glaxo Holdings Limited
Imperial Chemical Industries Limited
Lilly Research Centre Limited
May and Baker Limited
Merck Sharp and Dohme Limited
Miles Laboratories Limited
Ortho Pharmaceuticals Limited
Organon Laboratories Limited
Pfizer Limited
Philips-Duphar BV
Roche Products Limited
Sandoz Products Limited
Schering Chemicals Limited
G. D. Searle and Company Limited
E. R. Squibb and Sons Limited
The Wellcome Trust

Organised by a symposium committee consisting of:

M. Sandler (Chairman and Hon. Secretary)
T. J. Franklin
H. J. Smith
D. W. Straughan
K. F. Tipton

Symposium Contributors

Dr W. Aldridge, Toxicology Unit, Medical Research Council, Woodmansterne Road, Carshalton, Surrey SM5 4EF, UK

Dr M. Antonaccio, The Squibb Institute for Medical Research, Princeton, New Jersey 08540, USA

Dr M. Asaad, Department of Pharmacology, The Squibb Institute for Medical Research, Princeton, New Jersey 08540, USA

Dr A. Barrett, Molecular Pathology Department, Strangeways Laboratory, Cambridge CB1 4RN, UK

Dr H. Blaschko, Department of Pharmacology, University of Oxford, South Parks Road, Oxford, OX1, UK

Dr H. Cheung, The Squibb Institute for Medical Research, Princeton, New Jersey 08540, USA

Dr J. Coyette, Service de Microbiologie, Faculté de Medécine, Institut de Botanique, Université de Liège, Sart Tilman, B-4000 Liège, Belgium

Dr D. Cushman, The Squibb Institute for Medical Research, Princeton, New Jersey 08540, USA

Dr O. Dideberg, Laboratoire de Cristallographie approfondie, Institut de Physique, Université de Liège, Sart Tilman, B-4000 Liège, Belgium

Dr C. Duex, Service de Microbiologie, Faculté de Medécine, Institut de Botanique, Université de Liège, Sart Tilman, B-4000 Liège, Belgium

Dr J. Dusart, Service de Microbiologie, Faculté de Medécine, Institut de Botanique, University de Liège, Sart Tilman, B-4000 Liège, Belgium

Dr J. Elsworth, Bernhard Baron Memorial Research Laboratories and Institute of Obstetrics and Gynaecology, Queen Charlotte's Maternity Hospital, Goldhawk Road, London W6 0XG, UK

Dr J. Fisher, Department of Chemistry Harvard University, Cambridge, Massachusetts 02138, USA

Dr J. Frère, Service de Microbiologie, Faculté de Medécine, Institut de Botanique, Université de Liège, Sart Tilman, B-4000 Liège, Belgium

Dr J. Ghuysen, Service de Microbiologie, Faculté de Medécine, Institut de Botanique, Université de Liège, Sart Tilman, B-4000 Liège, Belgium

Dr V. Glover, Bernhard Baron Memorial Research Laboratories and Institute of Obstetrics and Gynaecology, Queen Charlotte's Maternity Hospital, Goldhawk Road, London W6 0XG, UK

Dr H. Goldenberg, Department of Pharmacology, The Squibb Institute for Medical Research, Princeton, New Jersey 08540, USA

Dr D. Harris, Department of Pharmacology, The Squibb Institute for Medical Research, Princeton, New Jersey 08540, USA

Dr G. Hitchings, The Wellcome Research Laboratories, Research Triangle Park, North Carolina 27709, USA

Dr R. John, Department of Biochemistry, University College, P.O. Box 78, Cardiff CF1 1XL, Wales

Dr M. Jung, Centre de Recherche Merrell International, 16 rue d'Ankara, 67084-Strasbourg Cedex, France

Dr J. Knoll, Semmelweis University of Medicine, Department of Pharmacology, 1445 Budapest, P.O.B. 370, Hungary

Dr J. Koch-Weser, Centre de Recherche Merrell International, 16 rue d'Ankara, 67084-Strasbourg Cedex, France

Dr J. Knowles, Department of Chemistry, Harvard University, Cambridge, Massachusetts 02138, USA

Dr J. Knox, Institute of Materials Science, University of Connecticut, Storrs, Connecticut 06268, USA

Dr R. Lewinsohn, Bernhard Baron Memorial Research Laboratories and Institute of Obstetrics and Gynaecology, Queen Charlotte's Maternity Hospital, Goldhawk Road, London W6 0XG, UK

Dr M. Leyh-Bouille, Service de Microbiologie, Faculté de Medécine, Institut de Botanique, Université de Liège, Sart Tilman, B-4000 Liège, Belgium

Dr G. Lienhard, Department of Biochemistry, Dartmouth Medical School, Hanover, New Hampshire 03755, USA

Dr S. Moncada, Wellcome Research Laboratories, Langley Court, Beckenham, Kent, BR3 4BS, UK

Dr M. Ondetti, The Squibb Institute for Medical Research, Princeton, New Jersey 08540, USA

Dr M. Phillips, Department of Pharmacology, The Squibb Institute for Medical Research, Princeton, New Jersey 08540, USA

Dr M. Reveley (Bernhard Baron Memorial Research Laboratories and Institute of Obstetrics and Gynaecology, Queen Charlotte's Maternity Hospital, Goldhawk Road, London W6 0XG, UK

Dr B. Roth, The Wellcome Research Laboratories, Research Triangle Park, North Carolina 27709, USA

Dr B. Rubin, The Squibb Institute for Medical Research, Princeton, New Jersey 08540, USA

Dr E. Sabo, The Squibb Institute for Medical Research, Princeton, New Jersey 08540, USA

Professor M. Sandler, Bernhard Baron Memorial Research Laboratories and Institute of Obstetrics and Gynaecology, Queen Charlotte's Maternity Hospital, Goldhawk Road, London W6 0XG, UK

Dr K. Schaper, Borstel Research Institute, 2061 Borstel, FDR

Dr J. Seydel, Borstel Research Institute, 2061 Borstel, FDR

Dr E. Shaw, Biology Department, Brookhaven National Laboratory, Upton, New York 11973, USA

Dr A. Sjoerdsma, Centre de Recherche Merrell International, 16 rue d'Ankara, 67084-Strasbourg Cedex, France

Dr K. Tipton, Department of Biochemistry, Trinity College, Dublin 2, Ireland

Dr R. Vane, Wellcome Research Laboratories, Langley Court, Beckenham, Kent, BR3 4BS, UK

Contents

1
Kinetics and enzyme inhibition studies

Keith F. Tipton (Department of Biochemistry,
Trinity College, Dublin 2, Ireland)

INTRODUCTION

Many pharmacologically important drugs act by inhibiting enzymes, and kinetic studies of their inhibitory effects can provide valuable information on their potency and mechanism of action which is, in many ways, complementary to that obtainable from direct chemical studies. The incorrect application of kinetic methods will, however, result, at best, in the loss of potentially important information and, at worst, in misleading data. This general survey will concentrate on the uses and limitations of kinetic approaches to analysing the effects of various types of inhibitors having pharmacological uses or potential, and the pitfalls which may result from ill-planned kinetic studies.

ENZYME ASSAYS

It is obviously essential to have a reliable method for measuring enzyme activity for the study of the effects of inhibitors; many potentially important studies have been rendered uninterpretable by failure to ensure the validity of the assay method used. The rate of an enzyme-catalysed reaction can be followed by determining the disappearance of substrate or the appearance of product with time (see Tipton, 1978, for a review of the different types of assay methods available). A time-course of the reaction determined in this way usually has an initial linear phase but then tends to curve off at longer times, as shown in figure 1.1. There are several possible causes for this departure from linearity, including a significant fall in substrate concentration as the reaction progresses, the reaction approaching equilibrium, inhibition of enzyme by the products formed, instability of the enzyme, a limitation of the assay method or changes in conditions as the reaction progresses. These effects have been discussed in greater detail by Tipton (1973, 1978) but it is clear from the discussion above that the causes of non-linearity may be complex, involving more than one factor. It is thus essential to measure the rate of the reaction during the initial linear phase (the initial rate—see figure

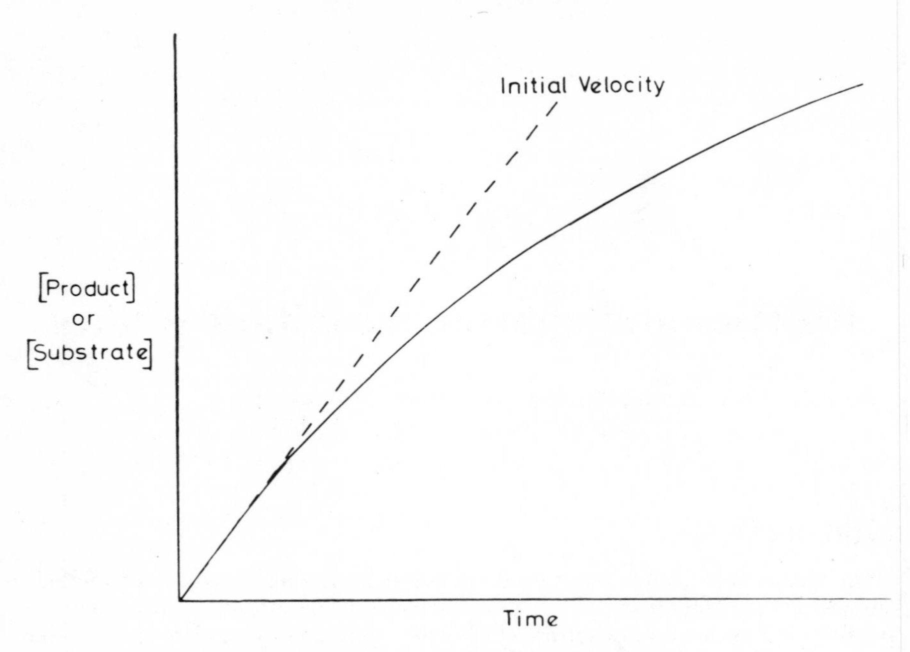

Figure 1.1 The progress curve of a typical enzyme-catalysed reaction.

1.1), when none of these factors has had any significant effect, if the results are to have any validity.

If it is possible to follow the time-course of the reaction continuously, for example by a direct assay in a recording spectrophotometer, it should be easy to identify the initial rate phase. However, in many cases it is either necessary or more convenient to use a procedure which involves stopping the reaction after fixed times before determining the change in substrate or product concentration, as in a radiochemical assay. In such a case, it is often tempting to determine the amount of product formed after a single convenient fixed time interval and to use this as a measure of the activity of the enzyme. This measurement will, however, only yield meaningful results if the time-course of the reaction is linear for the chosen time period and thus it is essential to carry out measurements after a series of increasing incubation times to ensure that the time interval chosen is within the linear initial rate period. It is also important to recognise that the period of linearity is likely to be affected by the assay conditions, necessitating repeated checks for the period of linearity each time the conditions, for example enzyme, substrate or inhibitor concentration, pH or temperature are changed. In experiments where enzyme or substrate concentrations are varied, it is frequently suffient to ensure that the reaction proceeds linearly for the time chosen at the lowest and highest concentrations to be used.

A frequently quoted procedure to ensure linearity is to keep the determinations down to a period in which less than 10 per cent of the total substrate is used. However, because of the multiple possible causes of departure from linearity, this approach may not be adequate and in cases where the initial substrate concentration is less than the K_m value of the enzyme for it, such a depletion could itself make an appreciable contribution to the curvature. In a few cases, enzyme-catalysed reactions may show initial bursts or lags in product formation before the linear steady-state rate is established. This may be an artefact of the assay method used (see Tipton, 1978) or, in the case of complex systems, it may be a genuine property of the reaction catalysed (see, for example, Frieden, 1970; Kurganov *et al.*, 1976). Such behaviour should be readily detectable if a continuous assay is used but can lead to serious errors with a stopped assay unless a time-course is used to ascertain the period of linearity.

Because enzymes act as catalysts, the initial velocity of the reaction can be expected to be proportional to enzyme concentration, provided it is always much less than concentration of the substrate, as is almost invariably the case in kinetic studies. Such behaviour is generally seen with a graph of initial velocity against enzyme concentration giving a straight line passing through the origin. Cases in which the dependence is not linear can arise through artefacts of the assay method used, the presence of a reversible inhibitor or activator in the enzyme solution, the existence of different polymeric states of the enzyme with different activities, or the presence of an irreversible inhibitor in the assay mixture (see Dixon and Webb, 1964; Tipton, 1978). Although such behaviour is relatively uncommon, it is important to check for departures from linearity and, if possible, to arrange the assay conditions so that such complications are eliminated, since the enzyme concentration is often allowed to vary during kinetic studies.

TYPES OF INHIBITION

A number of different types of inhibitor have found pharmacological and medical uses and the kinetic behaviour of some of them is summarised in table 1.1. Various kinetic approaches are required to obtain useful data and descriptions of the behaviour of these different systems; these will now be discussed in detail.

Table 1.1 Types of inhibitors

Type	Mechanism	Examples
Reversible	$E + I \rightleftharpoons EI$	Substrate analogues: competitive; tight binding; transition-state analogues; bisubstrate analogues
Non-specific irreversible	$E + I \rightarrow EI$	Group-specific reagents
Specific irreversible	$E + I \rightleftharpoons EI \rightarrow E\text{–}I$	Active-site directed; k_{cat} (suicide)

Reversible inhibitors

In the simplest case, combination of an inhibitor (I) with an enzyme (E) can be represented by the scheme

$$\mathrm{E} + \mathrm{I} \underset{k_{-1}}{\overset{k_{+1}}{\rightleftharpoons}} \mathrm{EI} \tag{1.1}$$

and the inhibitor has a relatively low affinity for the enzyme so that its concentration (i) is very much greater than that of the enzyme (e). Under these conditions the amount of inhibitor bound to the enzyme will be trivial compared with the total concentration of the inhibitor, so that this can be regarded as being unchanged by the association. Here the dissociation constant for the reaction, which is termed the inhibitor constant (K_i), is given by the relationships

$$K_i = \frac{k_{-1}}{k_{+1}} = \frac{(e - ei)\,i}{ei} \tag{1.2}$$

where ei represents the concentration of the enzyme–inhibitor complex.

It is usually characteristic of such a combination that the inhibition shows no time dependence as the reaction involves the formation of no covalent bonds between enzyme and inhibitor. Since the interaction is reversible, inhibition can be reversed by dialysis, gel-filtration or dilution, to reduce the inhibitor concentration. A simple test for reversibility is to compare the activity of the enzyme assayed in the presence of a fixed concentration of that inhibitor with that obtained when the enzyme is preincubated in a small volume with the same concentration of inhibitor before being diluted by addition to the remainder of the assay mixture. If inhibition is freely reversible, activity in the latter case should be greater than that in the former since the inhibitor will dissociate from the enzyme on dilution. If, for example, the initial inhibitor concentration were equal to the K_i value, equation (1.2) shows that one-half of the enzyme would be complexed with the inhibitor, ninefold dilution of the mixture would result in any one-tenth of the enzyme being complexed and only one-hundredth of the enzyme would be complexed if the dilution were 99-fold.

Reversible inhibitors are classified into a number of types, according to the effects they have on the constants in the simple Michaelis–Menten equation

$$v = \frac{Vs}{K_m + s} \tag{1.3}$$

in which v and V are the initial and maximum velocities, K_m is the Michaelis constant and s the substrate concentration. An inhibitor which increases the K_m value without affecting V is termed a competitive inhibitor, whereas a non-competitive inhibitor decreases V without affecting K_m. Inhibitors that cause an equal decrease in K_m and V are termed uncompetitive (or anti-competitive) and those that cause unequal changes in these two parameters are called mixed inhibitors. Some workers however, make no distinction between mixed and non-competitive inhibition and use the latter term to describe both types. The kinetic equations that describe these types are shown in table 1.2. The different types of inhibition can be distinguished by their effects on graphs of $1/v$ against $1/s$ according to the reciprocal form of equation (1.3):

$$\frac{1}{v} = \frac{K_m}{V} \cdot \frac{1}{s} + \frac{1}{V} \tag{1.4}$$

as shown in figure 1.2, and the K_i values can be determined from plots of the slopes and/or intercepts of these lines against inhibitor concentration, or from graphs of $1/v$ and s/v against inhibitor concentration (Segel, 1975; Cornish-Bowden, 1974; Dixon and Webb, 1979). The distinction between different types of inhibi-

Table 1.2 Equations describing simple reversible inhibition

Inhibition type	Kinetic equation	Effects
None	$v = \dfrac{V}{1 + K_m/s}$	
Competitive	$v = \dfrac{V}{1 + (K_m/s)(1 + i/K_i)}$	K_m increased V unchanged
Non-competitive	$v = \dfrac{V/(1 + i/K_i)}{1 + K_m/s}$	K_m unchanged V decreased
Uncompetitive	$v = \dfrac{V}{(1 + i/K_i) + K_m/s}$	K_m and V equally decreased
Mixed	$v = \dfrac{V}{(1 + i/K_i') + (K_m/s)(1 + i/K_i)}$	Unequal changes in K_m and V

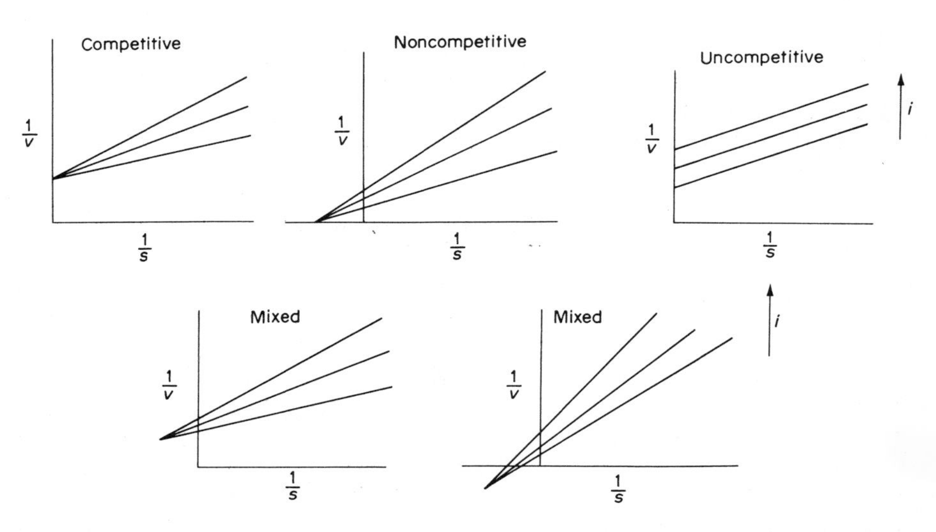

Figure 1.2 The effects of different types of reversible inhibitors on the double-reciprocal plots of simple enzyme-catalysed reactions.

tor is made solely in terms of the effects that they have on the Michaelis parameters without any reference to the mechanism of interaction between the inhibitor and the enzyme. The different mechanisms which can result in a given type of inhibition have been considered in detail by Frieden (1964) (see also Segel, 1975) who has shown that, even in the simplest case of an enzyme reaction involving only a single substrate, more than one mechanism can give the same effects. In addition, when an enzyme catalyses a reaction between two different substrates, the type of inhibition shown towards each substrate is generally different (see for example Cleland, 1971; Segel, 1975; Tipton, 1974).

The most common types of simple reversible inhibitors to have been used medically or pharmacologically are competitive inhibitors based on analogues of the true substrate for the enzyme. Some examples are shown in table 1.3. An

Table 1.3 Some 'useful' competitive inhibitors

Enzyme	Inhibitor	K_i (μM)	Reference
Acetylcholinesterase	Decamethonium	25	Ellman *et al.* (1961)
Alcohol dehydrogenase	Pyrazole	0.2	Theorell and Yonetani (1969)
Folate synthetase	Sulphanilamide	~ 0.1†	Blakley (1969)
Folate reductase	Pyrimethamine	~ 0.01	Blakley (1969)
Aconitase	Fluorocitrate, *Trans*-aconitate	1.0, 290‡ 140	Glusker (1971)

†May also act as a competitive substrate.
‡Value depends on the source of the enzyme; inhibition becomes irreversible on prolonged incubation.

advantage of simple reversible inhibitors of this type is that their effects should be very rapid when they are in contact with the enzyme. The effects of a competitive inhibitor will, however, be counteracted by any increase in substrate concentration. This can be seen from the equation for this case shown in table 1.2 and also from the simple formation of the competitive case shown below:

$$\begin{array}{ccccc} E & \underset{}{\overset{S}{\rightleftharpoons}} & ES & \longrightarrow & E + \text{Products} \\ {\scriptstyle I}\Big\downarrow\!\Big\uparrow & & & & \\ EI & & & & \end{array} \qquad (1.5)$$

which indicates that at a fixed concentration of inhibitor, it would be possible to displace all the inhibitor from the enzyme by increasing substrate concentration. This effect could result in the *in vivo* potency of a competitive inhibitor being less than that observed in experiments with purified enzymes since, in the former system, inhibition of an enzyme would be expected to result in an increase in the steady-state concentration of substrate.

The combination of an inhibitor with an enzyme can be more complicated, occurring in several steps, for example

$$\mathrm{E} + \mathrm{I} \underset{k_{-1}}{\overset{k_{+1}}{\rightleftharpoons}} \mathrm{EI} \underset{k_{-2}}{\overset{k_{+2}}{\rightleftharpoons}} (\mathrm{EI})^* \qquad (1.6)$$

where the second step may represent a conformational change in enzyme–inhibitor complex, perhaps allowing the inhibitor to resemble more closely the substrate transition state. The total concentration of enzyme–inhibitor complexes will be given by

$$ei + (ei)^* = \frac{(e - ei - (ei)^*)i}{K_\mathrm{i}} \left(1 + \frac{1}{K_\mathrm{i}'}\right) \qquad (1.7)$$

where $K_\mathrm{i}' = k_{-2}/k_{+2}$, and the proportions of the two forms will depend upon the values of two inhibitor constants. Such inhibitors will behave in the same way as the simpler cases. For example, if the inhibitor is competitive, the kinetic equation will become

$$v = \frac{V}{1 + [1 + (1 + 1/K_\mathrm{i}')\,(i/K_\mathrm{i})]\ K_\mathrm{m}/s} \qquad (1.8)$$

and the K_i value calculated in the normal way would be equal to $K_\mathrm{i}(1 + K_\mathrm{i}')$. If the rate constants describing the two steps were significantly different, it might be possible to detect them by rapid mixing or perturbation kinetics (see, for example, Gutfreund, 1972; Halford, 1974) and if the second step were very slow there might be a detectable time-dependence for the achievement of full inhibition.

Some compounds which have been used as competitors are alternative substrates for the enzyme in question rather than simple inhibitors. Some examples are given in table 1.4. Ethanol administration may be used to counter the effects of allyl alcohol which may be converted to a toxic aldehyde by alcohol dehydrogenase (see for example McLean, 1971). Compounds such as α-methyldopa were

Table 1.4 Some competitive substrates

Enzyme	Substrates
Alcohol dehydrogenase	Allyl alcohol: ethanol
Dopa decarboxylase	Dopa: α-methyldopa
Aldehyde dehydrogenase	Aryl aldehydes: acetaldehyde

initially used as antihypertensive drugs because of their known ability to inhibit dopa decarboxylase, but subsequent work has shown that their effects are largely due to further metabolism to certain 'false transmitters', α-methyldopamine and α-methylnoradrenaline in the case of α-methyldopa (see for example Kopin, 1971). The effects of ethanol ingestion on the peripheral metabolism of biogenic amines may be, at least in part, due to its metabolite, acetaldehyde, acting as a competing substrate for aldehyde dehydrogenase (see for example Tipton *et al.*, 1977). In such cases, the action of the enzyme itself, rather than other metabolism and elimination systems, will result in depletion of inhibitor concentration with time. Kinetic treatment of the simplest case in which the enzyme is presented with its normal substrate (S) and a competing substrate (X) which give rise to the products P and Z respectively:

$$\begin{array}{ccccc} & & ES & \xrightarrow{k_{+2}} & E + P \\ & k_{+1} \nearrow\!\!\swarrow k_{-1} & & & \\ E & & & & \\ & k_{+3} \searrow\!\!\nwarrow k_{-3} & & & \\ & & EX & \xrightarrow[k_{+4}]{} & E + Z \end{array} \tag{1.9}$$

results in a steady-state equation identical to that given by the simple competitive inhibitor case (table 1.2) except that K_i is replaced by the Michaelis constant for X ($K_m^X = (k_{-3} + k_{+4})/k_{+3}$).

Reversible inhibitors with high affinity for enzyme

Some inhibitors, including the transition-state analogues (see for example Wolfenden, 1976), bind to enzymes very tightly indeed. The high affinity necessitates the use of very low concentrations of the inhibitor in kinetic studies and, because of this, the assumption that the free inhibitor concentration is essentially unaffected by that removed in complexing with the enzyme will not be valid. Thus, equation (1.2) must be modified to take the depletion of the inhibitor concentration into account:

$$K_i = \frac{(e - ei)(i - ei)}{ei} \tag{1.10}$$

If it were possible to determine the free inhibitor concentration ($i_f = i - ei$) this equation could be expressed in the same form of equation (1.2) if the free concentration of inhibitor were substituted for i. It is, however, not possible to determine free inhibitor concentration for use in this equation since only total enzyme and inhibitor concentrations are known. Two graphical methods for analysing such cases have been found to be of particular value. That of Henderson (1972) involves the derivation of equations for linear graphs. A general equation relating the initial velocity of the reaction in the presence of the inhibitor (v_i) to that in its absence (v) can be written as

$$v_i = \frac{v}{(1 + K_m/s) + (i/s)(N/K_i)} \tag{1.11}$$

where N is a complex value able to contain substrate concentration and inhibitor and Michaelis constants. Using the relationships between free and total enzyme and inhibitor concentrations

$i = i_f + ei$ and $e = e_f + ei$

this equation can be rearranged to give

$$\frac{i}{(1 - v_i/v)} = e + \left(\frac{K_m + s}{N/K_i} \frac{v}{v_i}\right) \tag{1.12}$$

Thus a graph of $i/(1 - v_i/v)$ against v/v_i will be linear and will intersect the vertical axis at a value of e which represents total enzyme concentration in terms of inhibitor binding sites. The expressions representing the type of inhibition, which are identical to those shown in table 1.2, can be substituted for N/K_i in equation (1.12). Table 1.5 shows the expression for N for these simple cases and the full kinetic equations that will result. If a series of graphs are plotted each at a different fixed substrate concentration, the effects seen will depend on the type of inhibition. In the non-competitive case, table 1.5 shows that the line obtained will be

Table 1.5 Analysis of high-affinity reversible inhibitors†

Type of inhibition	N	Kinetic equation ($i/(1 - v_i/v) =$)
Competitive	K_m	$e + K_i \left(\frac{s + K_m}{K_m}\right) \frac{v}{v_i}$
Uncompetitive	s	$e + K_i \left(\frac{s + K_m}{s}\right) \frac{v}{v_i}$
Non-competitive	$s + K_m$	$e + K_i \left(\frac{v}{v_i}\right)$
Mixed	$\frac{s K_i}{K_i'} + K_m$	$e + \left(\frac{s + K_m}{K_m/K_i + s/K_i'}\right) \frac{v}{v_i}$

†The values of N in equation (1.12) and the full equations are shown for each type of inhibition.

independent of substrate concentration and will have a slope of K_i. In the competitive case, the slopes of the lines obtained will increase with increasing substrate concentration, and a graph of the slopes of these lines against substrate concentration will have a slope of K_i/K_m and will intersect the vertical axis at a value corresponding to K_i. Table 1.5 shows that, with uncompetitive inhibitors, the slope of the lines will decrease with increasing substrate concentration obeying the equation

$$\text{Slope} = K_i + \frac{K_i K_m}{s} \tag{1.13}$$

Thus, a replot of these slopes against substrate concentration will decrease hyberbolically towards a value of K_i as s tends to infinity. A graph of the slope against $1/s$ will, however, be linear, with a slope of K_iK_m and a vertical axis intercept of K_i. The mixed case, which behaves as a mixture of the uncompetitive and competitive cases, will show a more complicated picture. The method of Henderson (1972) may be extended to cases of inhibition that are more complicated than those discussed here and it has been adapted for use in computer-fitting procedures which allow standard errors of the kinetic parameters to be determined (Henderson, 1973).

Dixon (1972) has derived a geometrical method which, applied to curves of velocity against total inhibitor concentration, allows K_i value and total enzyme concentration to be determined in the cases of competitive and non-competitive inhibition. In the absence of inhibitor, the initial velocity (v_0) will be given by

$$v_0 = \frac{V}{1 + K_m/s} \tag{1.14}$$

whereas in the presence of a competitive inhibitor an equation of the form

$$\frac{i}{V - v(1 + K_m/s)} = \frac{K_i}{v} \frac{s}{K_m} + \frac{e}{V} \tag{1.15}$$

may be derived, where v represents velocity in the presence of inhibitor. A graph of v against i will give a curve such as that shown in figure 1.3. If a point n is taken anywhere on the curve, the velocity at this point (v') will be given by

$$v' = \frac{v_0}{n} = \frac{V}{n(1 + K_m/s)} \tag{1.16}$$

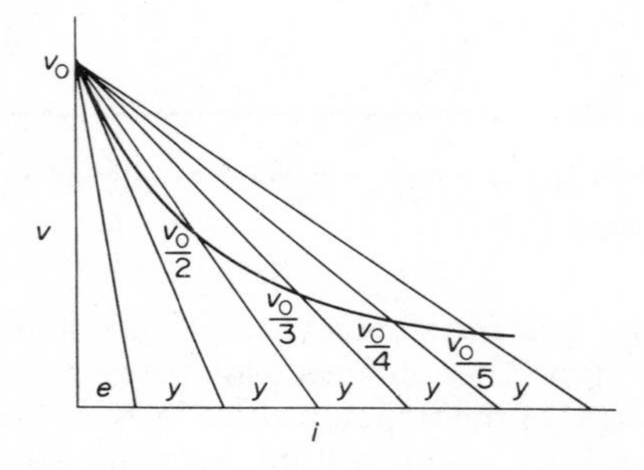

Figure 1.3 Kinetic analysis of the effects of high-affinity inhibitors by the method of Dixon (1972). Details of the constructions used are given in the text. The distances, y, between intersection points are defined in equation (1.19) for competitive inhibition.

In terms of equation (1.15) this point will represent

$$\frac{i}{V - v'(1 + K_m/s)} = \frac{i'}{v(1 - 1/n)} = \frac{i'}{V} \frac{n}{n - 1} = K_i \frac{s}{K_m} \frac{n}{v_0} + \frac{e}{V} \tag{1.17}$$

where i' represents the value of i_t at the point n.

A straight line drawn from the point v_0 through the point n on the curve will, by geometry, cut the base-line at a point x given by

$$x = \left(\frac{n}{n-1}\right) i' \tag{1.18}$$

and thus

$$x = nK_i \frac{s}{K_m} \frac{V}{v_0} + e = nK_i \left(1 + \frac{s}{K_m}\right) + e$$

Thus if a series of points (n) are chosen at values of $v_0/2, v_0/3, v_0/4$, etc., lines from the point v_0 drawn through them will intersect the base-line at equally spaced points, where the distance between them (y) will be given by

$$y = K_i \left(1 + \frac{s}{K_m}\right) \tag{1.19}$$

Thus K_i can be calculated if K_m is known or alternatively a graph of y against s will have a slope of K_i/K_m and a vertical axis intercept of K_i. A tangent to the initial part of the curve (corresponding to $n = v_0/1$) will cut the base line at a distance of y from the intersection point of the $n = v_0/2$ line and of $y + e$ from the vertical axis, allowing e to be calculated. In the case of competitive inhibitors which do not bind tightly to the enzyme, where K_i value is very much greater than enzyme concentration, the method will give similar results except that e will be negligible.

Details of the application of this method to the behaviour of non-competitive inhibitors and its modification to allow the concentrations of enzyme-substrate and enzyme-inhibitor complexes and free inhibitor and enzyme to be determined are given by Dixon (1972).

Binding studies

In some cases inhibitor binding is studied directly rather than through its effects on the velocity of the enzyme-catalysed reaction. Such binding studies can be carried out directly by determining the amount of complex formed when the enzyme has been equilibrated with a fixed concentration of inhibitor by using techniques such as equilibrium dialysis (Klotz *et al.*, 1946; Moe and Hammes, 1974), flow dialysis (Colowick and Womack, 1969), gel-filtration (Hummel and Dryer, 1962) or ultrafiltration (Sophianopoulos *et al.*, 1978). Alternatively they may be carried out indirectly by observing changes in a property of the enzyme or inhibitor, such as fluorescence, absorbance, optical rotatory dispersion or electron spin resonance, when binding occurs, although in these cases it is necessary to assume that the change observed is proportional to the degree of saturation of the enzyme.

When the inhibitor does not bind very tightly to the enzyme, binding data can be analysed by the method of Scatchard (1949). Equation (1.2) may be rearranged into the form

$$ei = e - \frac{K_i}{i}\ ei \tag{1.20}$$

In this equation e represents the enzyme concentration expressed in terms of its binding sites; that is, if the enzyme has n binding sites per molecule, the value of e in equation (1.20) will be n times its molar concentration. If the molar concentration of the enzyme is designated by E, dividing equation (1.20) by this gives

$$\frac{ei}{E} = \frac{e}{E} - \frac{K_i}{i}\frac{ei}{E} \tag{1.21}$$

Thus a graph of ei/E against $ei/E.i$ will give a straight line with a slope of $-K_i$ and an intercept on the vertical axis of e/E, which represents the number of inhibitor binding sites per mole of enzyme (figure 1.4). If the enzyme contains more than one type of binding site with different affinities for inhibitor, the Scatchard plot will give a curve, as shown in figure 1.4, which will tend to the slope representing the binding to the lower affinity sites at very high values of $ei/E.i$.

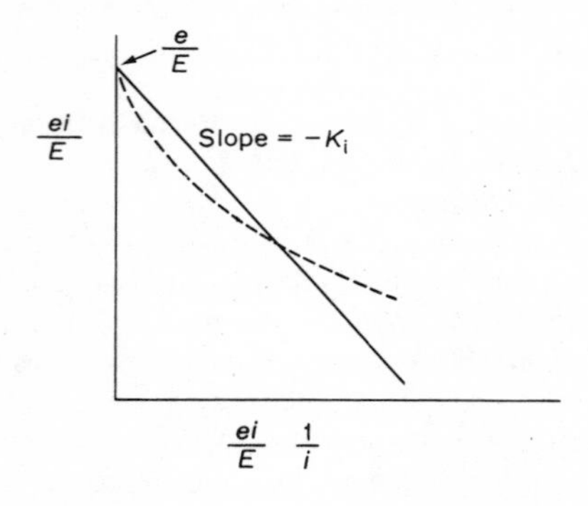

Figure 1.4 Determination of the dissociation constant and number of binding sites from inhibitor binding data by means of a Scatchard plot. The broken line shows the type of curve that will be obtained if the enzyme contains sites with different affinities for the inhibitor. The symbols are defined in the text.

In the case where i is not much greater than e, equation (1.10) must be used. Dixon (1972) has devised a graphical method for treating such cases which is similar to that discussed in the preceding section. In this case a curve of ei versus i is plotted and the value of ei obtained at saturating concentrations of i (ei_{max}) is marked as a horizontal line as shown in figure 1.5. By reasoning similar to that used in the previous case, Dixon showed that a tangent to the curve will cut the ei_{max} line at a distance of $e + K_i$ from the vertical axis, and that a line from the origin drawn through a point on the curve at $ei_{max}/2$ will intersect the ei_{max} line at a point at a distance of K_i from the intersection point of the tangent. Similarly, lines drawn through $ei_{max}/2$, $3ei_{max}/4$, $4ei_{max}/5$, etc., will each be separated from the previous one by a distance of K_i where it intersects the ei_{max} line.

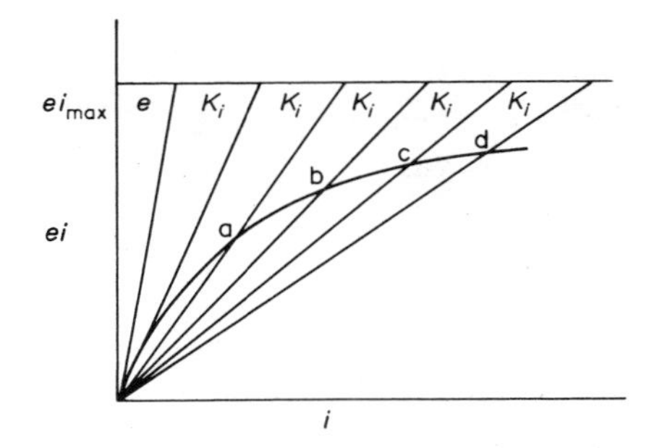

Figure 1.5 Analysis of the binding of high-affinity inhibitors by the method of Dixon (1972). Details of the constructions used are given in the text. The intersection points marked a–d represent points on the curve corresponding to $ei_{max}/2$, $2ei_{max}/3$, $3ei_{max}/4$ and $4ei_{max}/5$, respectively.

Bisubstrate analogues

In enzymes that catalyse reactions between two substrates that proceed by way of a ternary complex in which both substrates are bound to the enzyme, compounds which are analogues of both substrates can prove to be effective inhibitors with K_i values much lower than the dissociation constants of either substrate from the enzyme and also lower than the product of the dissociation constants of the two substrates (Byers, 1978). The affinity is dependent upon the kinetic mechanism obeyed by the enzyme and the orientations of the two substrate portions in the analogue, being greatest when this resembles that in the ternary complex most closely. The structures of some of these analogues are shown in table 1.6. Such compounds may, of course, be so designed that they are analogues of the transition state of the enzyme-catalysed reaction, in which case they would be expected to bind very tightly indeed (Wolfenden, 1976, 1977).

Systems may be designed in which the analogue is formed due to reaction between a substrate and the inhibitor on the enzyme surface; for example, Chase and Tubbs (1969) have shown that bromoacetylcarnitine (I) and coenzyme A (CoA) react on the surface of the enzyme, carnitine acetyltransferase, to produce the analogue shown in table 1.6 (E–I.CoA) in a process which can probably be represented by the kinetic mechanism

$$\begin{array}{ccccc} & \overset{K_a}{\nearrow\!\!\swarrow} & \text{E.CoA} & \overset{K_i'}{\searrow\!\!\nwarrow} & \\ \text{E} & & & & \text{E–I.CoA} \xrightarrow{k} \text{E.I–CoA} \\ & \underset{K_i}{\nwarrow\!\!\searrow} & \text{E.I} & \underset{K_a}{\swarrow\!\!\nearrow} & \end{array} \qquad (1.22)$$

In such systems, the inhibition will be time-dependent, and the mechanism shown in equation (1.22) will give rise to a kinetic equation describing the initial rate formation of the inhibitor (v_i) which takes the form

Table 1.6 Some bisubstrate analogues

Enzyme	Substrate	K_m (μM)	Analogue	K_i (nm)	Reference
Carnitine acetyl-transferase	Acetyl CoA $(CH_3)_3N^+CH_2CHOHCH_2COO^-$ (carnitine)	34 120	O (=) above C $(CH_3)_3\overset{+}{N}CH_2CHOCCH_2S{-}CoA$ ‖ CH_2COO^-	<12	Chase and Tubbs (1969)
Aspartate transcarbamylase	Carbamoyl phosphate Aspartate	27 11 000	$NHCOCH_2PO_3^{2-}$ | $CHCOO^-$ | CH_2COO^-	27	Collins and Stark (1971)
Adenylate kinase	MgATP AMP	100 625	O O O O O (= above each P) Ado-OPOPOPOPOPO Ado.Mg^{2+} $O^-O^-O^-O^-O^-$ (below each P)	2.5	Leinhard and Secemski (1973)
Lactate dehydrogenase	NADH Pyruvate	24 140	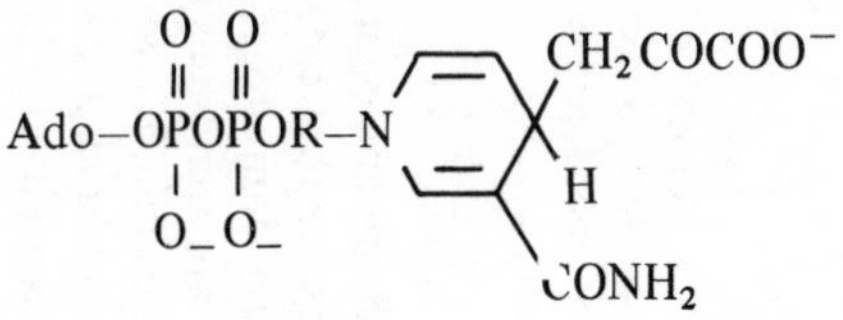 O O (= above each P) Ado–OPOPOR–N (dihydropyridine ring bearing CH_2COCOO^-, H and $CONH_2$) O^- O^- (below each P)	<1	Arnold and Kaplan (1974)

Ado = Adenosine, R = Ribose

$$v_i = \frac{ke}{K'_a/CoA + K'_i/i + K_iK'_a/CoA.i} \tag{1.23}$$

Let it be assumed that the rate of reaction to form the bisubstrate analogue is relatively slow so that the substrate and inhibitor binding steps remain in thermodynamic equilibrium and i and CoA represent inhibitor and CoA concentrations, respectively. Similar kinetic equations can be derived for cases in which inhibitor and substrate bind to the enzyme in compulsory order. Equation (1.23) shows that the rate of formation of the inhibitor will obey Michaelis (saturation) kinetics with respect to the concentrations of both I and CoA, double-reciprocal plots of $1/v_i$ against either $1/i$ or $1/CoA$ being linear and allowing all the kinetic constants to be determined in the manner normally used to treat equations of this type (see for example Segel, 1975; Tipton, 1974). Complications can occur, however, if there is an appreciable non-enzymic rate of formation of the inhibitor

$$\text{I} + \text{CoA} \xrightarrow{k} \text{I–CoA} \tag{1.24}$$

which will be described by a second-order rate equation of the integrated form

$$\frac{1}{i - CoA} \ln \frac{CoA\,(i - x)}{i\,(CoA - x)} = kt \tag{1.25}$$

where i and CoA represent the initial concentrations of these two species, x is the concentration of the I-CoA species and t is time.

Irreversible inhibitors

Irreversible inhibition can be represented by the equation

$$\text{E} + \text{I} \xrightarrow{k} \text{E–I} \tag{1.26}$$

The inhibition should not be reversible by dialysis, gel-filtration or dilution. In kinetic studies, however, it is difficult to make an absolute distinction between inhibitors which bind so tightly to the enzyme that they only dissociate very slowly indeed and those which are truly irreversible. Thus it is necessary to use an *operational* definition of irreversibility which defines an inhibitor as being irreversible if it does not dissociate significantly from the enzyme during the time period of the kinetic studies. Since inhibitors which are truly irreversible involve the formation of a covalent bond with the enzyme, it would be expected that such inhibition is time-dependent. The time-course of inhibition for systems obeying equation (1.26) would be described by a second-order equation of the form of equation (1.25), but if it is possible to have the concentration of the inhibitor so high that it is essentially unchanged during the reaction, *pseudo*-first-order conditions will apply, described by the integrated equation

$$\ln e - \ln (e - x) = k.i.t = k't \tag{1.27}$$

where e represents the initial concentration of enzyme, x is the concentration of E–I and i is that of the inhibitor.

Such inhibitors can be termed non-specific inhibitors since they do not form any kinetically significant non-covalent complex with the enzyme before the irreversible reaction, although they may react specifically with a specific residue, or group of residues in the enzyme. They have proved to be of great use in studies of enzyme structure and function (see for example Means and Feeny, 1971; Freedman, 1979 for reviews), but their lack of enzyme specificity has restricted their pharmacological and medical uses. In the latter type of application, inhibitors which are specific for individual enzymes or types of enzyme have, however, been of great value. Such inhibitors rely on the specificity of the enzyme to bind the inhibitor before the reaction to form the covalent enzyme–inhibitor complex occurs, a process which may be represented by the equation

$$\mathrm{E} + \mathrm{I} \underset{k_{-1}}{\overset{k_{+1}}{\rightleftharpoons}} \mathrm{EI} \xrightarrow{k_{+2}} \mathrm{E{-}I} \tag{1.28}$$

Equation (1.28) can represent the reaction of an active-site-directed inhibitor (see for example Baker, 1967) or a k_{cat} (suicide) inhibitor (Rando, 1974; Abeles and Maycock, 1976; Seiler *et al.* 1978) with the enzyme, since these two types of inhibition cannot be distinguished by steady-state kinetics. If it is assumed that the rate of formation of the covalent product is relatively slow, so that the initial binding equilibrium is not perturbed by its formation, the rate of formation of the irreversibly inhibited enzyme will be given by (Kitz and Wilson, 1962)

$$\frac{\mathrm{d}x}{\mathrm{d}t} = \frac{k_{+2}e}{1 + K_i/i} \tag{1.29}$$

where x represents the concentration of the E–I product and K_i is the inhibitor dissociation constant, k_{-1}/k_{+1}. Equation (1.29) may be integrated to give

$$k't = \ln e - \ln (e - x) \tag{1.30}$$

where

$$k' = \frac{k_{+2}}{1 + K_i/i} \tag{1.31}$$

Thus the first-order rate constant determined from graphs of ln $(e - x)$ against time (t) will show a hyperbolic dependence upon the inhibitor concentration. Equation (1.31) may be written in reciprocal form as

$$\frac{1}{k'} = \frac{K_i}{k_{+2}} \frac{1}{i} + \frac{1}{k_{+2}} \tag{1.32}$$

and thus both K_i and k_{+2} may be evaluated from a double-reciprocal plot as shown in figure 1.6.

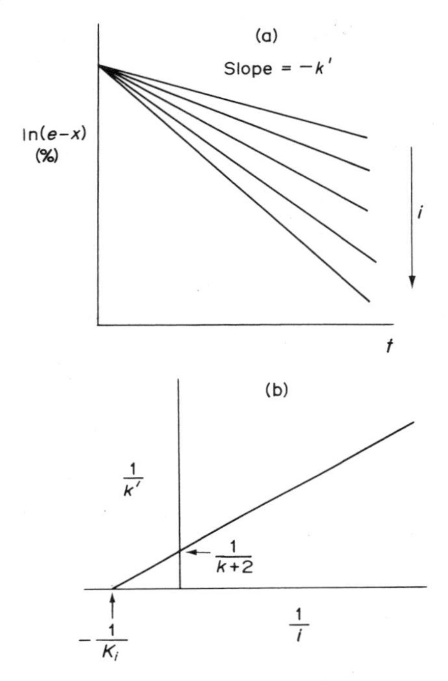

Figure 1.6 Determination of the kinetic parameters for irreversible inhibitor obeying equations (1.28)–(1.32). (a) A semi-logarithmic plot of the rate of reaction at a series of inhibitor concentrations (i). (b) The variation of the reciprocal of the apparent first-order rate constant ($1/k'$) on $1/i$.

Active-site-directed and k_{cat} inhibitors are assumed to differ from each other in that, after binding the inhibitor non-covalently, the enzyme plays a passive role in the former case and an active-site-directed inhibitor is assumed to react with a group at or near the active site by virtue of its high local concentration, whereas a k_{cat} inhibitor is converted to a reactive form by the catalytic action of the enzyme. Obviously such distinctions may not always be quite so clear-cut in practice, but this difference would suggest the existence of more than a single complex before the formation of the irreversibly-inhibited enzyme. Distinction between the mechanism shown in equation (1.28) and one involving additional complexes in the inhibition sequence is, however, not possible by steady-state kinetics alone, although rapid mixing or perturbation studies may allow this.

Protection by substrates

Because active-site-directed and k_{cat} inhibitors are assumed to form an initial non-covalent complex with the active site of the enzyme, one would expect that the presence of substrate would slow down the rate of inhibition by displacing some of this inhibitor. In the presence of a substrate or competitive inhibitor, equation (1.28) may be written as

$$\begin{array}{ccccc} & k_{+1}\ \mathrm{I} & & k_{+2} & \\ \mathrm{E} & \rightleftharpoons & \mathrm{EI} & \longrightarrow & \mathrm{E\text{-}I} \\ & k_{-1} & & & \\ k_{+3}\ \mathrm{S} \downarrow\uparrow k_{-3} & & & & \\ \mathrm{ES} & & & & \end{array} \tag{1.33}$$

which will change the value of k' in equations (1.30) and (1.31) to

$$k' = \frac{k_{+2}}{1 + (K_i/i)(1 + s/K_s)} \tag{1.34}$$

where $K_s = k_{-3}/k_3$. This is of the same form as the equation for simple competitive inhibition (table 1.2) and the constants may be determined in the same ways as are used in that case by using the reciprocal form

$$\frac{1}{k'} = \frac{K_i}{k_{+2}}\left(1 + \frac{s}{K_s}\right)\frac{1}{i} + \frac{1}{k_{+2}} \tag{1.35}$$

Thus a graph of $1/k'$ against $1/i$ will intersect the vertical axis at a value of $1/k_{+2}$. If a series of such graphs are plotted at different concentrations of substrate, a graph of the slopes of these lines against s will have a slope of $K_i/K_s k_{+2}$ and an intercept on the vertical axis of K_i/k_{+2}. This line will also extrapolate to cut the base-line at a value corresponding to $-K_s$.

In the case of non-specific inhibitors, substrate protection may not occur, although it will do if they inhibit by reacting with a group at the active site. In this case the reaction may be represented as

$$\begin{array}{ccc} & k\ \mathrm{I} & \\ \mathrm{E} & \longrightarrow & \mathrm{E\text{-}I} \\ k_{+1}\ \mathrm{S} \downarrow\uparrow k_{-1} & & \\ \mathrm{ES} & & \end{array} \tag{1.36}$$

Thus the presence of the substrate will reduce the amount of free enzyme available to react with inhibitor. In the presence of substrate, the free enzyme concentration (e_f) is reduced to

$$e_f = \frac{e}{1 + s/K_s} \tag{1.37}$$

where $K_s = k_{-1}/k_{+1}$. Thus k' in equation (1.27) must be replaced by k'_{app}, where

$$k'_{app} = \frac{k'}{1 + s/K_s} \tag{1.38}$$

Thus the value of the apparent first-order rate constant (k'_{app}) will decrease hyperbolically with increasing substrate concentration. Equation (1.38) may be written in reciprocal form as

$$\frac{1}{k'_{app}} = \frac{s}{K_s k'} + \frac{1}{k'} \tag{1.39}$$

and thus a graph of $1/k'_{app}$ against s will be a straight line intersecting the vertical axis at a value of $1/k'$ and with a slope of $1/K_s k'$. This line may be extrapolated to cut the base-line at a value corresponding to $-K_s$. In both these cases K_s should be replaced by K_m if the enzyme-catalysed reaction occurs (see section entitled 'Reversible inhibitors' above).

EXPRESSION OF RESULTS

In the case of reversible inhibitors the inhibitor constant (or constants in the case of mixed inhibitors), K_i, and the type of inhibition should be sufficient to allow its effectiveness to be predicted at any substrate concentration, although in the case of enzyme reactions involving two or more substrates, a knowledge of the kinetic mechanism obeyed in the absence of inhibitor may be necessary to avoid ambiguities (see for example Segel, 1975; Tipton, 1974). The I_{50} value, defined as the inhibitor concentration that gives 50 per cent inhibition is not an absolute value since it will depend upon the substrate concentration and type of inhibition (see Tipton, 1973) and therefore it has little practical value. The expression of percentage inhibition given at a single fixed concentration of substrate is also of little use and in some cases it can be misleading, as in the cases where different degrees of inhibition when different substrates are used have been cited as evidence that more than one enzyme is involved (see Tipton, 1973).

Irreversible inhibitors obeying equations (1.26) and (1.28) may be distinguished by the effects of varying the inhibitor concentration on the value of the apparent first-order rate constant for inactivation. In the case of systems obeying equation (1.26), only the first-order rate constant (or the second-order constant at very low inhibitor concentrations) is necessary to define the potency of the inhibitor whereas in the case of those obeying equation (1.28), the enzyme-inhibitor dissociation constant is also required.

It will be seen from table 1.2 that a non-competitive inhibitor acts to decrease the maximum velocity of an enzyme-catalysed reaction without affecting the K_m value, and a similar effect can be produced by irreversibly inhibiting a portion of the enzyme. Thus, irreversible inhibitors can give rise to patterns of double-reciprocal plots which can be confused with those of non-competitive inhibition; but in this case any K_i value calculated is meaningless since it will depend upon enzyme concentration and on preincubation time. Before the equations applicable to reversible inhibitors are applied it is important to show that inhibition is truly reversible within the time scale of the kinetic experiments which are to be used. In the analysis of systems obeying equation (1.28), there are two factors responsible for inhibition, the reversible competitive effect due to formation of the initial non-covalent complex and the irreversible effect resulting from the subsequent reaction. In order to measure the latter effect without interference from the former, it is necessary to assay at very high substrate concentrations so that the reversibly-bound inhibitor is all displaced or to dilute the enzyme–inhibitor mixture into the assay mixture so that the reversible complex dissociates.

I_{50} values—or sometimes their negative logarithm, pI_{50}—have frequently been used to express the potency of irreversible inhibitors, but it should be remembered that, in the absence of competing reactions, an equimolar concentration of the inhibitor would be expected to inhibit the enzyme completely, given a sufficiently long reaction time. Thus it is necessary to ensure that the enzyme and inhibitor are incubated until no further inhibition occurs before measurements are made. In addition, unlike the rate and affinity constants given earlier, the I_{50} value is not an absolute value since it will vary with enzyme concentration.

COMPARISONS AND COMPLICATIONS

As discussed earlier, reversible inhibitors generally have the advantage that, once they have reached the site of the enzyme, they normally act rapidly whereas irreversible inhibitors can take longer to exert their effects. The duration of effectiveness of a reversible inhibitor will depend upon the action of metabolic and elimination systems which tend to remove it, whereas in the case of competitive substrates the target enzyme will also be involved in its removal. In both cases, complications can occur if the product of the reaction has pharmacological actions of its own, as illustrated by the metabolism of α-methyldopa to produce false transmitters. The effectiveness of competitive inhibitors *in vivo* may be reduced by the rise in substrate concentration that will follow inhibition.

Irreversible inhibitors may be removed from the body by metabolism and elimination systems, but once an enzyme is inhibited the duration of the effect will depend only on the rate of turnover of the enzyme (see for example Schimke and Doyle, 1970).

If an enzyme is irreversibly inhibited, the rate of return of its activity will be given by its rate of synthesis minus its rate of degradation:

$$\frac{de}{dt} = k_s - k_d e \qquad (1.40)$$

where k_s is the zero-order rate constant for its rate of synthesis, k_d is the first-order rate constant for enzyme breakdown and e represents the enzyme concentration. Under steady-state conditions, the rate of synthesis and degradation are equal ($de/dt = 0$) and thus

$$k_s = k_d e_{ss} \qquad (1.41)$$

where e_{ss} is the steady-state concentration of active enzyme, which would be obtained at infinite time after administration of the irreversible inhibitor. Thus equation (1.40) may be rewritten as

$$\frac{de}{dt} = k_d\,(e_{ss} - e) \qquad (1.42)$$

which shows that the rate of regain of activity after inhibition will depend only upon its rate of degradation.

Equation (1.42) can be integrated to give

$$\ln(e_{ss} - e) = -k_d t + \text{constant} \tag{1.43}$$

Thus a graph of $\ln(e_{ss} - e)$ or $2.303 \log(e_{ss} - e)$ against time should give a straight line with a slope of $-k_d$, and the half-life of the enzyme ($T_{1/2}$) can be calculated from the relationship $T_{1/2} = 0.693/k_d$. Such an approach has been used to determine the rate of degradation for a number of enzymes. In the case of monoamine oxidase, for example, the method has been employed to calculate the rate of degradation and half-life of the enzyme after inhibition by pargyline (Neff and Goridis, 1972) and to study changes in the rates of degradation occurring with age in rat heart (Della Corte and Callingham, 1977). The half-lives of individual enzymes *in vivo* are very different, ranging from a few minutes in the case of ornithine decarboxylase (Russell and Snyder, 1969) to several weeks, and thus the effectiveness of a single dose of an irreversible inhibitor will be affected by the stability of the target enzyme *in vivo*.

The effectiveness of irreversible inhibitors which contain reactive groupings is likely to be reduced by their reaction with components other than the target enzyme. Such reactions might also give rise to toxic effects, for example the reaction of chloromethylketone derivatives with glutathione, potentially resulting in dangerous falls in tissue glutathione levels, has recently been discussed by Smith (1978). The k_{cat} inhibitors which do not contain highly reactive groupings in their native state should be of particular value in reducing these undesired side effects. In a recent study, Fuller and Hemrick (1978) have combined the rapid effects and superior controllability of reversible inhibitors to modulate the response of monoamine oxidase to irreversible selective inhibitors.

This brief survey has concentrated on the behaviour of relatively simple systems. More complicated cases of reversible inhibition can occur if the inhibitor only partially inhibits the enzyme or if the binding of more than one molecule of inhibitor contributes to the observed effect. Such behaviour does not, however, affect the patterns shown in figure 1.2, although it can lead to non-linearity in graphs plotted as a function of inhibitor concentration (see for example Segel, 1975; Dixon and Webb, 1979). Cases in which the product of an enzyme-catalysed reaction is a powerful inhibitor may either be analysed by the initial rate method discussed here or by allowing the reaction to proceed well past the linear phase (see figure 1.1) and analysing the progress curve by the use of integrated rate equations (see for example Orsi and Tipton, 1979) although here it is essential to ensure that inhibitor formation is the only cause of the departure from linearity. In the case of irreversible inhibition, it is possible to envisage more complicated mechanisms than those discussed here but they should not be invoked unless the data are incompatible with the simpler systems.

REFERENCES

Abeles, R. H. and Maycock, A. L. (1976). *Accounts chem. Res.*, 9, 313

Arnold, L. J. and Kaplan, N. O. (1974). *J. biol. Chem.*, 249, 652

Baker, B. R. (1967). *Design of Active-Site-Directed Irreversible Inhibitors*, John Wiley and Sons, New York
Blakley, R. L. (1969). *The Biochemistry of Folic Acid and Related Pteridines*, North-Holland, Amsterdam
Byers, L. J. (1978). *J. theor. Biol.*, **74**, 501
Chase, J. F. A. and Tubbs, P. K. (1969). *Biochem. J.*, **111**, 225
Cleland, W. W. (1971). In *The Enzymes*, vol. 1 (ed. P. D. Boyer), Academic Press, New York, p. 1–65
Collins, K. D. and Stark, G. R. (1971). *J. biol. Chem.*, **246**, 6599
Colowick, S. P. and Womack, F. P. (1969). *J. biol. Chem.*, **244**, 774
Cornish-Bowden, A. (1974). *Biochem. J.*, **137**, 143
Della Corte, L. and Callingham, B. A. (1977). *Biochem. Pharmac.*, **26**, 407
Dixon, M. (1972). *Biochem. J.*, **129**, 197
Dixon, M. and Webb, E. C. (1964). *Enzymes*, 2nd ed., Longman, London
Dixon, M. and Webb, E. C. (1979). *Enzymes*, 3rd ed., Longman, London, in the press
Ellman, G. L., Courtney, K. D., Andres, V. and Featherstone, R. M. (1961). *Biochem. Pharmac.*, **7**, 88
Freedman, R. B. (1979). In *Techniques in the Life Sciences*, vol. B1/1 (ed. H. L. Kornberg, J. C. Metcalfe, D. H. Northcote, C. I. Pogson and K. F. Tipton), Elsevier/North-Holland, Amsterdam, B117, pp. 1–23
Frieden, C. (1964). *J. biol. Chem.*, **239**, 3522
Frieden, C. (1970). *J. biol. Chem.*, **245**, 5788
Fuller, R. W. and Hemrick, S. K. (1978). *Life Sci.*, **22**, 1083
Glusker, J. P. (1971). In *The Enzymes*, vol. 5 (ed. P. D. Boyer), Academic Press, New York, pp. 413–39
Gutfreund, H. (1972). *Enzymes: Physical Principles*, Wiley-Interscience, London
Halford, S. E. (1974). In *Companion to Biochemistry* (ed. A. T. Bull, J. R. Lagnado, J. O. Thomas and K. F. Tipton), Longman, London, pp. 197–226
Henderson, P. J. F. (1972). *Biochem. J.*, **127**, 321
Henderson, P. J. F. (1973). *Biochem. J.*, **135**, 101
Hummel, J. P. and Dryer, W. J. (1962). *Biochim. biophys. Acta*, **62**, 530
Kitz, R. and Wilson, I. B. (1962). *J. biol. Chem.*, **237**, 3245
Klotz, I. M., Walker, F. M. and Pivan, R. B. (1946). *J. Am. chem. Soc.*, **68**, 1486
Kopin, I. J. (1971). *Fedn Proc. Fedn Am. Socs exp. Biol.*, **30**, 904
Kurganov, B. I., Dorozhko, A. I., Kagan, Z. S. and Yakovlev, V. A. (1976). *J. theor. Biol.*, **60**, 247, 271 and 287
Leinhard, G. E. and Secemski, I. I. (1973). *J. biol. Chem.*, **248**, 1121
McLean, A. E. M. (1971). In *Mechanisms of Toxicity* (ed. W. N. Aldridge), Macmillan, London, pp. 219–28
Means, G. E. and Feeny, R. E. (1971). *Chemical Modification of Proteins*, Holden-Day, San Francisco
Moe, O. and Hammes, G. G. (1974). *Biochemistry*, **13**, 2547
Neff, N. H. and Goridis, C. (1972). *Adv. Biochem. Psychopharmac.*, **5**, 307
Orsi, B. A. and Tipton, K. F. (1979). In *Methods in Enzymology*, vol. 63A (ed. D. L. Purich), Academic Press, New York, in the press
Rando, R. R. (1974). *Science, N.Y.*, **185**, 320
Russell, D. and Snyder, S. H. (1969). *Molec. Pharmac.*, **5**, 253
Scatchard, G. (1949). *Ann. N.Y. Acad. Sci.*, **51**, 460
Schimke, R. T. and Doyle, D. (1970). *A. Rev. Biochem.*, **39**, 929
Segel, I. H. (1975). *Enzyme Kinetics*, John Wiley, New York
Seiler, N., Jung, M. J. and Koch-Weser, J. (eds) (1978). *Enzyme-Activated Irreversible Inhibitors*, Elsevier/North-Holland, Amsterdam
Smith, H. J. (1978). *J. theor. Biol.*, **73**, 531

Sophianopoulos, J. A., Durham, S. J., Sophianopoulos, A. J., Ragsdale, H. L. and Cropper, W. P. (1978). *Arch. Biochem. Biophys.*, **187**, 132
Theorell, H. and Yonetani, T. (1969). *Acta chem. scand.*, **23**, 255
Tipton, K. F. (1973). *Biochem. Pharmac.*, **22**, 2933
Tipton, K. F. (1974). In *Companion to Biochemistry* (ed. A. T. Bull, J. R. Lagnado, J. O. Thomas and K. F. Tipton), Longman, London, pp. 227–51
Tipton, K. F. (1978). In *Techniques in the Life Sciences*, vol. B1/11 (ed. H. L. Kornberg, J. C. Metcalfe, D. H. Northcote, C. I. Pogson and K. F. Tipton), Elsevier/North-Holland, Amsterdam, B112, pp. 1–56
Tipton, K. F., Houslay, M. D. and Turner, A. J. (1977). *Essays Neurochem. Neuropharmac.*, **1**, 103
Wolfenden, R. (1976). *A. Rev. Biophys. Bioengng.*, **5**, 271
Wolfenden, R. (1977). In *Methods in Enzymology*, vol. 46 (ed. W. B. Jakoby and M. Wilchek), Academic Press, New York, p. 15

2
Design of irreversible inhibitors

Elliott Shaw (Biology Department, Brookhaven National Laboratory
Upton, New York 11973, USA)

GENERAL DESCRIPTION OF AFFINITY LABELLING

The development of irreversible inhibitors was a natural goal of studies aimed at obtaining increasingly effective ones. The attempt to label a serotonin receptor using a structural analogue with alkylating potential states clearly the essential idea of affinity labelling although covalency, in this early try, was not achieved (Shaw and Woolley, 1954). Subsequent applications to enzyme function were more successful but the initial motivation was drug design (Shaw and Woolley, 1954; Baker, 1967). Affinity labelling has now become well established as a means of studying the mechanism by which enzymes act, since it has been fruitful in identifying functional amino acid residues at active centres of numerous enzymes (Jakoby and Wilchek, 1977). The 'labelling' consists of covalent bond formation between an amino acid side chain and a substrate-like reagent designed to promote stable bond formation. The stability of the covalent bond permits degradative protein chemistry to be carried out on the inhibited enzyme to determine the site of structural alteration. An early example of this approach was the inactivation of chymotrypsin by reaction with tosylphenylalanine bromo- or chloromethyl ketone (Schoellmann and Shaw, 1962, 1963) a structure which, because it contains phenylalanine, was expected to bind at the active centre of chymotrypsin since this enzyme hydrolyses phenylalanine bonds in proteins and also tosylphenylalanine esters. The chloromethyl ketone portion of the structure provides an alkylating potential near the carbonyl bond of the substrate to which functional groups at the active centre of chymotrypsin are directed to promote peptide bond hydrolysis. The role of serine-195 had already been established. Tosylphenylalanine chloromethyl ketone combined with chymotrypsin in a stoichiometric fashion with inactivation. Histidine-57 was found to be alkylated at N-3 (Ong *et al.*, 1965). When the three-dimensional structure of chymotrypsin was elucidated, His-57 and Ser-195 were found to be in close juxtaposition (Matthews *et al.*, 1967). Chymotrypsinogen, which lacks substrate binding capacity, is inert to the reagent.

This specific modification of a side chain of chymotrypsin by alkylation which, with a closely related reagent, took place 10^6 times more rapidly than the reaction of a simple histidine derivative (Shaw and Ruscica, 1971), demonstrated the great rate enhancement that could be provided by enzymes for their own inactivation, facilitating selective reactions that would not otherwise be possible. Coming as it did in a period of intensive effort to determine enzyme sequence, mechanism, and structure by crystallography, this development stimulated wide application to enzymes of all types.

Although substrates of intermediary metabolism vary considerably in chemical nature, proteins have a limited number of side chains which can be considered functional in a chemical sense. There are *N*-terminal amino groups, the ϵ-amino group of lysyl residues, guanidino group of arginyl residues, β- or γ-carboxyl group of aspartyl or glutamyl residues or *C*-terminal carboxyl group, phenolic group of tyrosyl residues, hydroxyl group of seryl or threonyl residues, imidazole group of histidyl residues, sulphydryl group of cysteinyl residues, methylmercapto group of methionyl residues, and possibly indole ring of tryptophyl residues. Many other side chains may be involved in creating specific binding regions but these (for example, the alkyl or aryl side chains of alanyl, valyl, leucyl, isoleucyl and phenylalanyl residues) are not smoothly linkable by covalent bands (although brave attempts have been made). In view of the limited group of protein side chains available, a relatively small number of reactive groupings have been used with a variety of substrates. In fact, an alkylating function, for example

$$\mathrm{X{-}CH_2\overset{\overset{\displaystyle O}{\|}}{C}{-}}$$

has been very widely employed and, depending on the remainder of the structure (substrate), in one enzyme may alkylate histidine, in others esterify glutamic acid, etherify serine, or alkylate methionine. Thus the groupings (table 2.1) are not in themselves diagnostic for a given type of side chain but, when attached to a substrate molecule, may become positioned close to a functional side chain and the chemical reaction which follows is directed by the geometry within the enzyme–inhibitor complex (Shaw, 1970). In fact, there is a fair possibility that the reactive grouping does not become positioned near a susceptible side chain and therefore the inhibitor remains a competitive inhibitor. There is undoubtedly a track of failures which reflect our ignorance of active centres or exaggerated expectations.

Unreactive substrate analogues which become reactive after being acted on by enzymes represent, in certain cases, a different approach from that briefly summarized above. Some very interesting irreversible inhibitors have been obtained with enzymes promoting proton abstraction (Abeles and Maycock, 1976; Rando, 1977; and authors in this volume). This approach should diminish side reactions. However, it may be limited to a mechanistically narrow group of enzymes. In addition, inherent in the more general approach to irreversible enzyme inhibition described above is the possible use of unreactive groups since a proximity effect within enzyme–inhibitor complexes may have a tremendous rate-enhancing action. We may expect to see some ingenious chemical approaches in either case.

Table 2.1 Some chemical groupings used to achieve covalent interactions with enzymes. Selective side chain modification depends on enzyme and reagent design.

Grouping	Substrate analogue	Enzyme	Modified side chain	Reference
$-C(=O)-CH_2X$	Ribulose 1,5-bisphosphate	ribulosephosphate carboxylase	Lys	Hartman *et al.* (1973)
	Amino acid amide	chymotrypsin	His	Schoellman and Shaw (1962)
	Amino acid amide	trypsin	Ser	Schroeder and Shaw (1971)
	Galactosamine	β-galactosidase	Met	Naider *et al.* (1972)
	Amino acid amide	papain	CySH	Whitaker and Perez-Villasenor (1968)
	Nucleotide	Staphylococcal nuclease	Tyr	Cuatrecasas *et al.* (1969)
$-C(=O)-CHN_2$	Glutamine	formalglycinamide ribotide amidotransferase	CySH	David *et al.* (1963)
	Glutamine	glutaminase	Thr	Holcenberg *et al.* (1978)
	Peptide	papain	CySH	Leary *et al.* (1977)
	Peptide	pepsin	COOH	Rajagopalan *et al.* (1966)
epoxide $-\underset{H}{C}-\underset{H}{C}-$ (bridged by O)	Amino acid	pepsin	Asp, Met	Tang, (1971); Cheng and Tang (1972)
	Glycosides	lysozyme	Asp	Eshdat *et al.* (1973)
	Glucopyranoside	β-glucosidase	Asp	Bause and Legler (1974)
$Aryl\text{-}N_2^+$	Nucleotides	staphylococcal nuclease	Tyr, His or Trp	Cuatrecasas and Wilchek (1977)

PROBLEM OF SELECTIVITY

To produce a therapeutically useful inhibitor, it has long been recognized that the major challenge to the biochemist is to obtain selectivity. Depending on the system under study, this may mean selective toxicity to a pathogen or tumour cell or, as in the work to be described in detail below, selective inactivation of one enzyme among a group of closely related ones.

There are at least two ways that selectivity may be achieved by affinity labelling. In the first way, the affinity of the reagent may be greater for the target enzyme than for other enzymes. In this case, the approach to selectivity is not different from that of a reversible inhibitor. However, irreversible inactivation is viewed as a distinct advantage as a means of achieving a prolonged inactivation, combatting dilution by physiological circumstances. In the second situation, selectivity may be achieved in the step involving covalent bond formation. Here, bond formation is *independent* of affinity and results from a geometric difference in enzyme active centres which permit bond formation in the target enzyme. This outcome is possible uniquely with irreversible reagents.

Results described below demonstrate both types of selectivity.

RECENT RESULTS WITH SOME TRYPSIN-LIKE ENZYMES

Selectivity controlled by affinity

The trypsin-like enzymes represent an interesting challenge since these proteases cleave at arginine or, to a lesser extent, lysine residues in proteins and utilise a common mechanism consisting of an active centre serine for acyl-enzyme formation. An active centre histidine facilitates proton transfer in the hydrolytic steps. An ionised aspartic acid residue in the primary specificity site serves to bind the substrate -Arg- residue in a productive orientation assisted by a region of peptide bond which positions an analogous stretch of the substrate.

Characterization of proteases in blood coagulation, fibrinolysis, kinin production, complement action and fertilization reveals that we are dealing with a family of homologous proteases acting very selectively in nature. They represent an attractive group for studies of drug design since there are opportunities for a variety of potentially useful actions. At the same time, the similarities in specificity suggest that it might be very difficult to produce synthetic inhibitors discriminating enough to be useful.

Recent sequence studies of the sites of cleavage in physiological substrates of trypsin-like enzymes provide information which can be incorporated into inhibitors (table 2.2). Since peptides of lysine chloromethyl ketone (Coggins *et al.*, 1974) and arginine chloromethyl ketone (Kettner and Shaw, 1977, 1978, 1979*a,b*) can be synthetised with defined sequence, it was hoped that selective inhibitors could be obtained for a given protease by utilising the appropriate substrate sequence.

In the case of thrombin there are at least four recognised sites of cleavage, the release of the two fibrinopeptides from fibrinogen, a cleavage of prothrombin that could have regulatory significance and the activation of factor XIII for fibrin clot cross-linking (table 2.2). Chloromethyl ketone derivatives corresponding to the sequences in three of the thrombin substrates were synthesised and compared with respect to their ability to inactivate thrombin.

Table 2.2 Specificity of some trypsin-like enzymes

Protease	Substrate	Sequence at site of cleavage	Reference
Thrombin	Fibrinogen A-chain	–Gly–Gly–Val–Arg–Gly–Pro	Iwanaga *et al.* (1969)
	Fibrinogen B-chain	Phe–Ser–Ala–Arg–Gly–Pro	Iwanaga *et al.* (1969)
	Prothrombin	Val–Ile–Pro–Arg–Ser–Gly	Magnusson *et al.* (1975)
	Factor XIII	Val–Pro–Arg–Gly–Val	Takagi and Doolittle (1974)
Factor Xa	Prothrombin	Ile–Glu–Gly–Arg–Ile–Val	Magnusson *et al.* (1975)
Factors IXa and VIIa	Factor X	Gln–Val–Val–Arg–Ile–Val	Titani *et al.* (1975)
Kallikreins	Kininogen	Ser–Pro–Phe–Arg–	Pisano (1975)
Plasminogen Activators	Plasminogen	Cys–Pro–Gly–Arg–Val–	Sottrup-Jensen *et al.* (1975)

Irreversible enzyme inhibitors are time-dependent in their action, in contrast with reversible inhibitors. The rate of enzyme inactivation is dependent on the concentration of enzyme inhibitor complex which, in turn, is dependent on inhibitor concentration:

$$E + I \rightleftharpoons EI \xrightarrow{k_2} \text{Inactive enzyme}$$

Dilutions of inhibitor which avoid saturation effects are generally chosen for study in a range in which the experimentally observed rate of inactivation is proportional to inhibitor concentration. This makes it possible to derive a second-order rate constant for inactivation (last column, table 2.3) which provides the usual basis for comparison of inhibitors. The Pro–ArgCH_2Cl derivatives were an order of magnitude superior to the fibrinopeptide A analogue, Gly–Val–ArgCH_2Cl (table 2.3). From these rates, we conclude that Val–Pro–ArgCH_2Cl is 54/1.9 = 28 times as effective as Gly–Val–ArgCH_2Cl in inactivating thrombin. This difference is due to the greater affinity of the former, with K_i = 0.6 μM versus 13 μM for the latter. In contrast, the complex of each reagent with thrombin has a similar rate constant for the formation of inactive enzyme, namely 0.31 and 0.30 min^{-1} (Kettner and Shaw, 1977).

Table 2.3 Relative effectiveness of chloromethyl ketones derived from thrombin substrates in the inactivation of thrombin (Kettner and Shaw, 1977)

Affinity label	Concentration of inhibitor (M)	$t_{1/2}$† (min)	k_{app}/I‡ (min^{-1} M^{-1} × 10^{-4})
Gly–Val–ArgCH_2Cl	2.0 × 10^{-6}	18	1.9
Val–Pro–ArgCH_2Cl	7.5 × 10^{-8}	17	54
Ile–Pro–ArgCH_2Cl	7.5 × 10^{-8}	22	42
Val–Ile–Pro–ArgCH_2Cl	7.5 × 10^{-8}	13	73

†$t_{1/2}$ is the half-time for the pseudo-first-order inactivation of thrombin at pH 7.0 and 25 °C.
‡k_{app}/I is the ratio of the apparent, first-order rate constant for inactivation of thrombin to the concentration of affinity label.

As shown in figure 2.1, an effective thrombin-directed reagent, Ile–Pro–ArgCH_2Cl is less effective in inactivating plasmin and plasma kallikrein. Figure 2.1 illustrates the typical activity-decay 'curves' obtained through the use of irreversible inhibitors, the action of which is time- and concentration-dependent. The individual points represent rate assays measured on aliquots of a reaction mixture. From graphical treatment of rates obtained at varied inhibitor concentrations, one may calculate K_i and k_2 (Kitz and Wilson, 1962; Shaw, 1970).

Work in this laboratory (Shaw, 1975) and elsewhere indicated that a D-amino acid residue in the P^3-position (two amino acid residues prior to the primary specificity site in the *N*-terminal direction) provided enhanced effectiveness for binding to thrombin. D-Phenylalanine was apparently of value in this regard (Bajusz *et al.*, 1975). Consequently D-Phe–Pro–ArgCH_2Cl was synthesised and proved to be a remarkably powerful inactivator of thrombin, so rapid in its

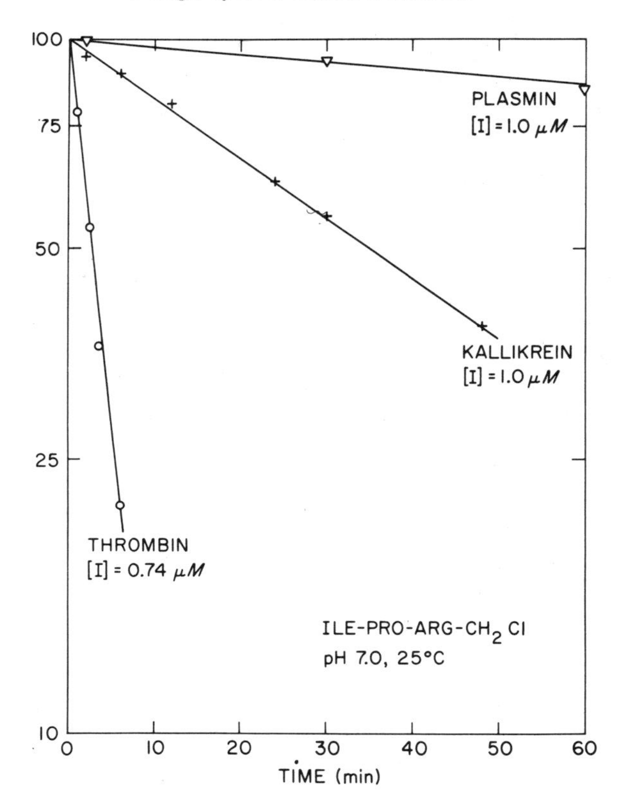

Figure 2.1 Comparison of the susceptibility of thrombin, plasmin, and plasma kallikrein to inactivation by Ile–Pro–Arg–CH_2 Cl at pH 7.0 and 25 °C.

action that, even at high dilution, it combines quickly with the enzyme (figure 2.2). It has not been possible with our present methods to measure the affinity of this reagent for thrombin, but this problem is still under study. The selectivity of D-Phe–Pro–ArgCH_2 Cl for thrombin is evident from the results in figure 2.3 (Kettner and Shaw, 1979*a*). Other trypsin-like enzymes are susceptible to the reagent at concentrations that would normally be considered indicative of reactivity; however, the ratio of second-order rate constants provides a measure of the selectivity for thrombin (table 2.4).

The above work was carried out on purified enzymes. To determine whether or not it is possible to inhibit thrombin under more physiological conditions, for example during its formation and in the presence of fibrinogen, the effect of D-Phe–Pro–ArgCH_2 Cl on the partial thromboplastin time of plasma was measured and a doubling of coagulation time was produced with 10^{-7} M reagent pointing to its ability to act in competition with macromolecular substrates.

Plasma kallikrein has been subjected to the same approach. Pro–Phe–ArgCH_2 Cl was synthesised as a reagent expected to conform to the specificity of plasma kallikrein (table 2.2) and therefore to inactivate it preferentially. This inhibitor

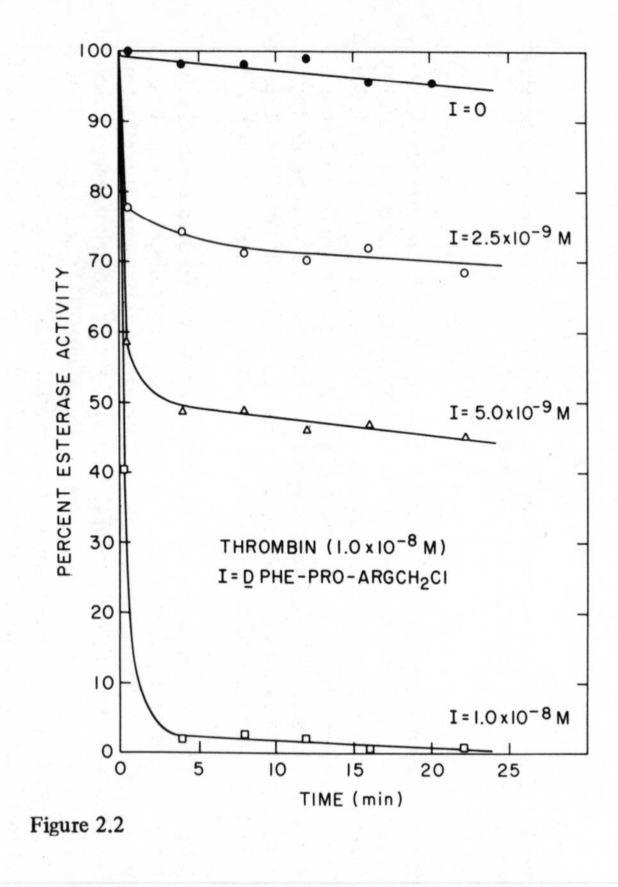

Figure 2.2

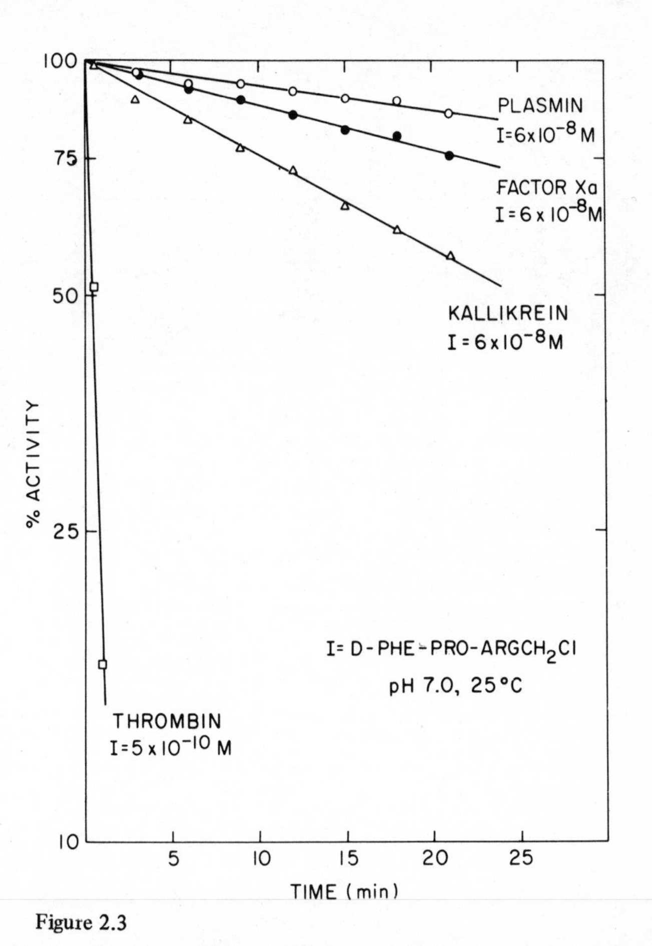

Figure 2.3

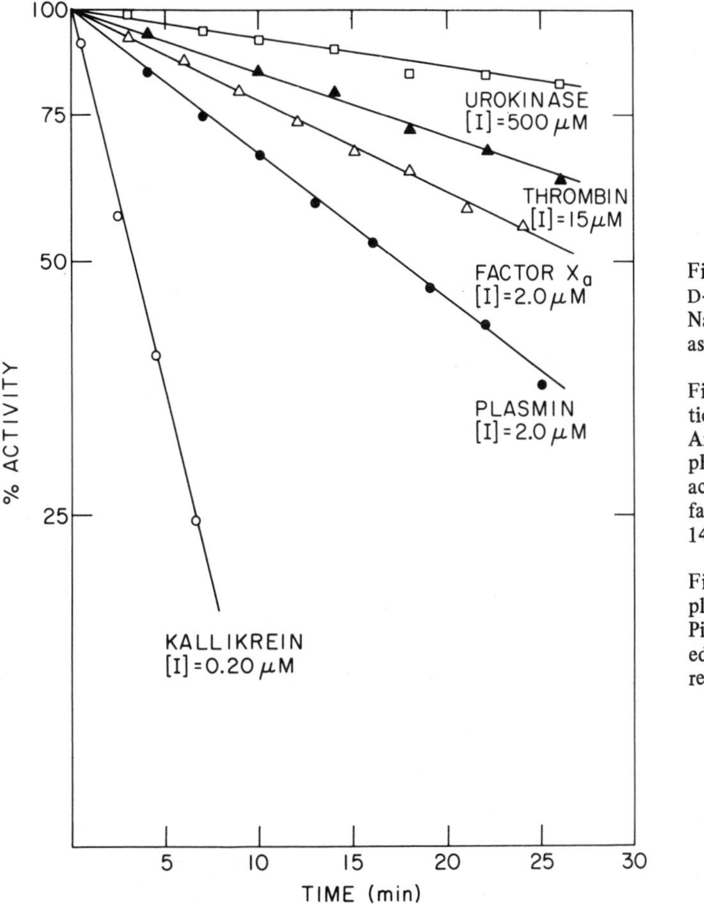

Figure 2.4

Figure 2.2 Stoichiometry of the inactivation of thrombin by D-Phe–Pro–ArgCH$_2$Cl in 50 mM Pipes buffer, pH 7.0, 0.20 M in NaCl, and 25 °C. Portions were removed at timed intervals and assayed for esterase activity.

Figure 2.3 Selectivity of D-Phe–Pro–ArgCH$_2$Cl in the inactivation of thrombin. The proteases were incubated with D-Phe–Pro–ArgCH$_2$Cl at the indicated concentrations in 50 mM Pipes buffer, pH 7.0; aliquots were removed and assayed for residual esterase activity. The initial concentrations of thrombin, plasma kallikrein, factor Xa, plasmin and urokinase (not shown) were 0.23, 2.5, 14, 1.8 and 13 nM, respectively.

Figure 2.4 Selectivity of Pro–Phe–ArgCH$_2$Cl in the activation of plasma kallikrein. The proteases were incubated at 25 °C in 50 mM Pipes buffer (pH 7.0) containing Pro–Phe–ArgCH$_2$Cl at the indicated concentration. Inactivation of the protease was monitored by removing timed aliquots and assaying for esterase activity.

Table 2.4 Selectivity of D-Phe-Pro-$ArgCH_2Cl$ as a protease inhibitor (Kettner and Shaw, 1979a)

Protease	Concentration of affinity label (μM)	$t_{1/2}$ (min)	k_{app}/I (M^{-1} min^{-1} × 10^{-4})	Relative rate of inactivation
Thrombin	5×10^{-10}	1.2	115 000	185 000
Plasma kallikrein	6×10^{-8}	24.3	47.5	77
Factor Xa	2×10^{-7}	23.25	14.9	24
Plasmin	5×10^{-7}	32.3	4.30	6.9
Urokinase	5×10^{-6}	22.3	0.62	1

is in fact extremely effective in inactivating plasma kallikrein (figure 2.4) (Kettner and Shaw, 1978). Other trypsin-like enzymes are susceptible to higher concentrations of the reagent. The actual rates differ by 10^5 from the most susceptible (kallikrein) to the least susceptible (urokinase). In the case of plasmin, the rate is about 1/40th that of plasma kallikrein. The basis for this selectivity has been explored by a more complete kinetic analysis of three reagents, two of which, Pro-Phe-$ArgCH_2Cl$ and Ala-Phe-$ArgCH_2Cl$ are more effective against plasma kallikrein than plasmin, whereas the third, Ala-Phe-$LysCH_2Cl$ has a slight preference for plasmin. It was found (table 2.5) that the rate, k_2, of formation of the alkylated enzyme (analogous to V_{max} for ES breakdown) was relatively constant for each protease, 0.35 min^{-1} for plasma kallikrein and 0.18 min^{-1} for plasmin. On the other hand, the affinity of each reagent in the reversible phase, K_i, varied with structure, accounting for the differences in selectivity.

A sizeable group of peptides of arginine chloromethyl ketone have now been synthesised and evaluated with a number of trypsin-like enzymes of physiological importance. The details cannot all be reviewed here. However one additional class of this family should be considered because of current interest, namely plasminogen activators. These have been known for some time by investigators of fibrinolysis, as reviewed by Astrup (1975). More recently, evidence for their increased production by transformed cells and the association of this property with oncogenicity (Unkless *et al.*, 1973; Reich, 1975) makes it important to try to obtain selective inhibitors for plasminogen activators in order to deduce their role in neoplasia. Since the activation of plasminogen involves the cleavage at a particular Pro-Gly-Arg sequence (Sottrup-Jensen *et al.*, 1975), the corresponding chloromethyl ketone was prepared and compared with a number of other reagents in ability to inactivate urokinase, a plasminogen activator of kidney origin, and also an activator obtained from HeLa cells. The results with urokinase (table 2.6) (Kettner and Shaw, 1979*b*) show the importance of structure in determining reactivity since the rates have a range of 10^4 among the arginine tripeptide derivatives. Pro-Gly-$ArgCH_2Cl$ is near the top of the list. The most effective reagents have a Gly-$ArgCH_2Cl$ sequence. The results with the plasminogen activator from HeLa cells were similar (Coleman *et al.*, 1979) but differences exist with urokinase which demonstrate the ability of the reagents to discriminate between closely related enzymes.

Table 2.5 Comparison of kinetic constants in the inactivation of plasma kallikrein and plasmin by arginine and lysine chloromethyl ketones (Kettner and Shaw, 1978)

Affinity label	Plasma kallikrein			Plasmin		
	K_i† (μM)	k_2† (min^{-1})	k_2/K_i (M^{-1} min^{-1} × 10^{-3})	K_i (μM)	k_2 (min^{-1})	k_2/K_i (M^{-1} min^{-1} × 10^{-3})
Ala-Phe-ArgCH_2Cl‡	0.078	0.35	4400	1.3	0.18	140
Ala-Phe-LysCH_2Cl§	4.9	0.35	72	0.83	0.18	210
Pro-Phe-ArgCH_2Cl¶	0.24	0.36	1500	4.2	0.15	37

†The reversible dissociation constants of the substrate-like complex, K_i, and the first-order rate constants of the irreversible steps of the reaction, k_2, were determined by the method of Kitz and Wilson (1962) at pH 7.0 and 25 °C.
‡The concentration range of Ala-Phe-ArgCH_2Cl over which values of k_{app} were measured in determining the kinetic constants was 0.0083–0.025 μM for plasma kallikrein and 0.17–0.76 μM for plasmin.
§The concentration range of Ala-Phe-LysCH_2Cl was 0.47–1.4 μM for plasma kallikrein and 0.12–0.50 μM for plasmin.
¶The concentration range of Pro-Phe-ArgCH_2Cl over which values of k_{app} were measured in determining the kinetic constants was 0.16–0.05 μM for plasma kallikrein and 0.55–2.0 μM for plasmin.

Table 2.6 Comparison of the reactivity of urokinase with peptides of arginine and lysine chloromethyl ketones† (Kettner and Shaw, 1979b)

Affinity label	Concentration (μM)	$t_{1/2}$‡ (min)	k_{app}/I§ (M^{-1} min^{-1} × 10^{-4})
Glu–Gly–ArgCH_2Cl	0.20	17.0	20.4
DNS–Glu–Gly–ArgCH_2Cl	0.50	28.9	4.80
Ac–Gly–Gly–ArgCH_2Cl	0.50	54.0	2.57
Pro–Gly–ArgCH_2Cl	2.0	44.0	0.79
Val–Pro–ArgCH_2Cl	10.0	12.8	0.54
Ile–Pro–ArgCH_2Cl	10.0	17.8	0.39
Phe–Ala–ArgCH_2Cl	10.0	23.9	0.29
Val–Ile–Pro–ArgCH_2Cl	20.0	19.6	0.18
Phe–Ala–LysCH_2Cl	100.0	28.7	0.024
Gly–Val–ArgCH_2Cl	100.0	41.3	0.017
Ile–Leu–ArgCH_2Cl	200.0	23.9	0.014
Ala–Phe–ArgCH_2Cl	500.0	23.6	0.0059
Pro–Phe–ArgCH_2Cl	500.0	91	0.0015
Ala–Phe–LysCH_2Cl	2500	83	0.0003
Tos–LysCH_2Cl	50 000	85	0.000 02

†Inactivations were conducted at 25°C in 50 mM Pipes buffer, pH 7.0, containing 0.20 M NaCl. The initial concentration of urokinase was 13 nM.

‡$t_{1/2}$ is the half-time for the first-order inactivation of urokinase at the indicated concentration of the affinity labelling reagent.

§k_{app}/I is the ratio of the apparent, first-order rate constant for inactivation to the concentration of the affinity label. These values were calculated from the values of $t_{1/2}$ using the relationship, $k_{app} = \ln 2/t_{1/2}$, and the indicated concentration of the affinity label.

Although this work has provided effective inhibitors of plasminogen activators, selectivity is lacking. The better reagents also inactivate other trypsin-like enzymes. In this case, further work will be necessary to obtain selectivity. Although it is not clear how this may be achieved, some clues may be found in the following section.

The specific interaction of trypsin-like enzymes with their physiological substrates may well involve other structural features of the substrate in addition to the sequence at the cleavage site. Even within this limited region of interaction, the conformation may be a decisive feature which is not duplicated by a peptide even of substantial length. However, the importance of sequence has been demonstrated by the foregoing results, particularly in contrast with those for trypsin and acrosin which are less specific in their action (Kettner *et al.*, 1978). With these proteases, the spread of reactivity among a large group of tripeptides of arginine chloromethyl ketone is only 100-fold in contrast with the 10^4–10^5 range of rates obtained with plasma kallikrein and urokinase.

Selectivity in the step of covalent bond formation

Observations with a group of inhibitors chemically distinct from the above provide evidence for a different origin for selectivity, one which is not related to affinity. Catalysis by serine proteinases involves the transient formation of an acyl-enzyme. It became evident from the effectiveness of benzamidine and phenylguanidine as competitive inhibitors (Mares-Guia and Shaw, 1965) that analogues of arginine could be obtained with an aromatic ring in place of the side chain methylene groups (Mares-Guia *et al.*, 1967). Esters of some of these act as substrates. However, ethyl *p*-guanidinobenzoate inactivated trypsin due to formation of a relatively stable acyl-enzyme (Mares-Guia and Shaw, 1967). In this case, we are not dealing with an irreversible inhibitor in the strictest sense but with a long-lived intermediate. The situation is analogous to the temporary inactivation of acetylcholinesterase by carbamates (Aldridge and Reiner, 1972). One should not get bogged down in terminology since an agent which produces a covalent inactive derivative of an enzyme with a life-time measurable in hours rather than microseconds can have powerful physiological actions. From the point of view of selectivity it is of interest that *p*-guanidinobenzoyl-plasmin is considerably more stable (200-fold) than the thrombin derivative although those enzymes are homologous (Chase and Shaw, 1969). Because of these observations a more extended study of aromatic analogues has been carried out. Both sulphonic esters and carboxylic esters have been examined with various positively charged groups intended to function as the guanidino group in arginine does, namely, by initial interaction with the ionized aspartic acid conserved in the primary specificity-site of trypsin-like enzymes.

Sulphonic acid analogues are of most recent interest. If sulphonylation of the active centre serine takes place, the protease is permanently inactivated since the deacylation does not take place, in contrast to carboxylate acylation. (Apparently

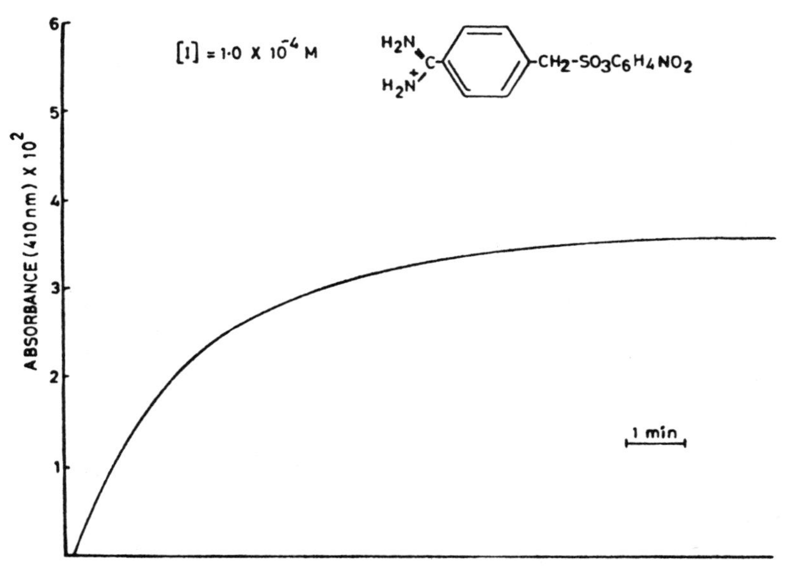

Figure 2.5 Action of thrombin (3×10^{-6} M) on *p*-nitrophenyl-*p*-amidinophenylmethanesulphonate (I), pH 7.4, 25 °C. Stoichiometric nitrophenol release accompanying inactivation. (Reproduced with permission from Wong and Shaw, *Arch. Biochem. Biophys.*, **161**, 536–43 (1974) by courtesy of Academic Press.)

the binding region of the active centre for a water molecule to function in hydrolysis is not available.)

p-Nitrophenyl-*p*′-amidinophenylmethane sulphonate provides an interesting example of selective action since, among a group containing plasmin, trypsin, urokinase, plasma kallikrein, and thrombin, only the latter is inactivated (Wong and Shaw, 1974). A stoichiometric amount of nitrophenol is released in the process and activity is not recoverable (figure 2.5). The *meta*-amidino isomer does not inactivate thrombin. However, both reagents have good affinity for the enzymes examined, acting as competitive inhibitors with K_i values in the 10^{-4}-10^{-5} M range. The only susceptible enzyme has, in fact, lower affinity than several of the others. In this case it is clear that there are geometrical differences in the active

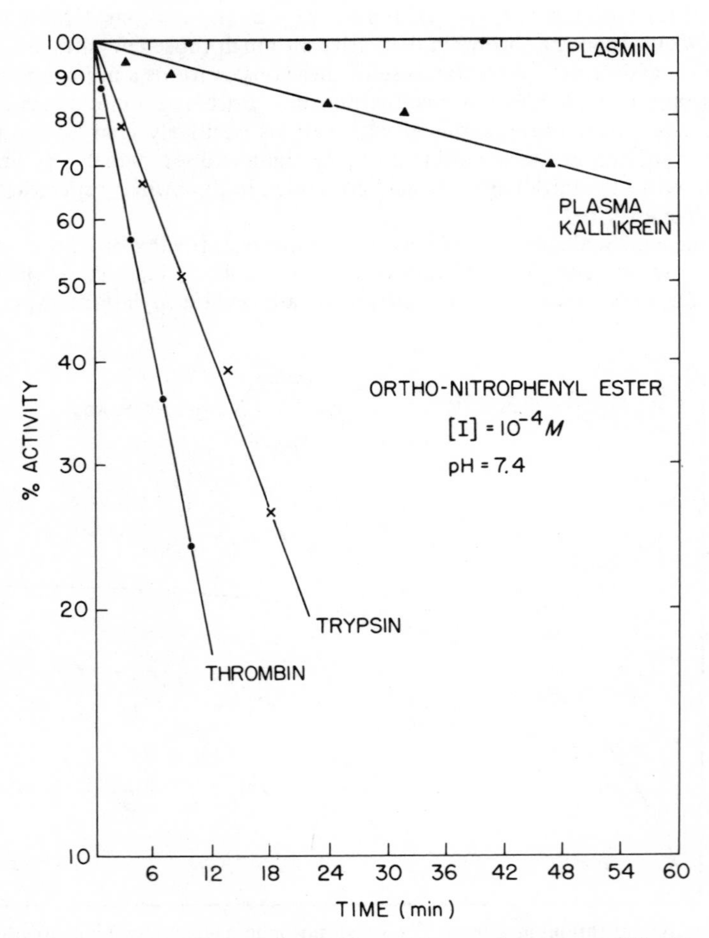

Figure 2.6 Relative rates of inactivation of thrombin, trypsin, plasma kallikrein, and plasmin by *o*-nitrophenyl-*p*-amidinophenylmethanesulphonate at pH 7.4. (Reproduced with permission from Wong and Shaw, *Arch. Biochem. Biophys.*, **176**, 113–18 (1976) by courtesy of Academic Press.)

centres of these trypsin-like enzymes such that within the enzyme–inhibitor complex, suitable juxtaposition for bonding to the active centre serine is provided only in the case of thrombin. This is probably due in part to the rigidity of the inhibitor in contrast with the flexibility of the natural amino acid side chains. The departing group, *p*-nitrophenol, plays some role in the steric restriction which has to be dealt with by the proteases since the *ortho* isomer is less selective. Not only thrombin but, at a slower rate, now trypsin and plasma kallikrein are inactivated (figure 2.6). This indicates some difference in the departing group region of these enzymes (Wong and Shaw, 1976).

In addition to being relatively rigid, the esters studied have other differences from natural substrates. One important variable is the distance from the positive charge to the atom involved in acyl- (sulphonyl-) enzyme formation. The normal substrate conformation in the active centre of trypsin is considered to be in an extended form (Kreiger *et al.*, 1974) and consequently a comparison is made of various inhibitor lengths as in figure 2.7, which summarises a variety of structure-

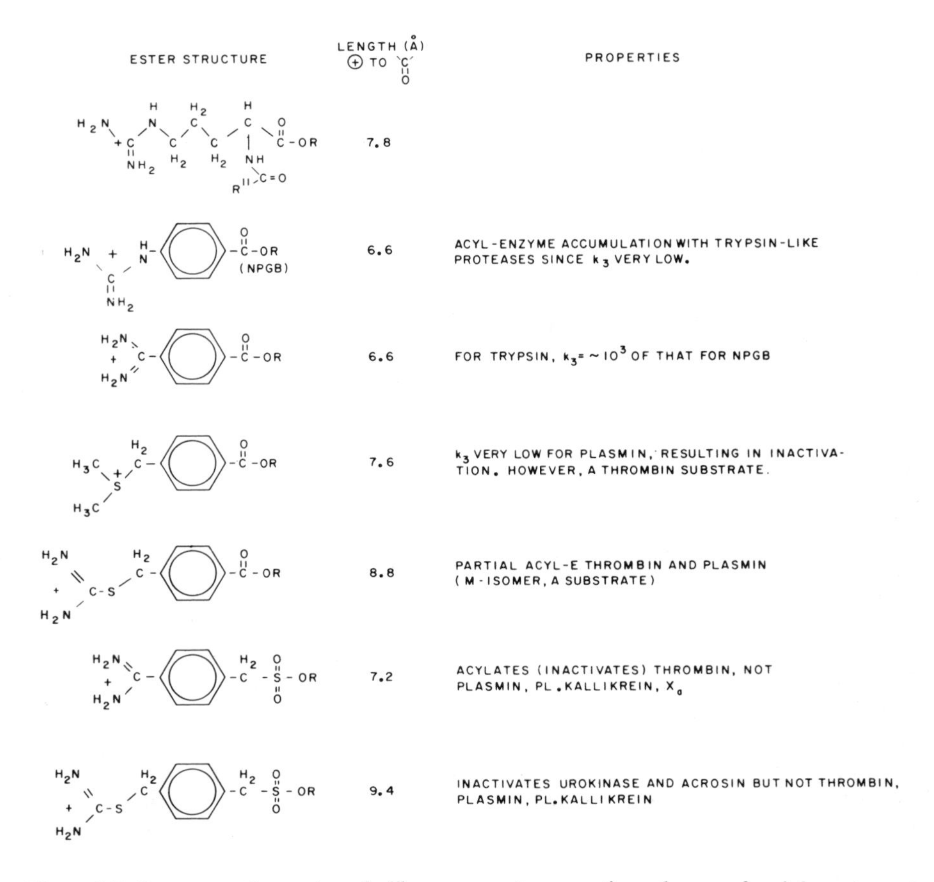

Figure 2.7 Response of some trypsin-like enzymes to aromatic analogues of arginine esters.

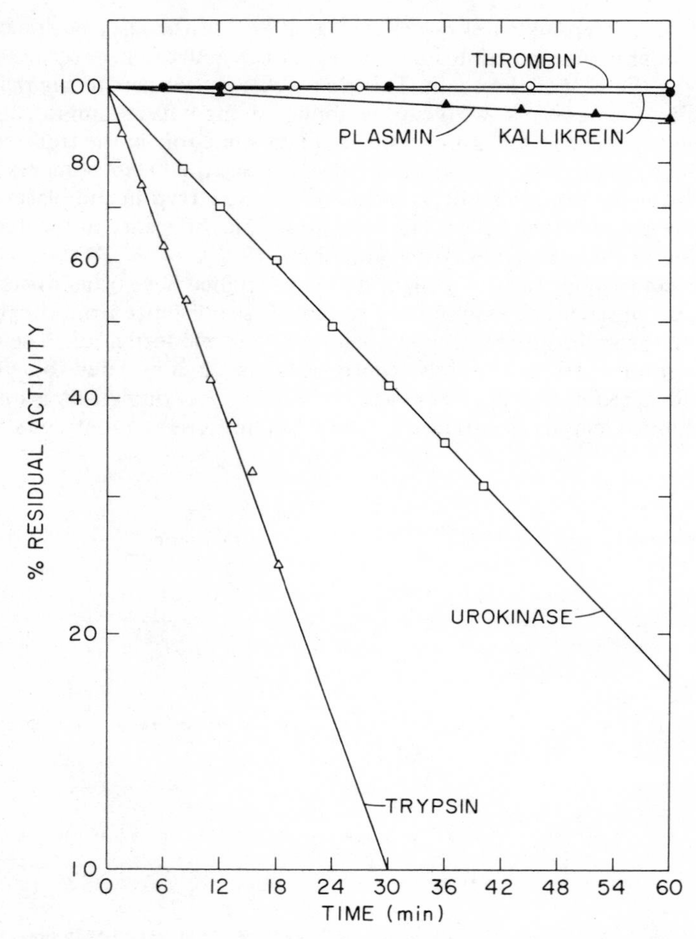

Figure 2.8 The action of the isothiouronium ester (figure 2.7, last compound) at 5×10^{-5} M on various trypsin-related proteases at pH 7.4 in 0.1 M Tris chloride, pH 7.4; enzyme concentration $\times 10^{7}$ M for bovine trypsin, 6.4; human plasmin, 4.8; human plasma kallikrein, 2.0; bovine thrombin, 16; and human urokinase, 51.

activity findings. The fact that inhibitors of a different length than arginine can form acyl- or sulphonyl-enzymes is consonant with the idea that the active centres have some flexibility in their primary specificity sites, a conclusion supported by the hydrolysis of esters of ornithine and homoarginine which are, respectively, shorter and larger than arginine in charge to carbonyl distance.

Among the esters studied, the nitrophenyl ester of the isothiouronium sulphonate provided another example of selectivity. Although essentially inert relative to thrombin, plasmin and plasma kallikrein, it readily inactivates urokinase (figure 2.8) (Wong *et al.*, 1978).

DISCUSSION

Since there are at least two approaches to selectivity, as demonstrated above, it would seem profitable for further progress to exploit both in a single class of reagents to take advantage of the geometrical differences existing in the primary specificity site of these closely related enzymes. Affinity labelling is very sensitive to geometry. Furthermore, the departing group region has hardly been utilized in this work. It seems likely that great selectivity can be achieved with relatively small synthetic inhibitors and that these may be therapeutically useful. The dangers of non-specific and toxic side reactions should diminish with inhibitors which are effective at micromolar concentrations or even higher dilutions.

The utilisation of amino acid sequence formation is obvious in drug design for selective protease inhibitors. The structural bases for other differences that may be utilised are not evident now, although evidence for their existence can be obtained as described above. Exploration of these differences is somewhat empirical at the present time but this may change. A more favourable situation would be to have available three-dimensional structures of the active centres of the proteases of interest for model building. Technological improvements permitting rapid determination of protein structures from small crystals through the use of synchrotron radiation are in the making. Interactive display systems that permit a chemist to have access to protein structures in a way conducive to understanding topographical detail are already at hand. Departure from known structures for the design of new irreversible ligands has not yet become a way of life (see, however, Chen and Plapp, 1978) but should become an essential part of inhibitor design.

REFERENCES

Abeles, R. H. and Maycock, A. L. (1976). *Accounts chem. Res.*, **9**, 313
Aldridge, W. N. and Reiner, E. (1972). *Enzyme Inhibitors as Substrates*, North-Holland, London
Bajusz, S., Barabas, E., Szell, E. and Bagdy, D. (1975). In *Peptides: Chemistry, Structure and Biology* (ed. R. Walter and J. Meienhofer), Ann Arbor Science Publishers, Ann Arbor, Mich., p. 603
Astrup, T. (1975). In *Proteases in Biological Control* (ed. E. Reich, D. B. Rifkin and E. Shaw), Cold Spring Harbor Press, Cold Spring Harbor, N.Y., pp. 343–55
Baker, B. R. (1967). *Design of Active-Site-Directed Irreversible Enzyme Inhibitors*, John Wiley, New York
Bayley, H. and Knowles, J. R. (1977). In *Methods in Enzymology*, Vol. 46 (ed. W. B. Jakoby and M. Wilchek), Academic Press, New York, pp. 69–114
Bause, E. and Leger, G. (1974). *Z. physiol. Chem.*, **355**, 438
Chase, T., Jr and Shaw, E. (1969). *Biochemistry*, **8**, 2212
Chen, W.-S. and Plapp, B. V. (1978). *Biochemistry*, **17**, 4916
Chen, K. C. S. and Tang, J. (1972). *J. biol. Chem.*, **247**, 2566
Coggins, J. R., Kray, W. and Shaw, E. (1974). *Biochem. J.*, **138**, 579
Coleman, P., Kettner, C. and Shaw, E. (1979). *Biochim. biophys. Acta*, **569**, 41
Cuatrecasas, P. and Wilchek, M. (1977). In *Methods in Enzymology*, Vol. 46 (ed. W. B. Jakoby and M. Wilchek), Academic Press, New York, pp. 358–62
Cuatrecasas, P., Wilchek, M. and Avfinsen, C. B. (1969). *J. biol. Chem.*, **244**, 4316
Dawid, J. B., French, T. C. and Buchanan, J. M. (1963). *J. biol. Chem.*, **238**, 2178
Eshdot, Y., McKelvy, J. F. and Sharon, N. (1973). *J. biol. Chem.*, **248**, 5892
Hartman, F. C., Welch, M. H. and Norton, I. L. (1973). *Proc. natn. Acad. Sci. U.S.A.*, **70**, 3721

Holcenberg, J. S., Ericsson, L. and Roberts, J. (1978). *Biochemistry*, **17**, 411
Iwanaga, S., Wallen, P., Grondahl, N. J., Henschen, A. and Blombach, B. (1969). *Eur. J. Biochem.*, **8**, 189
Jakoby, W. B. and Wilchek, M. (eds) (1977). *Methods in Enzymology*, Vol. 46, Academic Press, New York
Kettner, C. and Shaw, E. (1977). In *Chemistry and Biology of Thrombin* (ed. R. L. Lundblad, K. G. Mann and J. W. Fenton), Ann Arbor Science Publishers, Ann Arbor, Mich. pp. 129–43
Kettner, C. and Shaw, E. (1978). *Biochemistry*, **17**, 4778
Kettner, C. and Shaw, E. (1979*a*). *Thrombosis Res.*, **14**, 969
Kettner, C. and Shaw, E. (1979*b*). *Biochim. biophys. Acta*, **569**, 31
Kettner, C., Springhorn, S., Shaw, E., Muller, W. and Fritz, H. (1978). *Z. physiol. Chem.*, **359**, 1183
Kitz, R. and Wilson, I. B. (1962). *J. biol. Chem.*, **237**, 3245
Krieger, M., Kay, L. M. and Stroud, R. M. (1974). *J. molec. Biol.*, **83**, 209
Leary, R., Larsen, D., Watanabe, H. and Shaw, E. (1977). *Biochemistry*, **16**, 5857
Magnusson, S., Petersen, T. E., Sottrup-Jensen, L. and Claeys, H. (1975). In *Proteases and Biological Control* (ed. E. Reich, D. B. Rifkin and E. Shaw), Cold Spring Harbor Press, Cold Spring Harbor, N.Y., pp. 123–49
Mares-Guia, M. and Shaw, E. (1965). *J. biol. Chem.*, **240**, 1579
Mares-Guia, M. and Shaw, E. (1967). *J. biol. Chem.*, **242**, 5782
Mares-Guia, M., Shaw, E. and Cohen, W. (1967). *J. biol. Chem.*, **242**, 5777
Matthews, B. W., Sigler, P. B., Henderson, R. and Blow, D. M. (1967). *Nature, Lond.*, **214**, 652
Naider, F., Bohok, Z. and Yariv, J. (1972). *Biochemistry*, **11**, 3202
Ong, E. B., Shaw, E. and Schoellmann, G. (1965). *J. biol. Chem.*, **240**, 694
Pisano, J. J. (1975). In *Proteases and Biological Control* (ed. E. Reich, D. B. Rifkin and E. Shaw), Cold Spring Harbor Press, Cold Spring Harbor, N.Y., pp. 199–222
Poulos, T. L., Alden, R. A., Freer, S. T., Birktoft, J. J. and Kraut, J. (1976). *J. biol. Chem.*, **251**, 1097
Rajagopalan, T. G., Stein, W. H., and Moore, S. (1966). *J. biol. Chem.*, **241**, 4295
Reich, E. (1975). In *Proteases in Biological Control* (ed. E. Reich, D. B. Rifkin and E. Shaw), Cold Spring Harbor Press, Cold Spring Harbor, N.Y., pp. 333–41
Rando, R. R. (1977). In *Methods in Enzymology*, Vol. 46 (ed. W. B. Jakoby and M. Wilchek), Academic Press, New York, pp. 28–41
Schoellmann, G. and Shaw, E. (1962). *Biochem. biophys. Res. Commun.*, **7**, 36
Schoellmann, G. and Shaw, E. (1963). *Biochemistry*, **2**, 252
Schroeder, D. D. and Shaw, E. (1971). *Arch. Biochem. Biophys.*, **142**, 340
Shaw, E. (1970). *Physiol. Revs*, **50**, 244
Shaw, E. (1975). In *Proteases and Biological Control* (ed. E. Reich, D. B. Rifkin and E. Shaw), Cold Spring Harbor Press, Cold Spring Harbor, N.Y., pp. 455–65
Shaw, E. and Ruscica, J. (1971). *Arch. Biochem. Biophys.*, **145**, 484
Shaw, E. and Woolley, D. W. (1954). *J. Pharm. exp. Therap.*, **111**, 43
Sottrup-Jensen, L., Zajdel, M., Claeys, H., Petersen, T. E. and Magnusson, S. (1975). *Proc. natn. Acad. Sci., U.S.A.*, **72**, 2577
Takagi, T. and Doolittle, R. F. (1974). *Biochemistry*, **13**, 750
Tang, J. (1971). *J. biol. Chem.*, **246**, 4510
Titani, K., Fujikawa, K., Enfield, D. L., Ericsson, L. J., Walsh, K. A. and Neurath, H. (1975). *Proc. natn. Acad. Sci. U.S.A.*, **72**, 3082
Unkless, J. C., Tobia, A., Ossowski, L., Quigley, J. P., Rifkin, D. B. and Reich, E. (1973). *J. exp. Med.*, **137**, 85
Whitaker, J. R. and Perez-Villasenor, J. (1968). *Arch. Biochem. Biophys.*, **124**, 70
Wong, S.-C. and Shaw, E. (1974). *Arch. Biochem. Biophys.*, **161**, 536
Wong, S.-C. and Shaw, E. (1976). *Arch. Biochem. Biophys.*, **176**, 113
Wong, S.-C., Green, G. D. J. and Shaw, E. (1978). *J. med. Chem.*, **21**, 456

3
Transition-state analogues

Gustav E. Lienhard (Department of Biochemistry, Dartmouth Medical School, Hanover, New Hampshire 03755, USA)

INTRODUCTION

This article is divided into two parts. In the first, the application of the transition-state theory of kinetics to enzymic catalysis is described. This application forms the basis for defining the class of compounds known as transition-state analogues and of understanding why they should bind very strongly to enzymes. The treatment is a brief one which largely follows that of Wolfenden (1976). More extensive descriptions of the applications of transition state theory to enzymic catalysis and discussions of the implications thereof are given in articles by Wolfenden (1972, 1976, 1978) and Lienhard (1973).

In the second part, guidelines for the design of transition-state analogues are presented and illustrated with examples from the author's own work. Readers who wish to familiarise themselves with the large number of analogues that are now known for a variety of enzymes should consult the review articles by Lienhard (1972), Lindquist (1975) and Wolfenden (1976). A review article about transition-state analogues as drugs will appear shortly (Stark *et al.*, 1979).

THEORY

The uncatalysed transformation of a reactant S into a product P will proceed predominantly by the pathway with the lowest energy barrier. In the transition-state theory of kinetics, the structure of highest energy on this pathway is referred to as the transition state (figure 3.1, $S^{\ddagger}$ + E). An enzyme which catalyses the same reaction must do so by providing a pathway with a transition state of lower energy (figure 3.1, $ES^{\ddagger}$). Consequently, the hypothetical reaction in which the transition state of the enzymic reaction ($ES^{\ddagger}$) dissociates to form the transition state of the non-enzymic reaction ($S^{\ddagger}$) and free enzyme (E) must have a low value for its equilibrium constant. In other words, $S^{\ddagger}$ must bind strongly to the enzyme.

An estimate of the affinity of the enzyme for $S^{\ddagger}$ can be made in the follow-

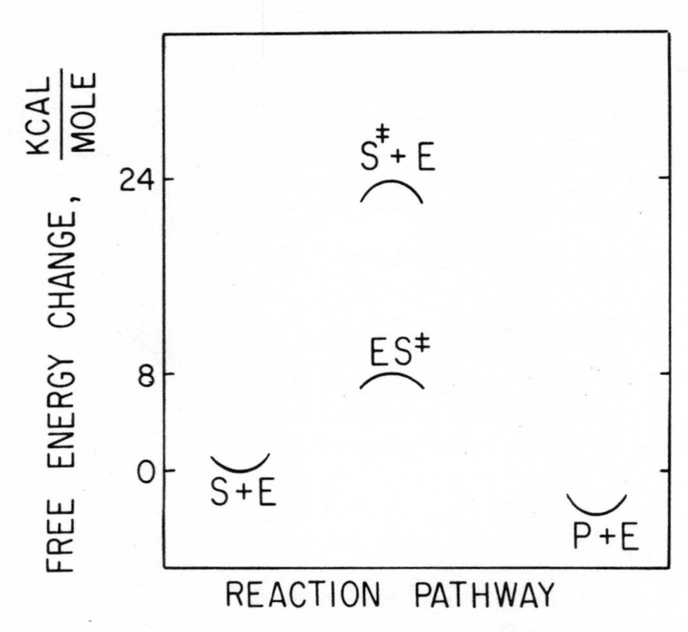

Figure 3.1 Free energy changes for an enzymic reaction and the corresponding uncatalysed reaction. k_n and k_{cat}/K_m (see text) were taken as 10^{-5} s^{-1} and 10^7 M^{-1} s^{-1} at 25 °C, respectively.

ing way. According to transition-state theory, the transition state is in equilibrium with the ground state of the reactant(s). Moreover, its breakdown to product(s) occurs at the same rate for all reactions; the universal rate constant for this process is equal to $\bar{k}T/\bar{h}$, where $\bar{k}$ is the Boltzman constant and $\bar{h}$ the Planck constant. Thus, the first-order rate constant for the uncatalysed reaction (k_n) is directly proportional to the equilibrium constant for formation of $S^‡$ ($K_n^‡$) (scheme 3.1). The second-order rate constant for the enzymic reaction, which

$$S \underset{}{\overset{K_n^‡}{\rightleftharpoons}} S^‡ \xrightarrow{\bar{k}T/\bar{h}} P$$

$$\text{Rate} = \frac{\bar{k}T}{\bar{h}} K_n^‡ [S]$$

$$S \xrightarrow{k_n} P$$

$$\text{Rate} = k_n [S]$$

$$k_n = \frac{\bar{k}T}{\bar{h}} K_n^‡$$

Scheme 3.1

is equal to the ratio of the catalytic constant (k_{cat}) to the Michaelis constant (K_m), is related in the same way to the equilibrium constant for formation of ES‡ ($K_e^‡$) (scheme 3.2). The combination of these two reactions gives an overall scheme (scheme 3.3) that includes the equilibrium for dissociation of ES‡ to S‡ and E and shows that the dissociation constant for this equilibrium (K_T) is equal to the ratio of the rate constant for the non-enzymic reaction to that for the enzymatic reaction.

$$E + S \overset{K_e^{\ddagger}}{\rightleftharpoons} ES^{\ddagger} \xrightarrow{\bar{k}T/\bar{h}} E + P$$

$$E + S \xrightarrow{k_{cat}/K_m} E + P$$

$$\frac{k_{cat}}{K_m} = \frac{\bar{k}T}{\bar{h}} K_e^{\ddagger}$$

Scheme 3.2

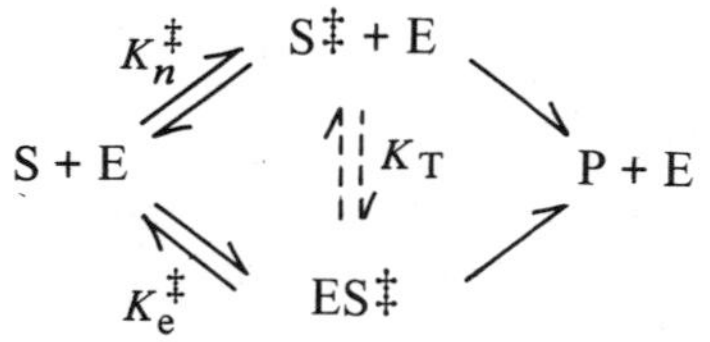

$$\begin{array}{ccccc} & K_n^{\ddagger} \nearrow\!\!\swarrow & S^{\ddagger} + E & \searrow & \\ S + E & & \updownarrow K_T & & P + E \\ & K_e^{\ddagger} \searrow\!\!\nwarrow & ES^{\ddagger} & \nearrow & \end{array}$$

$$K_T = K_n^{\ddagger}/K_e^{\ddagger} = k_n/(k_{cat}/K_m)$$

Scheme 3.3

Accurate values of K_T for the various enzymic reactions are generally not available simply because the rates of the corresponding non-enzymic reactions are too slow to allow determination of the values of k_n. For the few cases in which k_n has been estimated, the values of K_T range from 10^{-12} to 10^{-16} M (Jencks, 1969; Wolfenden, 1972). It seems likely that the value of K_T for most enzymic reactions will be at least as low as this range. Thus, the affinity of an enzyme for the transition state of the corresponding uncatalysed reaction is enormous.

An enzymic reaction involving two substrates can be described similarly (scheme 3.4). In this case, the rate constant for the corresponding uncatalysed reaction (k_n) is a second-order rate constant, and the rate constant for the enzymic reaction is a third-order one that is equal to the turnover number for the enzyme (k_{cat}) divided by the product of Michaelis constants for each substrate (K_{m1}, K_{m2}). Note that here the hypothetical dissociation of the enzymic transi-

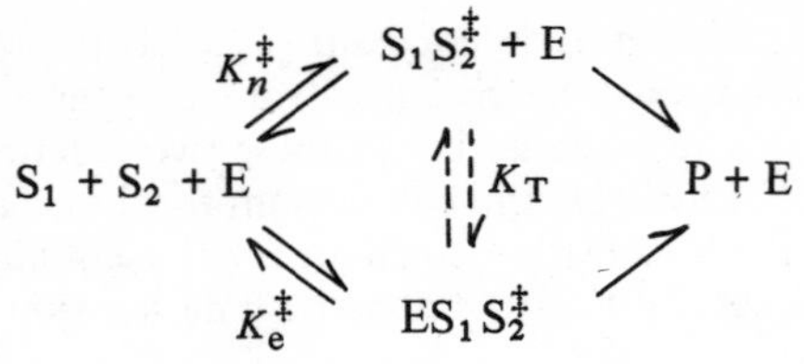

$$
\begin{array}{ccccc}
 & K_n^{\ddagger} \nearrow\!\!\swarrow & S_1S_2^{\ddagger} + E & \searrow & \\
S_1 + S_2 + E & & \uparrow\downarrow K_T & & P + E \\
 & K_e^{\ddagger} \searrow\!\!\nwarrow & ES_1S_2^{\ddagger} & \nearrow &
\end{array}
$$

$$K_T = K_n^{\ddagger}/K_e^{\ddagger} = k_n/(k_{cat}/K_{m1}K_{m2})$$

Scheme 3.4

tion state yields a non-enzymic transition state consisting of both substrates ($S_1S_2^{\ddagger}$).

Schemes 3.3 and 3.4 follow directly from the transition-state theory of kinetics without any assumptions. It should be recognised that, although the substrate portion of the transition state of the enzymic reaction is probably often similar in structure to the corresponding transition state for the uncatalysed reaction, the schemes do not require such similarity. For example, the enzymic and non-enzymic reactions may proceed by fundamentally different mechanisms. Alternatively, both the enzymic and non-enzymic reactions may proceed through similar transition states for the chemical interconversion; however, the rate-limiting step for the enzymic reaction, and thus the transition state associated with the value of k_{cat}/K_m, may be dissociation of the enzyme–product complex. In cases such as these, there are reasons, which have been presented elsewhere (Wolfenden, 1972, 1976, 1978; Lienhard *et al.*, 1971; Lienhard, 1973), to conclude that the value of the equilibrium constant for dissociation of that enzymic transition state in which the covalent changes are occurring to yield a non-enzymic transition state of similar structure for the substrate will generally be even smaller than the value of K_T from schemes 3.3 or 3.4.

PRACTICE

A transition-state analogue is a stable compound which resembles in structure the substrate portion of the enzymic transition state for chemical change. Because the affinity of the enzyme for this transition state is so large, even a crude analogue, the only type that can be synthesised, should bind to the enzyme very strongly.

The design of a transition-state analogue for an enzyme is based on knowledge of the mechanism of the enzymic reaction. Because the transition state itself possesses partially broken and/or made covalent bonds, it is generally impossible to design a stable compound which mimics the transition state closely. It is usually more productive to focus on the high-energy, metastable intermediates in the mechanism. Often, stable analogues of these can be envisaged. Since a metastable intermediate is closer in structure to the transition state than the substrate itself (Hammond, 1955), an analogue of such an intermediate is a crude transition-state analogue.

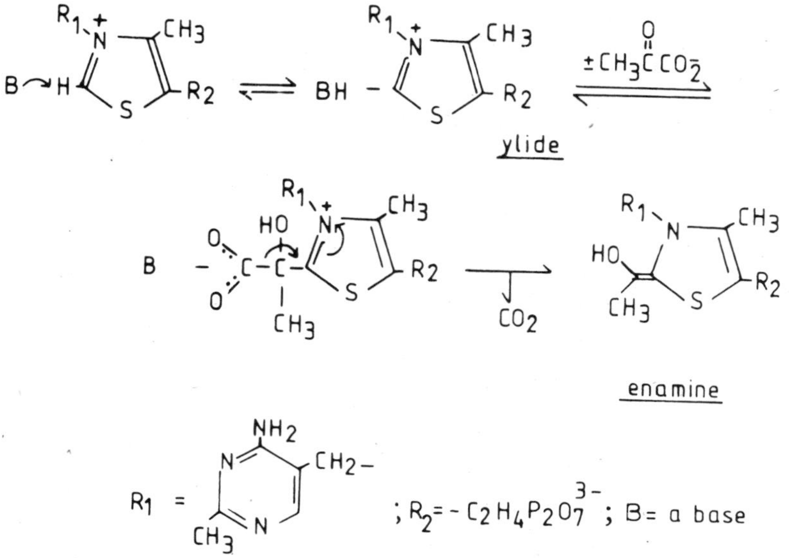

ylide

enamine

; $R_2 = -C_2H_4P_2O_7^{3-}$; B = a base

Figure 3.2 Initial steps in the thiamine pyrophosphate-dependent decarboxylation of pyruvate.

An example of this approach is provided by our design of an analogue for thiamine pyrophosphate-dependent enzymic reactions (Gutowski and Lienhard, 1976). The basic mechanisms for thiamine pyrophosphate-dependent enzymic reactions have been elucidated through investigations of both model reactions and the enzymic ones. In figure 3.2, the initial steps in the decarboxylation of pyruvate are shown. The metastable intermediates in this mechanism are known from model studies to be the ylide, formed by proton transfer from the thiazolium nucleus of thiamine pyrophosphate, and the enamine, which is the immediate product of the decarboxylation of pyruvate. Note that in contrast with the

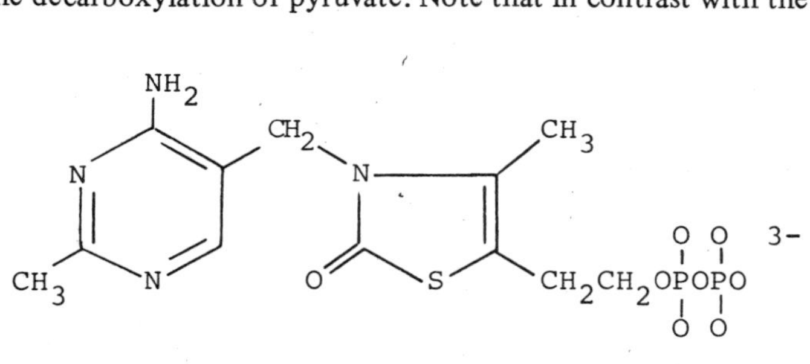

$$\frac{K_i}{K_s} = \frac{<5 \times 10^{-10}\ \text{M}}{10^{-5}\ \text{M}} < 5 \times 10^{-5}$$

Figure 3.3 Thiamine thiazolone pyrophosphate, a transition state analogue for thiamine pyrophosphate.

positively charged thiazolium ring of thiamine pyrophosphate and of its pyruvate adduct, the ring in these intermediates bears no net charge. This consideration led us to view thiamine thiazolone pyrophosphate as a stable analogue of these metastable intermediates (figure 3.3). We synthesised this compound and tested it as an inhibitor of the pyruvate dehydrogenase complex from *Escherichia coli*. The compound binds so tightly that it has only been possible to estimate an upper limit for its dissociation constant, of 5×10^{-10} M. By contrast, the value of the dissociation constant for thiamine pyrophosphate is 10^{-5} M. Thus, the analogue binds at least 20 000 times more strongly than the cofactor. Thiamine pyrophosphate slows the rate at which the thiazolone binds to the enzyme, and an analysis of this effect is consistent with the notion that both compounds bind to the same site.

Many enzymes catalyse reactions involving two substrates. Through analysis of the steady-state kinetics, it usually is possible to determine whether reaction occurs within a ternary complex of the enzyme and both substrates or through the sequential transfer of a portion of one substrate to the enzyme followed by reaction of the modified enzyme with the other substrate (Cleland, 1970). In the former case, a reasonable assumption is that the two substrates are bonded to one another in the transition state. Since the pathway to the transition state involves the bringing together of both substrates from dilute solution, a suitable analogue is simply a molecule which possesses, in an appropriate spatial relationship, the

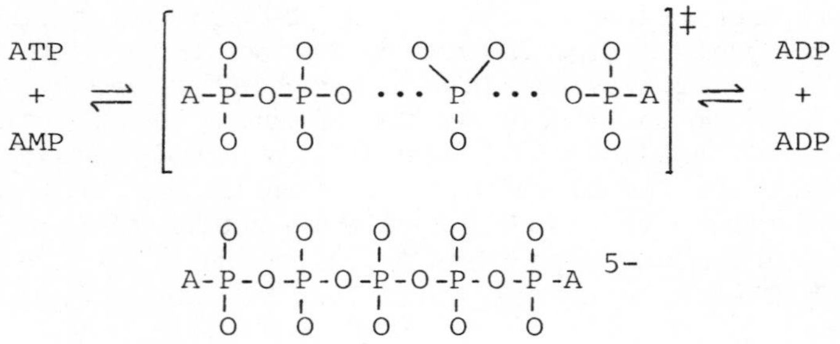

ATP + AMP ⇌ [A-P(O)(O)-O-P(O)(O)-O ··· P(O)(O)(O) ··· O-P(O)(O)-A]‡ ⇌ ADP + ADP

A-P(O)(O)-O-P(O)(O)-O-P(O)(O)-O-P(O)(O)-O-P(O)(O)-A $^{5-}$

$$K_i/K_{mATP} \cdot K_{mAMP} = 2 \times 10^{-9}\ M/10^{-4} \cdot 6 \times 10^{-4}\ M^2$$

Figure 3.4 The adenylate kinase reaction and P^1,P^5-di(adenosine-5′)pentaphosphate.

binding determinants for *both* substrates. Such analogues have also been termed 'multisubstrate' analogues (Wolfenden, 1972).

An example of an analogue of this type is one described for the enzyme adenylate kinase (Lienhard and Secemski, 1973). This enzyme catalyses the transfer of the terminal phosphoryl group of ATP to AMP; the reaction almost certainly proceeds by the direct displacement at the terminal phosphorus of ATP by a phosphoryl oxygen atom of AMP (figure 3.4). The compound P^1, P^5-di(adenosine-5′)pentaphosphate is a potent inhibitor of the enzyme. The value of its dissociation constant, 2.5×10^{-9} M, is about 10^{-4} times that for either substrate. This fact, and the finding that its inhibition is competitive with respect to both ATP and AMP, indicate that P^1, P^5-di(adenosine-5′)pentaphosphate binds simultaneously to the sites for the adenosine moieties of both substrates.

This example illustrates a convenient feature of multisubstrate analogues: it is not necessary to mimic the transition-state structure of the functional group undergoing reaction in order to achieve tight binding. Here, the phosphoryl group undergoing transfer presumably is pentavalent, with a trigonal bipyramidal geometry of oxygen atoms about it in the transition state (figure 3.4); the pyrophosphate group which substitutes for this transition-state function in the analogue bears no obvious detailed resemblance. The pyrophosphate group presumably serves simply as a linking function of the correct size and structure to allow the two adenosine rings to bind simultaneously. Because of this feature, the design of multisubstrate analogues does not require a detailed knowledge of the enzymic mechanism and is a relatively easy task. In many cases, it may be possible to synthesise an effective inhibitor simply by the condensation of appropriate derivatives of the two substrates themselves.

The above discussion has dealt with the design of transition-state analogues *de novo*. It is important to recognise that enzymic reactions fall into mechanistic classes, within each of which all the reactions proceed by the same fundamental mechanism, and that enzymic specificity among the enzymes within a class resides largely in interactions between enzyme and substrate which are distal to those between the enzyme and the reacting functional group in the transition state. Thus, once a transition-state analogue for one enzyme has been identified, analogues for other enzymes of the same class but with different specificities can probably be prepared by introducing appropriate substituents on the critical functional group that resembles and/or replaces the reacting functional group in the transition state. Since over 50 possible transition-state analogues for enzymes catalysing a variety of reaction types are now known (Lindquist, 1975; Wolfenden, 1976), there is a solid background on which to base this mimicry of mimicry.

In this regard, I would point out one area deserving of further investigation.

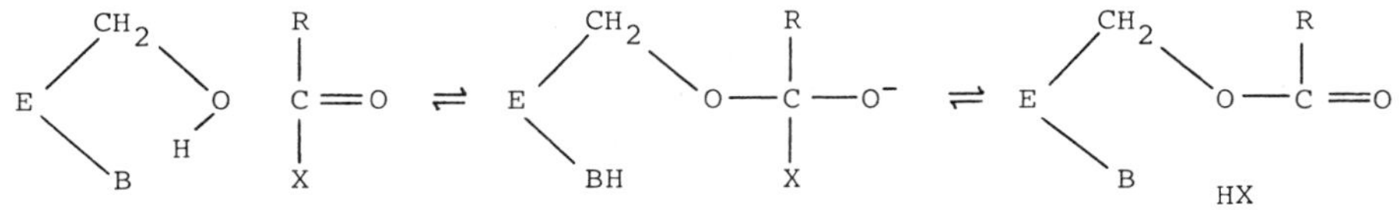

Figure 3.5 Mechanism for acylation and, in the reverse, deacylation of enzyme seryl residue (CH_2OH) with assistance by a basic group of the enzyme (B).

There are a number of enzymes which catalyse acyl transfer reactions via the intermediacy of an ester or thiol ester with a seryl or cysteinyl residue. The transition states for acylation and deacylation of these enzymes probably resemble the metastable tetrahedral intermediates expected to occur on the reaction pathway (figure 3.5). For some of these enzymes, boron acids have been found to be strong inhibitors (table 3.1). The explanation for the strong inhibition lies in the

Table 3.1 Boron acid inhibitors

Enzyme	Inhibitor	K_i/K_s†
Chymotrypsin‡	$C_6H_5CH_2CH_2-B(OH)_2$	7×10^{-3}
	$C_6H_5CONHCH_2-B(OH)_2$	10^{-3}
Subtilisin§	$C_6H_5-B(OH)_2$	5×10^{-3}
Acetylcholinesterase¶	$(CH_3)_3\overset{+}{N}CH_2CH_2CH_2-B(OH)-CH_3$	10^{-4}
Glutamyl transpeptidase‖	$^{-}O_2CCH(\overset{+}{N}H_3)CH_2O-B(OH)_2$	10^{-6}
Lipase††	$CH_3(CH_2)_{16}CH_2-B(OH)_2$	4×10^{-4}

†Ratio of the value of the dissociation constant for the boron acid to the value of the dissociation constant for the analogous substrate or amide inhibitor.
‡Koehler and Lienhard (1971); Lindquist and Nguyen (1977). The evidence for the structure of the boronic acid analogue of *N*-benzoylglycine is also consistent with the structure $C_6H_5C(NH)OCH_2-B(OH)_2$, an imido ester (private communication from D. Matteson, Washington State University).
§Lindquist and Terry (1974).
¶Koehler and Hess (1974).
‖Tate and Meister (1978). Here K_i has been calculated on the assumption that the value of the equilibrium constant for the reaction

$$\text{Serine} + B(OH)_3 \rightleftharpoons H_2O + \text{Serine ester of boric acid}$$

is 1. K_S is the dissociation constant for L-glutamine (Thompson and Meister, 1977).
††Garner (1979).

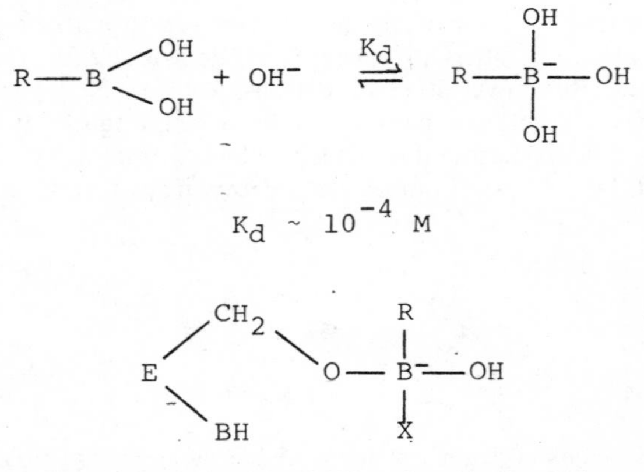

Figure 3.6 The equilibrium for ionization of a boronic acid; the stable structure analogous to the tetrahedral intermediate in figure 3.5.

fact that boron acids form stable tetrahedral adducts and thus bind to the nucleophilic residue at the active site of the enzyme in a structure resembling the tetrahedral intermediate (figure 3.6) (Koehler and Lienhard, 1971; Rawn and Lienhard, 1974; Matthews *et al.*, 1975). To date, only one very simple boron acid analogue of a peptide has been synthesised, and there is some doubt as to whether the

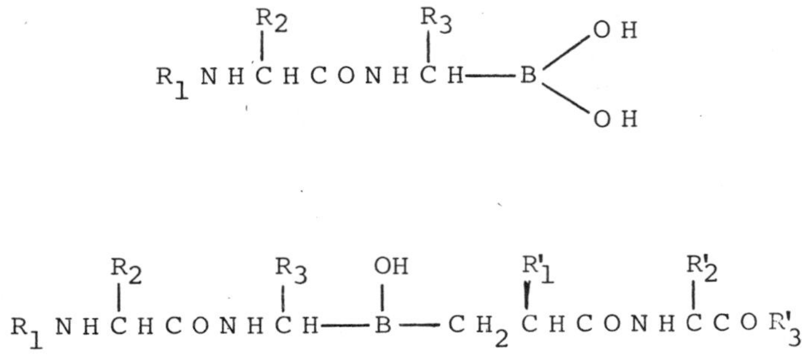

$$R_1NHCH(R_2)CONHCH(R_3)\text{—}B(OH)_2$$

$$R_1NHCH(R_2)CONHCH(R_3)\text{—}B(OH)\text{—}CH_2CH(R'_1)CONHCH(R'_2)COR'_3$$

Figure 3.7 Boronic (upper) and borinic (lower) acid analogues of peptides.

proposed structure is the correct one (table 3.1). Future work in this area might profitably be devoted to the synthesis and testing of boron acid analogues of amino acids and peptides (figure 3.7) with structures appropriate to the known specificities of the proteases in this class. It can be confidently predicted that such compounds will be extremely potent inhibitors.

REFERENCES

Cleland, W. W. (1970). In *The Enzymes*, 3rd ed., vol. 2 (ed. P. D. Boyer), Academic Press, New York, pp. 1–66
Garner, C. W. (1979). *Fedn Proc. Fedn Am. Socs exp. Biol.*, **38**, 729
Gutowski, J. A. and Lienhard, G. E. (1976). *J. biol. Chem.*, **251**, 2863
Hammond, G. S. (1955). *J. Am. chem. Soc.*, **77**, 334
Jencks, W. P. (1969). *Catalysis in Chemistry and Enzymology*, McGraw-Hill, New York, p.4
Koehler, K. A. and Hess, G. P. (1974). *Biochemistry*, **13**, 5345
Koehler, K. A. and Lienhard, G. E. (1971). *Biochemistry*, **10**, 2477
Lienhard, G. E. (1972). *A. Rep. med. Chem.*, **7**, 249
Lienhard, G. E. (1973). *Science, N.Y.*, **180**, 149
Lienhard, G. E. and Secemski, I. I. (1973). *J. biol. Chem.*, **248**, 1121
Lienhard, G. E., Secemski, I. I., Koehler, K. A. and Lindquist, R. N. (1971). *Cold Spring Harbor Symp. quant. Biol.*, **36**, 45
Lindquist, R. N. (1975). In *Drug Design*, Vol. 5 (ed. E. J. Ariens), Academic Press, New York, pp. 24–80
Lindquist, R. N. and Nguyen, A. C. (1977). *J. Am. chem. Soc.*, **99**, 6437
Lindquist, R. N. and Terry, C. (1974). *Arch. Biochem. Biophys.*, **160**, 135
Matthews, D. A., Alden, R. A., Birktoft, J. J., Freer, S. T. and Kraut, J. (1975). *J. biol. Chem.*, **250**, 7120
Rawn, J. D. and Lienhard, G. E. (1974). *Biochemistry*, **13**, 3124
Stark, G. R., Greenwell, P. and Johnson, R. (1979). *Pharmac. Therap.*, in the press
Tate, S. S. and Meister, A. (1978). *Proc. natn. Acad. Sci. USA.*, **75**, 4806
Thompson, G. A. and Meister, A. (1977). *J. biol. Chem.*, **252**, 6792
Wolfenden, R. (1972). *Accounts chem. Res.*, **5**, 10
Wolfenden, R. (1976). *A. Rev. Biophys. Bioengng*, **5**, 271
Wolfenden, R. (1978). In *Transition States of Biochemical Processes* (ed. R. D. Gandour and R. L. Schowen), Plenum, New York, pp. 555–78

4

Quantitative structure-activity relationships as applied to enzyme inhibitors

J. K. Seydel and K.-J. Schaper (Borstel Research Institute, 2061 Borstel, FDR)

INTRODUCTION

The greatest dream of every medicinal chemist is the 'rational' design of new drugs. We all know that this wish will probably remain unfulfilled for a long time. Nevertheless progress in our understanding of drug action and in drug development is proceeding rapidly, being stimulated especially by the methods applied in quantitative structure–activity relationship (QSAR) analysis. These computerised statistical methods try to explain the observed variance in biological effect of certain classes of compounds as a function of molecular changes caused by the substituents. There are two main applications of QSAR analysis:

(I) The predictive aspect

(a) the extrapolation of correlations found between structural parameters and biological activity to propose the synthesis of more active derivatives;

(b) to avoid the synthesis and testing of derivatives with the same 'equivalent' activity.

(II) The diagnostic aspect

(a) to answer the question of what is the rate-determining step;

(b) to support or reject a mode of action hypothesis.

Several approaches are used in QSAR studies (Seydel and Schaper, 1979; Martin, 1978). The essential procedures are the non-parametric methods such as that of Free and Wilson (1964), the parametric method of Hansch (1969), discriminant analysis (Martin, 1974) and the pattern recognition technique (Kirschner and Kowalski, 1979) which is essentially a substructure analysis. The choice of method is dependent on the quality of the biological data, number of compounds tested, degree of variance in the results, and ratio of the time needed for synthesis and biological testing respectively.

The most widely used approach continues to be the linear free energy-related model, the so-called Hansch approach (Hansch and Dunn, 1972), where the variance in biological effect (ΔBE) is explained by the variance of certain linear free energy-related substituent constants which describe the change in lipophilic/hydrophilic ($\Delta L/\Delta H$), electronic (ΔE), steric (ΔE_s) or other properties of the parent molecule induced by the substituents:

$$\Delta BE \equiv \log \frac{c_1}{c_2} = f(\Delta L/H; \Delta E, \Delta E_s) \tag{4.1}$$

The variance in electronic properties can be expressed by Hammett σ constants, pK_a values or spectroscopic data like chemical shift from NMR spectroscopy (p.p.m.), intensity and wavelength of signals from IR or UV spectra, etc. The change in lipophilicity can be described by R_m values from reversed-phase chromatography, by the partition coefficient $\log P$ or the substituent constant π which is defined as $\pi = \log P_X - \log P_H$, where X refers to the substituted derivative and H to the parent compound (Fujita *et al.*, 1964). The steric influence of the substituents can be expressed by the Taft steric constant E_s (Taft, 1956), the van der Waals volume, or the molar volume according to Bondi (1964) and Exner (1967). In a stepwise linear multiregression analysis, the physicochemical parameters which have the most statistical significance for the explanation of the observed biological variance undergo evaluation as follows:

$$\log BE = a + b\pi$$
$$\log BE = a + c\sigma$$
$$\log BE = a + dE_s$$
$$\log BE = a + b\pi + c\sigma$$

etc.

The major problem in such an analysis is the complexity of the biological system. Drug action is proceeding in several branched or consecutive reaction steps. Many equilibria can be involved and each of the steps may be rate limiting. Due to these complex events, the derivation of a significant quantitative structure-activity relationship is often difficult. Even if a statistically sound equation is obtained by regression analysis, it might still be difficult to interpret the results by physicochemical and biochemical reasoning. These difficulties can be reduced if a 'simple' biological model is selected, that is if we do not determine activity data for the drug from whole animal studies or isolated organs. In this case, the number of possible steps in which drug action may be involved is decreased. In most cases, also, the error in the biological activity data becomes smaller. Isolated enzymes are therefore very reliable systems in which to study inhibitory effects and derive quantitative structure–activity relationships; they may provide information about the rate-limiting step and the physicochemical forces involved in the interaction between drug and enzyme proteins. It is therefore not surprising that the QSAR techniques have been applied more and more to the description of the biological activity of enzyme inhibitors. A tabulated compilation is given by Seydel and Schaper (1979).

Instead of reviewing the several examples given in the literature, we wish to discuss the derivation of quantitative structure–activity relationships of some enzyme inhibitors which have been studied in great detail in our laboratories (Seydel, 1971, 1975, 1979; Miller *et al.*, 1972; Seydel *et al.*, 1973; Bock *et al.*, 1974). These are the sulphonamides (SA), sulphones and *p*-aminobenzoic acid derivatives known to inhibit the bacterial 7,8-dihydropteroate synthetase (E_2 in figure 4.3).

We believe that SA are amongst the most completely studied enzyme inhibitors using QSAR analysis. It was possible to derive separate QSAR for a cell-free folate-synthesising enzyme system (Miller *et al.*, 1972; Seydel *et al.*, 1973; Bock *et al.*, 1974; Seydel, 1975), for whole cell bacteria (*Escherichia coli*) (Seydel, 1971) and for animals and humans (Seydel *et al.*, 1973; Seydel, 1979). This enables us to understand the variation in biological activity in the various steps and systems arising as a function of the variation of the physicochemical properties of the substituents. QSAR studies make it possible to estimate the rate-limiting step and they allow a more rational synthesis of biologically active SA with certain pharmacokinetic properties.

If we limit the discussion to an *in-vitro* system—in the case of antibacterial enzyme inhibitors, to bacteria in a synthetic culture medium and to isolated enzyme proteins—the possible rate-determining steps may be reduced to (see figure 4.1):

(1) permeation of bacterial cell wall;
(2) binding to the specific receptor protein;
(3) reaction with the enzyme proteins or with substrate to form a product (after reversible interaction with the enzyme protein (2)).

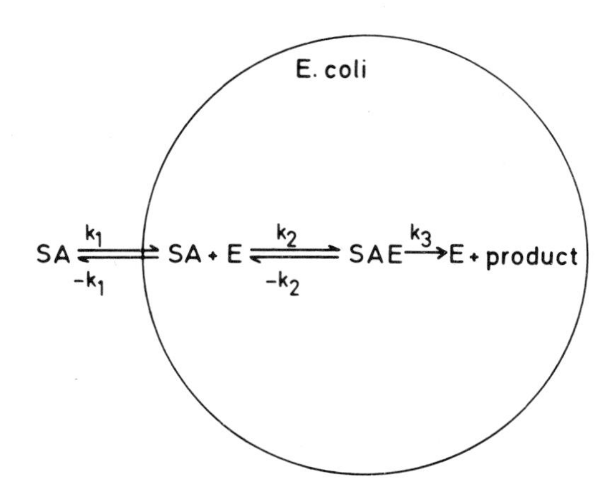

Figure 4.1 Possible rate determining steps in sulphonamide (SA) and diaminodiphenylsulphone (DDS, dapsone) action on *E. coli.*

WHOLE CELL INHIBITORY ACTIVITIES

Whole cell inhibitory activities have been determined for a series of substituted N^1-phenylsulphanilamides (**I**) and N^1-pyridylsulphanilamides (**II**) using serial dilution techniques and *E. coli* as a test organism (Seydel, 1971; Miller *et al.*, 1972).

H_2N–⟨ ⟩–SO_2–NH–⟨ ⟩–R $\qquad\qquad$ H_2N–⟨ ⟩–SO_2–NH–⟨ N ⟩–R

I $\qquad\qquad\qquad\qquad$ **II**

It has been proved that the minimum inhibitory concentrations (MIC) obtained are paralleled by activity or rate constants obtained from bacterial growth kinetic studies (Garrett *et al.*, 1969; Seydel, 1971). This is a necessary condition if MIC values are used in free energy-related equations where the log MIC is plotted against the pK_a value of the SA (or against the chemical shift of the amino group protons of the precursor amines; in this series substituted anilines are used for the synthesis of the N^1-phenylsulphanilamides (Seydel, 1971)). The results are summarized in figure 4.2, where log MIC is plotted against the pK_a and chemical

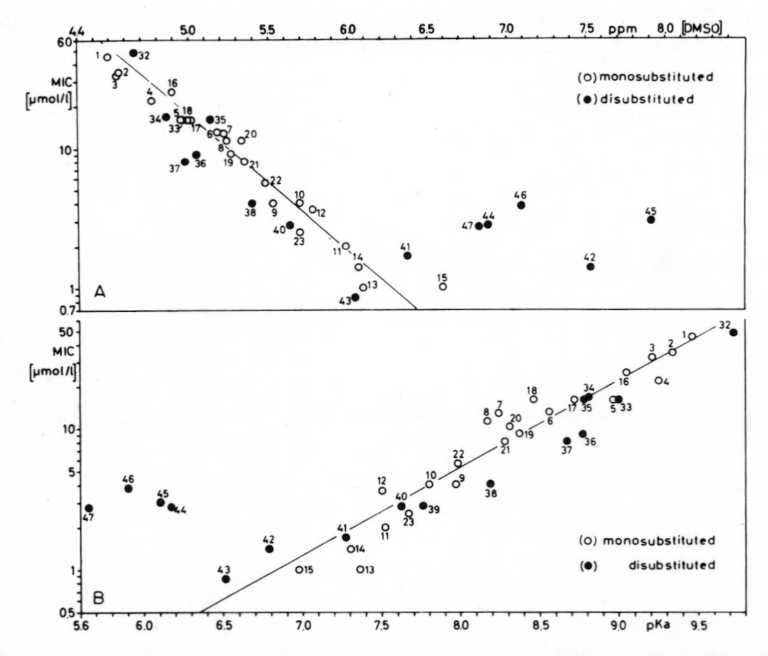

Figure 4.2 Correlation between (A) chemical shift (p.p.m., solvent dimethylsulphoxide (DMSO)) of the amino group protons of various substituted precursor anilines and (B) the acid pK_a of the corresponding SA and the log of the corresponding MIC (*E. coli*) of the SA.

shift (p.p.m., NMR) of the SA and amines respectively. As in many cases where relationships between physicochemical parameters and biological activites are observed (Ariens, 1969), an optimum is seen. After a certain threshold in pK_a ($\leqslant 6.7$) or in p.p.m. ($\leqslant 6.5$) is reached, a cut-off in increase in activity is observed. This phenomenon is well known for activities depending on lipid-water partition or for a change in activity with number of carbon atoms in aliphatic chains (Bray, 1950; Malcolm-Dyson and May, 1959; Hansch and Clayton, 1973). It has also been described for SA by Bell and Roblin (1942); a bell-shaped curve was obtained when the pK_a values of different SA were plotted against the logarithm of the *in-vitro* activities.

The parallelism between figure 4.2(A) and (B) is striking. The difference from the curve drawn by Bell and Roblin is the linearity in correlation between pK_a or p.p.m. respectively and log MIC (see also Cammarata, 1967). Previous optima in SA structure–activity relationships all resulted from the combination of several SA series (Bell and Roblin, 1942). The presented results, with a closely homologous series of SA, offer a convincing confirmation of the existence of such an optimum, but no bell-shaped curve is obtained (figure 4.2).

The results for QSAR studies discussed so far within a homologous series of SA and its precursor amines enable one to predict, within certain limits, the biological activity of a SA before its synthesis and thus to carry out a more rational design. The results, however, give no answer to questions about the rate-limiting step and the mechanism of action of this type of drug. There is a significant correlation between log MIC and pK_a or p.p.m., but this does not necessarily mean that the electron density at the N^1-nitrogen atom is directly involved in interaction with the receptor. It also does not explain the observed maximum or cut-off in activity. The reason might be a change in the rate-limiting step, for example limitation in permeability of the bacterial cell wall if the drug becomes increasingly ionised with decreasing pK_a. More information can only be obtained, therefore, by studying SA inhibitory activity directly at the enzyme proteins involved.

DETERMINATION OF CELL-FREE ACTIVITY (i_{50}) OF SA AND SULPHONES

Recognition of the importance of *p*-aminobenzoic acid (PABA) in folic acid metabolism and the ability of certain folate derivatives and/or products to reverse the effect of SA or sulphones has led to a general theory of SA action. In principle, this theory states that SA are competitive inhibitors of the enzymic incorporation of PABA into folic acid (Rogers *et al.*, 1964). The sequential pathway of folic acid synthesis has been evaluated by Jaenicke and Chan (1960), Brown (1962), Shiota *et al.* (1964) and Ortiz and Hotchkiss (1966) and is given in figure 4.3. It has also been shown that bacterial (Wolf and Hotchkiss, 1963; Ortiz and Hotchkiss, 1966), plant (Kazuo and Okinaka, 1968; Mitsuda and Suzuki, 1969) and plasmodial (Ferone, 1973) cell-free folate synthesising extracts are inhibited by SA.

Brown (1962) has measured inhibitory activities of several SA in such a cell-free folate-synthesising system. He reported that, in general, they are proportional to the antibacterial activities (MIC) of these compounds. The relatively small number and the heterogeneity of the compounds studied, however, have limited their usefulness for a detailed quantitative structure–activity analysis. Similarly, the

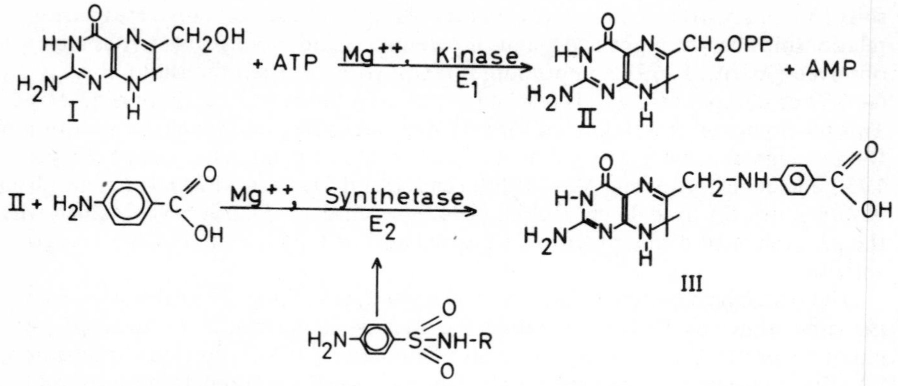

I + ATP —(Mg^{++}, Kinase, E_1)→ II + AMP

II + H_2N–C$_6$H$_4$–COOH —(Mg^{++}, Synthetase, E_2)→ III

↑ H_2N–C$_6$H$_4$–SO_2–NH–R

Figure 4.3 Reaction scheme of the biosynthesis of 7,8-dihydropteroic acid (**III**).

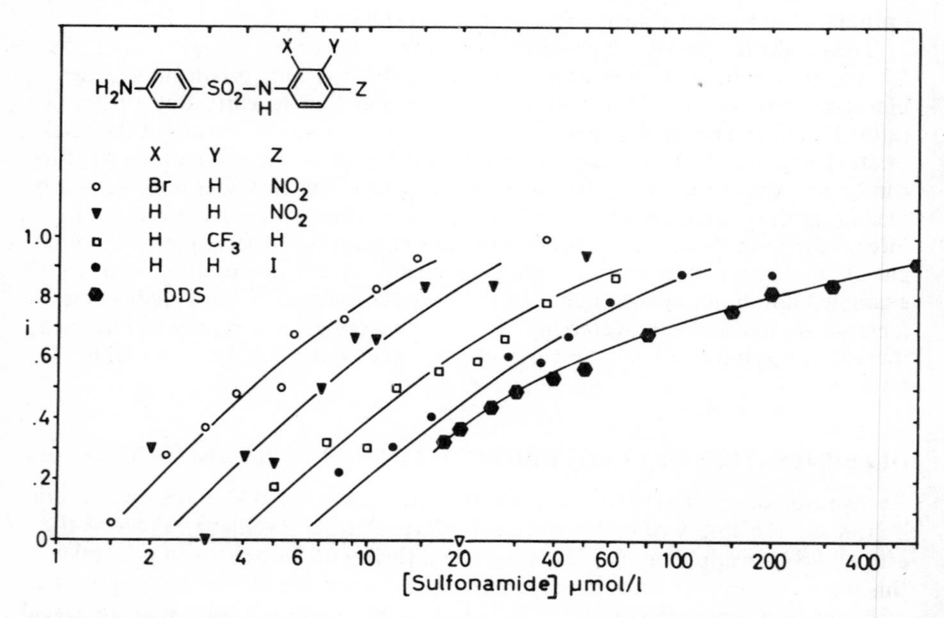

Figure 4.4 Determination of the inhibitory activity of several N^1-phenylsulphanilamides and DDS by the single-point method. Plots of the fraction of inhibition i of cell-free folate synthesis versus logarithm of the SA or DDS concentration were used to estimate i_{50} values. The conditions were as described by Miller *et al.* (1972).

data have limited utility in assessing the role of permeability in antibacterial activity.

To gain more information about structural requirements and the physicochemical forces which are directly related to the enzyme-SA or enzyme–diaminodiphenylsulphone (DDS) interaction, and to find an explanation for the observed optimum in whole cell activity (MIC), the inhibitory activity of a closely homologous series of SA and of DDS has been studied in our laboratories. Results of some typical (single-point) experiments conducted to determine the inhibitory activities of several N^1-phenyl (I) and N^1-pyridyl (II) sulphanilamides and DDS have been published (Miller *et al.*, 1972). An example is given in figure 4.4. It shows a plot of fractional inhibition of folate synthesis against inhibitory concentration used, that is a dose–response curve. Amount of folate synthesised was determined after a fixed time interval (3 or 5 h). From such plots, the SA or DDS concentration causing 50 per cent inhibition was determined. The tested SA (~ 50 in total) show a wide range in physicochemical properties (for example, lipophilicity (π, log P), electron distribution (pK_a), steric effects (E_s)) and exhibit an approximately 25-fold variation in inhibitory activity in the cell-free system.

COMPARISON OF WHOLE CELL INHIBITORY ACTIVITIES AND CELL-FREE SYSTEM INHIBITORY ACTIVITIES

If the inhibitory activity in whole cell systems (MIC) is plotted in a double-logarithmic system against the activity we have found in the cell-free system (i_{50}) (Miller *et al.*, 1972), a linear relationship is obtained (figure 4.5). It is apparent that, with the exception of four compounds (43–45, 47), all SA and dapsone can be fitted by a single line (Miller *et al.*, 1972).

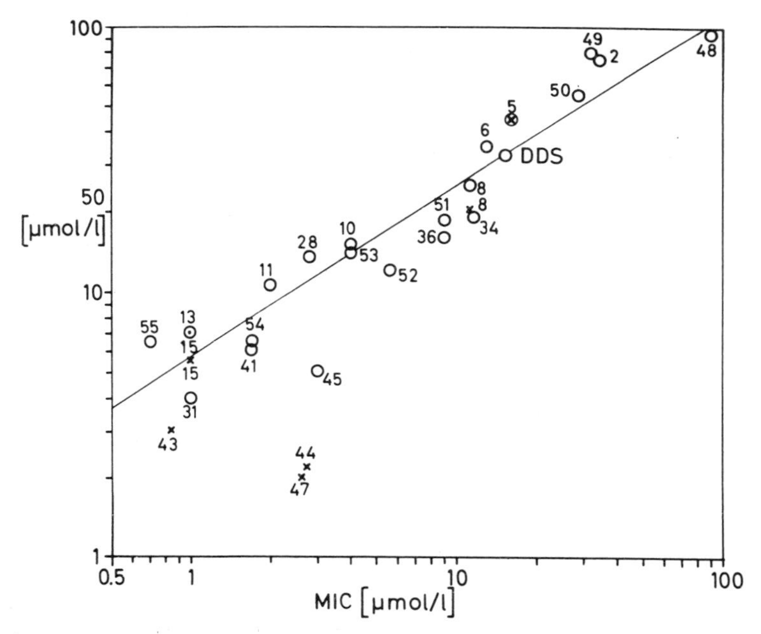

Figure 4.5 The relationship of cell-free inhibitory activities to whole cell inhibitory activities. A single relationship for N^1-pyridyl- and N^1-phenylsulphonamides and DDS is obtained. For discussion of the exceptions (compounds 43–45 and 47) see text. (•, Single-point experiments; ×, kinetic experiments.)

$$\log i_{50} = 0.64 \log \text{MIC} + 0.75$$
$$n = 22 \quad r = 0.95 \quad s = 0.13 \quad F = 187 \tag{4.2}$$

In these equations, n is the number of data points used in the analysis, r is the correlation coefficient, s the standard error of estimate, the F value is the decision statistic of the F test of significance.

This result indicates that the differences in MIC values are paralleled by differences in cell-free activities. Therefore these differences cannot be attributed to permeability factors but must be associated with the reactions of the SA and sulphones catalysed by the enzyme proteins of *E. coli*. It therefore seems most likely that permeability factors do not contribute substantially to the antibacterial activity of the tested compounds *in vitro*.

COMPARISON OF QSAR OF WHOLE CELL AND CELL-FREE ACTIVITIES

The observed variance in MIC values has been correlated and explained by changes in physicochemical properties of the SA. In this correlation especially, the pK_a value was of importance as indicator for changes in electron distribution (Seydel, 1971). The following equation was obtained:

$$\log \text{MIC}_{o,m,p} = (0.66 \pm 0.07)\, pK_a - 4.76$$
$$n = 18 \quad r = 0.92 \quad s = 0.17 \quad F = 96 \tag{4.3}$$

In view of the relationship found between log MIC and log i_{50}, we would expect all physicochemical parameters describing electronic influences of the various substituents which gave successful correlations with MIC data (equation (4.3)) to be equally successful with cell-free activity. In addition, exceptions (figure 4.2) should not be exceptional as far as cell-free activity is concerned. This does indeed appear to be the case.

A similar analysis to explain the observed variance in cell-free activity (i_{50}) resulted in the following equation:

$$\log (i_{50})_{o,m,p} = (0.44 \pm 0.07)\, pK_a - 2.50$$
$$n = 18 \quad r = 0.84 \quad s = 0.17 \quad F = 39 \tag{4.4}$$

The dependencies on structural changes are almost identical in both systems, thus supporting the assumption that interaction with enzyme protein is the rate-limiting step.

Close inspection of figure 4.6, however, reveals that compounds with pK_a values less than 7, that is compounds which are more than 90 per cent ionised, do not fit this relationship as well as the others. They are highly active in the cell-free system; nevertheless they are less active than expected (compounds 44, 47 with circle). These activities have been determined with the single-point method, that is inhibition of folate synthesis was determined after a certain time interval (5 h of incubation). An explanation of the observed smaller activity was found by examination of the kinetics of inhibition.

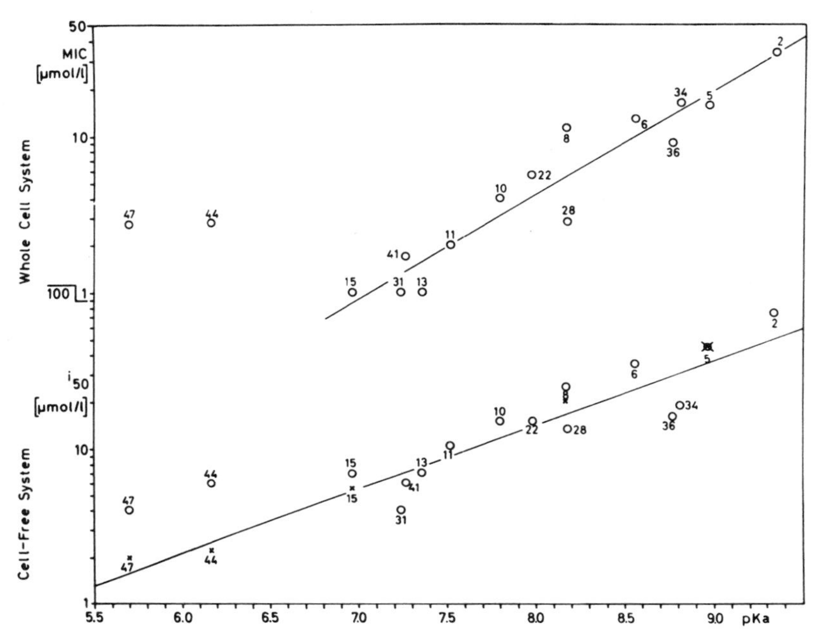

Figure 4.6 The relationship of inhibitory activities to pK_a values of the N^1-sulphonamides. Cell-free inhibitory activities (○, single-point method; ×, kinetic method) of substituted N^1-phenylsulphanilamides exhibit a normal linear free-energy relationship with the N^1- pK_a values. A similar relationship has previously been reported (Seydel, 1971) for whole cell activity. It will be seen that compounds with pK_a values less than approximately 7 (44, 47) do not fit this linear free-energy relationship (compounds 44 and 47 were excluded from regression analysis for log MIC versus pK_a (equation (4.3)).

KINETICS OF INHIBITION IN A CELL-FREE SYSTEM

For these compounds, the kinetics of inhibition was determined in a cell-free system (Miller *et al.*, 1972; Bock *et al.*, 1974). Figure 4.7 shows a typical kinetic experiment in which the amount of folate-like material produced in the reaction mixture as a function of time was determined for several SA concentrations. The concentration causing a 50 per cent inhibition in rate of folate synthesis (k_{50}) was derived from such plots (figure 4.7). The observed k_{50} values are in agreement with the observed i_{50} values (concentration needed to reduce by one-half the amount of folate produced after 5 h of incubation). For the very active SA, however, k_{50} was much smaller than i_{50} for the first 2-3 h interval. Later, the rate of folate synthesis returned to the rate of the control probe (figure 4.8). Figure 4.9 shows the results of kinetic experiments with N^1-4-nitrophenylsulphanilamide where the SA was preincubated with pteridine substrate and the extract prior to the addition of PABA. A degree of inhibition is found which is much higher than that found when all substrates and the inhibitor are combined simultaneously (figure 4.9(left)). If the amount of the pteridine substrate (7,8-dihydro-6-hydroxymethylpterin is 1.7 times higher (figure 4.9(right)), the degree of inhibition in the preincubation experiment is identical with the experiment where all substrates

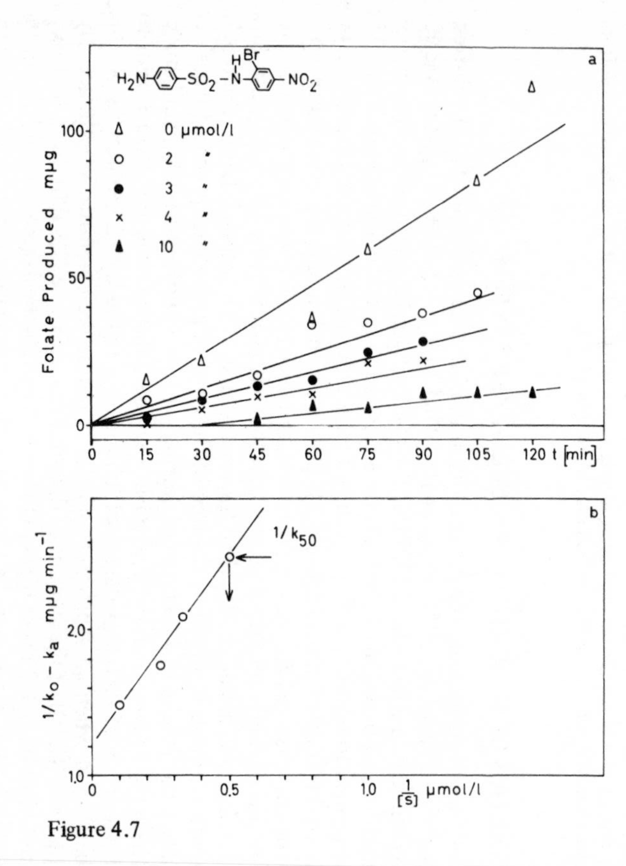

Figure 4.7

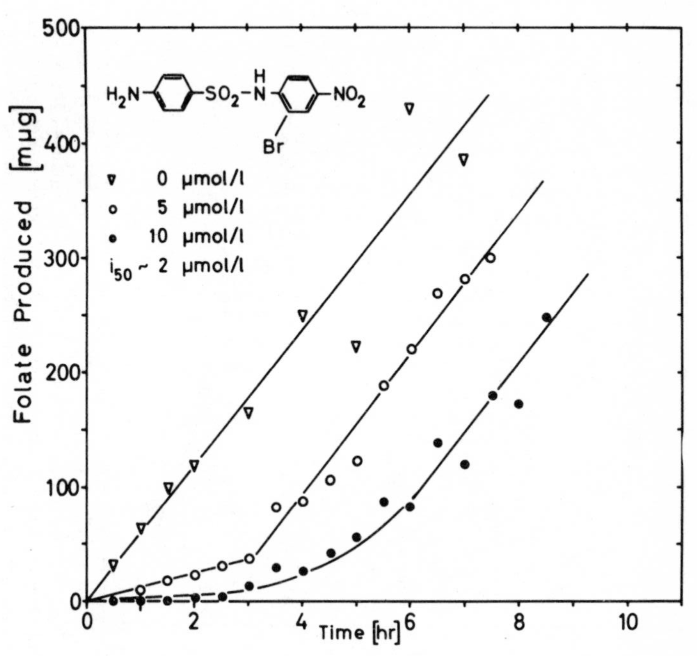

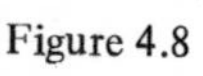
Figure 4.8

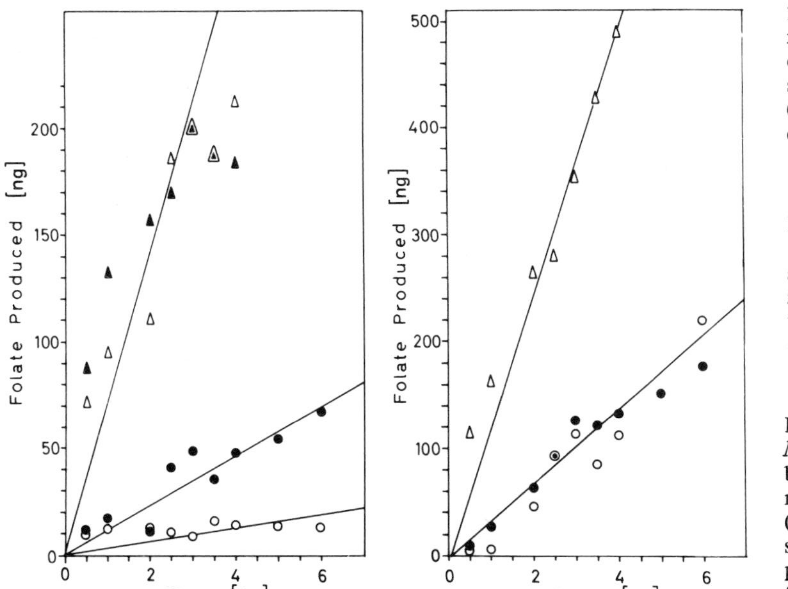

Figure 4.9

Figure 4.7 Determination of the inhibitory activity of 2-bromo-4-nitro-N^1-phenylsulphanilamide by the kinetic method. Rate constants for folate synthesis in the presence of several concentrations of 2-bromo-4-nitro-N^1-phenylsulphanilamide were obtained from kinetic experiments (a) and used in Lineweaver-Burke type plots (b) to obtain estimates of k_{50} (Miller *et al.*, 1972).

Figure 4.8 The kinetics of inhibition of folate synthesis by 2-bromo-4-nitro-N^1-phenylsulphanilamide. For this highly active compound, the kinetics of inhibition was of pseudo-zero order only for a short initial period. After this time, the rate of folate synthesis is similar to that seen in the control (Bock *et al.*, 1974).

Figure 4.9 Kinetics of inhibition of folate synthesis by N^1-4-nitro-phenylsulphonamide with and without preincubation ((a), left). Kinetics of inhibition where the sulphonamide (12 μM) and 7,8-dihydro-6-hydroxymethylpterin (32 μM) were preincubated for 3 h and PABA (20 μM) was subsequently added (○) and where sulphonamide (12 μM), pterin component (32 μM) and PABA (20 μM) were added together to the reaction mixture at zero time (●). The appropriate controls where no sulphonamide was added are (△) and (▲). ((b), right): Kinetics of inhibition by N^1-4-nitrophenylsulphonamide as under (a); however, the concentration of 7,8-dihydro-6-hydroxymethylpterin was 54 μM (○) and (●), control (△) (Bock *et al.*, 1974).

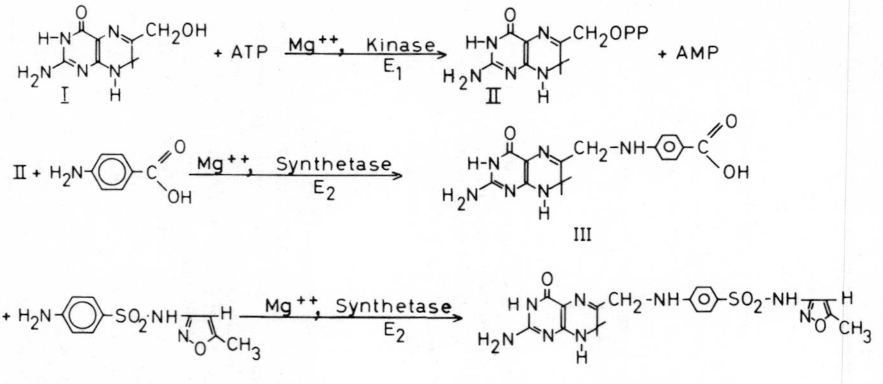

Figure 4.10 Reaction scheme for the biosynthesis of a 7,8-dihydropteroic acid analogue in the presence of sulphonamides (2-sulpha-3-methylisoxazole).

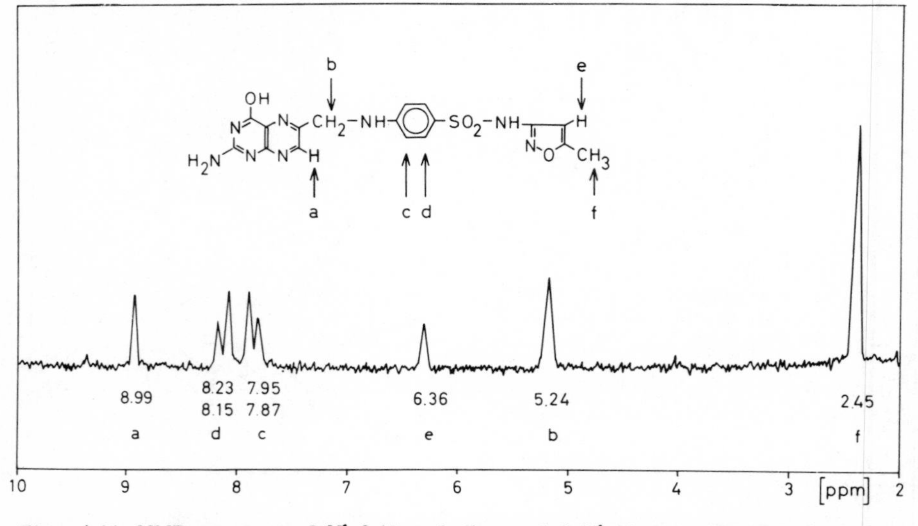

Figur 4.11 NMR spectrum of N^1-3-(5-methylisoxazolyl)-N^4-(6-pterinylmethyl)sulphanilamide in CF_3COOD, 100 MHz, tetramethylsilane being used as internal standard (Bock *et al.*, 1974).

and inhibitors are added simultaneously. This behaviour points to an exhaustion of inhibitor in a reaction with the substrate 7,8-dihydropteridine alcohol (figure 4.10). In the cases of SA and sulphones, we were able to demonstrate that such an analogue of 7,8-dihydropteroic acid is indeed produced (Bock *et al*, 1974). The NMR spectrum of a dihydropteroic acid analogue, where the PABA moiety is replaced by 3-sulpha-5-methylisoxazole, is given in figure 4.11.

If the k_{50} values are used in the regression analysis for the highly active SA

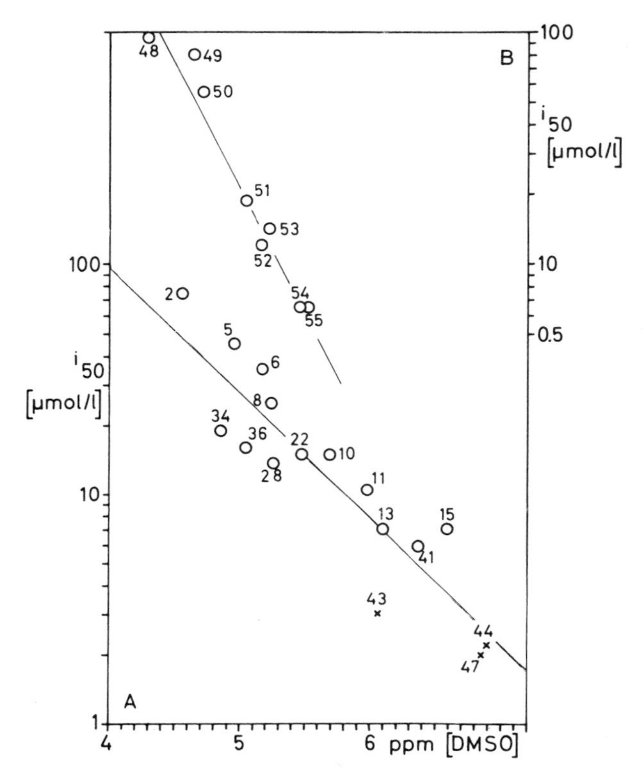

Figure 4.12 The relationship of cell-free inhibitory activities to chemical shift values of sulphonamide precursor amines. Chemical shift values of substituted anilines (A) or 3-amino-pyridines (B) were obtained in dimethylsulphoxide (DMSO) using tetramethylsilane (TMS) as a standard. The correlation of cell-free activities with these chemical shift values is linear over the entire p.p.m. scale. Previously reported correlations (Seydel, 1971) of these chemical shift values with whole cell activities are not linear at higher (> 6.5) p.p.m. values, in a similar manner to the correlation of these activities with pK_a values (figure 4.2).

instead of the i_{50} values, the correlation between the logarithm of the biological activity and the physicochemical parameters becomes more significant (equation (4.5) and figure 4.6) compared to an equation including i_{50} for compounds 44 and 47.

$$\log (i_{50})_{o,m,p} = (0.39 \pm 0.043)\, pK_a - 2.05 \qquad (4.5)$$

$$n = 20 \quad r = 0.90 \quad s = 0.17 \quad F = 82$$

Close inspection of figure 4.12 reveals, however, that compounds 28, 34, 36, 43–44 and 47 still deviate from the regression curve by a nearly constant factor. They are all more active than predicted. These compounds have in common the fact that they bear a substituent in the *o*- or *m*-position. It was realized that these compounds seem to react faster with the substrate to form products (Bock *et al.*, 1974; Seydel *et al.*, 1973). If this 'steric' influence of the substituents is represented by an indicator variable (I) (assigning the value 1 for o- and m-substituted and 0 for p-substituted derivatives), all SA derivatives (type I) can be combined in one equation with improved statistical significance.

$$\log (i_{50})_{o,m,p} = (0.48 \pm 0.03)\ \mathrm{p}K_\mathrm{a} - (0.33 \pm 0.035) I - 2.47 \tag{4.6}$$

$$n = 18 \quad r = 0.97 \quad s = 0.08 \quad F = 123$$

The corresponding correlation equation for whole cell activities (MIC) is

$$\log \mathrm{MIC}_{o,m,p} = (0.68 \pm 0.06)\ \mathrm{p}K_\mathrm{a} - (0.24 \pm 0.08) I - 4.78 \tag{4.7}$$

$$n = 18 \quad r = 0.95 \quad s = 0.14 \quad F = 67$$

THE RATE-LIMITING STEP IN FOLATE INHIBITION

The results lead to the question of whether the rate-limiting step in inhibition of folate synthesis by SA, sulphones and also substituted PABA derivatives is competition for binding sites or rate of formation of the 7,8-dihydropteroic acid analogue (see figures 4.1 and 4.10). The kinetics of analogue formation was therefore studied in the absence of PABA. Decrease in substrate concentration (7,8-

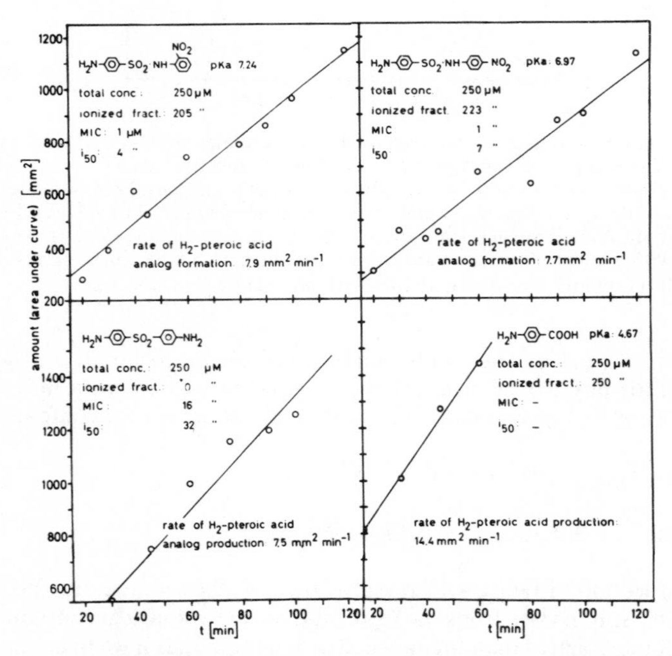

Figure 4.13 Rate of 7,8-dihydropteroic acid analogue formation in presence of various inhibitors and absence of PABA compared with the formation of 7,8-dihydropteroic acid in the presence of PABA and absence of inhibitors.

Table 4.1 Rate of H_2-pteroic acid analogue formation by various N^1-phenylsulphanilamides (SA) in a cell-free system in the absence of PABA at pH 7.71, 37 °C

H_2N-C$_6$H$_4$-$SO_2 \cdot NH$-C$_6$H$_4$-R	SA conc. (μM)		MIC	i_{50}	Rate	pK_a	π	D
	Total	Ionised fraction	(μM)	(μM)	(mm^2 min^{-1})			
4–CH_3	5239	223	21.8	–	6.50	9.25	0.44	0
H	2840	223	16.0	45.0	7.40	8.97	0	0
4–Cl	1242	223	13.0	35.0	7.58	8.56	0.83	0
4–Cl	273	49	13.0	35.0	7.90	8.56	0.83	0
2–Cl	991	223	2.8	13.5	6.35	8.18	0.37	1
2–Cl	341	117	2.8	13.5	6.65	8.18	0.37	1
2–NO_2	250	205	1.0	4.0	7.90	7.24	0.12	1
4–NO_2	250	223	1.0	7.0	7.70	6.97	0.59	0
PABA	250	250	–	–	14.00	4.67	–	–
DDS	250	0	16.0	32.0	7.50	–	log P 0.9	0

Table 4.2 Rate of H_2-pteroic acid analogue formation by various substituted PABA derivatives (5.2×10^{-4} M) in a cell-free system in the absence of PABA, pH 7.7, 37 °C

H_2N–C₆H₃(R)–COOH	Rate × 10^{10} (mol min^{-1})	pK_a	π
H	3.34 ± 0.34	4.67	0
2–NH_2	4.45 ± 0.56†	5.81	−1.14
2–CH_3	3.76 ± 0.35	4.83	0.31
2–NO_2	3.35 ± 0.27	2.78	0.38
2–F	3.31 ± 0.26	4.13	0.46
2–Br	3.75 ± 0.56	3.69	0.66
2–J	3.28 ± 0.26	3.81	0.82
2–OCH_3	3.03 ± 0.27	5.04	−0.38
2–OC_3H_7	1.31 ± 0.07	5.25	1.24
2–OC_6H_{13}	0.92 ± 0.14	5.37	2.00
PABA-ethylester	no formation		

†Insufficient separation on TLC plates

dihydro-6-hydroxymethylpterine) and formation (increase) of analogue was followed by quantitative thin layer chromatography (Bock *et al.*, 1978). The results are given in figure 4.13 and tables 4.1 and 4.2. All SA derivatives studied, including *o*-substituted SA and DDS, show the same rate of analogue formation (figure 4.13), despite the large variance in their inhibitory power (MIC, k_{50}, i_{50}). (Only PABA itself shows a faster rate of incorporation into 7,8-dihydropteroic acid.)

Similar results are obtained for the substituted PABA derivatives which are all totally ionised under the conditions of the experiment (table 4.2). The substituents do not influence rate of incorporation. However, if the substituent in an *o*-position to the carboxy group becomes too bulky, the rate of analogue formation decreases independently of the pK_a or π value (Seydel and Butte, 1977). Therefore the rate-limiting step in folate inhibition by these inhibitors is not the rate of formation of dihydropteroic acid analogue, but must be the different affinity of the inhibitory SA, sulphones and PABA derivatives for the enzyme, dihydropteroic acid synthetase, compared with the affinity of the natural metabolite, PABA.

CONTRIBUTION OF IONISED AND UNIONISED FORMS TO THE INHIBITORY EFFECT

Experiments were performed to determine the molar concentration of PABA (K_m) needed to reduce the inhibitory effect of SA (at constant concentration of its ionised form) on the rate of folate production to one-half of that obtained in the absence of the inhibitors. Some of the results are given in table 4.3. It is obvious that the ionised form of SA is a more powerful inhibitor of folate synthesis than the unionised form. To our surprise, however, a contribution of the unionised form of the SA to the inhibitory action can be derived if the K_m values obtained are considered. There is a significant increase in K_m values, despite the fact that the ionised

Table 4.3

Substance $H_2N-C_6H_4-SO_2\cdot NH-C_6H_4-R$	pK_a	pH at 37 °C	$[SA^-]$ ionised fraction (μmol)	$[SA^n]$ neutral fraction (μmol)	K_m PABA (μmol)
4-Nitro	6.97	7.65	10	2.089	44.4
4-Acetyl	7.52	7.65	10	7.413	43
4-Iodo	8.17	7.73	10	27.55	55.5
4-Chloro	8.56	7.65	10	81.283	69.2
H	8.97	7.65	10	208.914	97.24
4-Methoxy	9.34	7.65	10	489.750	164.35
4-Methyl	9.39	7.65	10	549.597	144.15
H	8.97	7.73	2.72	47.28	24.06
4-Chloro	8.56	7.16	1.92	48.08	23.9
4-Iodo	8.17	7.16	4.45	45.55	28.1
4-Methoxy	9.34	8.30	4.18	45.82	26.1

fraction in this series of substituted N^1-phenylsulphanilamides is kept constant. We tried to quantify, by multiregression analysis, the contribution to inhibition arising from the two molecular species. The results for all compounds studied, including N^1-disubstituted SA and sulphone, are given in equations (4.8) and (4.9).

$$K_{[\mathrm{PABA}]} = (6.61 \pm 0.41)\,[\mathrm{SA}^-] + (0.169 \pm 0.038)\,[\mathrm{SA}^{\mathrm{n}}] - 1.87 \qquad (4.8)$$

$$n = 23 \quad r = 0.96 \quad s = 25.5 \quad F = 131$$

If *ortho* substitution is accounted for by an indicator variable, equation (4.9) is obtained, which is statistically more significant on the 95-97 per cent level:

$$K_{[\mathrm{PABA}]} = (5.76 \pm 0.53)\,[\mathrm{SA}^-] + (0.19 \pm 0.035)\,[\mathrm{SA}^{\mathrm{n}}] + (41.9 \pm 18.5)I - 2.36$$

$$n = 23 \quad r = 0.97 \quad s = 23.2 \quad F = 107 \qquad (4.9)$$

It can be derived from these equations that the inhibitory power of the ionised form is approximately 30 times greater than that of the unionised form; that is, the contribution of the neutral form is less than 5 per cent and could not therefore be detected in a plot where log i_{50} or log MIC was plotted against pK_a (figure 4.2 and equation (4.3)) and a linear correlation obtained. The results of the competition experiments also make it possible to include unionised (disubstituted) SA and sulphones in the general regression equation. It is not possible, however, from the data available, to decide whether all unionised forms, regardless of substituent, have the same molar activity or not. It would seem better to assume an identical and constant contribution of all ionised species. This would help to explain the observed cut-off in increase in activity if total ionisation is approached. The completely ionised molecule, however, cannot penetrate the bacterial cell wall, as is strikingly demonstrated by the total inactivity of sulphanilic acid in whole cell cultures and its very high inhibitory action in cell-free systems (i_{50} = 5 μmol l^{-1}). The cut-off observed in MIC correlations might also be due to exhaustion of SA during the observation period, since only small amounts of the very active SA were present.

The results support the theory of Bell and Roblin (1942), derived from experiments on whole cell activity, suggesting that both forms are active; however, the ionised form is significantly more active. The findings, however, also support the hypothesis of Brueckner (1943) and Cowles (1942) that the unionised form only can permeate the bacterial cell wall.

The results described demonstrate the usefulness of QSAR analysis in enzymic systems in order to explain quantitatively the observed variance in biological activity, to detect exceptions, to determine the binding forces involved (electrostatic, no hydrophobic interaction) and to evaluate the mode of action of enzyme inhibitors.

REFERENCES

Ariens, E. J. (1969). *Il Farmaco*, **24**, 3
Bell, P. H. and Roblin, R. D. (1942). *J. Am. chem. Soc.*, **64**, 2905
Bock, L., Miller, G. H., Schaper, K.-J. and Seydel, J. K. (1974). *J. med. Chem.*, **17**, 23

Bock, L., Butte, W., Richter, M. and Seydel, J. K. (1978). *Anal. Biochem.*, **86**, 238
Bondi, A. (1964). *Phys. Chem.*, **68**, 441
Bray, H. G. (1950). *Biochem. J.*, **47**, 294
Brown, G. M. (1962). *J. biol. Chem.*, **257**, 536
Brueckner, A. H. (1943). *Yale J. Biol. Med.*, **15**, 813
Cammarata, A. (1967). *J. pharm. Sci.*, **56**, 640
Cowles, P. B. (1942). *Yale J. Biol. Med.*, **14**, 599
Exner, O. (1967). *Coll. Czech. Chem. Commun.*, **32**, 1
Ferone, R. (1973). *J. Protozool.*, **20**, 459
Free, S. M. and Wilson, J. W. (1964). *J. med. Chem.*, 7, 395
Fujita, T., Iwasa, J. and Hansch, C. (1964). *J. Am. chem. Soc.*, **86**, 5175
Garrett, E. R., Mielck, J. M., Seydel, J. K. and Kessler, H. J. (1969). *J. med. Chem.*, **12**, 740
Hansch, C. (1969). *Accounts chem. Res.*, **2**, 232
Hansch, C. and Dunn, W. J. III (1972). *J. pharm. Sci.*, **61**, 1
Hansch, C. and Clayton, J. M. (1973). *J. pharm. Sci.*, **62**, 1
Jaenicke, L. and Chan, P. C. (1960). *Angew. Chem.*, **72**, 752
Kazuo, I. and Okinaka, O. (1968). *J. Vitaminol.*, **14**, 170
Kirschner, G. L. and Kowalski, B. (1979). In *Drug Design*, vol. VIII (ed. E. J. Ariens), Academic Press, New York and London, p. 73
Malcolm-Dyson, G. and May, P. (1959). In *Chemistry of Synthetic Drugs* (ed. P. May), Longmans Green, London, p. 526
Martin, Y. C. (1974). *J. med. Chem.*, **17**, 409
Martin, Y. C. (1978). *Quantitative Drug Design*, Marcel Dekker, New York
Miller, G. H., Doukas, P. H. and Seydel, J. K. (1972). *J. med. Chem.*, **15**, 700
Mitsuda, H. and Suzuki, Y. (1969). *Biochem. biophys. Res. Commun.*, **36**, 1
Ortiz, P. J. and Hotchkiss, R. D. (1966). *Biochemistry*, **5**, 67
Rogers, E. F., Clark, R. L., Becker, H. J., Pessolano, A. A., Leanza, W. J., McManus, E. C., Andriuli, F. J. and Cuckler, A. C. (1964). *Proc. Soc. exp. Biol. Med.*, **117**, 488
Seydel, J. K. (1971). *J. med. Chem.*, **14**, 724
Seydel, J. K. (1975). In *Topics in Infectious Diseases*, vol. 1 (ed. J. Drews and F. E. Hahn), Springer, Wien and New York
Seydel, J. K. (1979). *Infection,* 7, suppl. 4, 313
Seydel, J. K. and Butte, W. (1977). *J. med. Chem*, **20**, 439
Seydel, J. K. and Schaper, K.-J. (1979). Chemische Struktur und biologische Activitat von Wirkstoffen, Methoden der Quantitativen Struktur-Wirkung-Analyse, Verlag Chemie, Weinheim
Seydel, J. K., Miller, G. H. and Doukas, P. H. (1973). In *Medicinal Chemistry* (ed. P. Pratesi), Butterworths, London, p. 139
Seydel, J. K., Wempe, E. and Richter, M. (1980). *Int. J. Leprosy*, in press
Shiota, T., Disraely, M. N. and McCann, M. P. (1964). *J. biol. Chem.* **239**, 2259
Taft, R. W. (1956). In *Steric Effects in Organic Chemistry* (ed. M. S. Newman), Wiley, New York, p. 556
Wolf, B. and Hotchkiss, R. D. (1963). *Biochemistry*, **2**, 145

5

Enzyme-induced inactivation of pyridoxal phosphate-dependent enzymes: Approaches to the design of specific inhibitors

Robert A. John (Department of Biochemistry,
University College, P.O. Box 78, Cardiff CF1 1XL, Wales)

INTRODUCTION

Ideally, compounds intended to act as drugs by selectively inhibiting a particular metabolic step should show absolute specificity towards the enzyme catalysing that step. One approach to this ideal is to prepare a compound that not only bears a strong structural resemblance to the normal substrate but also remains unreactive as an irreversible inhibitor until one or more steps of the target enzyme's normal catalytic mechanism have introduced modifications in it. The appeal of this mechanism is that it incorporates two tiers of specificity. Not only must the compound complement the active site of the enzyme but at least part of the normal catalytic mechanism must occur before the irreversible step takes place.

The potential value of the use of this type of inactivation *in vivo* has been pointed out by several authors (Fasella and John, 1969; Rando, 1973; Abeles and Walsh, 1973). Rather unfortunately, different names have been used by the several reviewers of the field. Rando (1974*a*) named them 'k_{cat} inhibitors' whereas Abeles and Maycock (1976) used the term 'suicide enzyme inactivators' and Walsh (1977) 'suicide substrates'. Whilst there is little analogy between this inhibitory mechanism and the voluntary taking of one's own life, the instant appeal of the term 'suicide' seems to make it the most frequently used. This article uses a derivative of the prosaic but informative description used as a title for a recent symposium on the subject, 'Enzyme-activated irreversible inhibitors' (Seiler *et al.*, 1978).

The article will include within its scope any compound where the inactivating mechanism results from changes induced in the substrate by the operation of part of the normal catalytic mechanism. The term 'irreversible' will, inevitably, be imprecisely defined. A functional definition related to the time scale of the experiments must be adopted. Irreversible inhibition or inactivation will be considered to have taken place if, after separation from the compound originally responsible, no significant enzyme activity returns within a matter of days. This definition is preferred to one which requires covalent bond formation between enzyme protein and the compound since the latter would exclude potent and potentially valuable irreversible inhibitors in which the generation of a very tightly bound coenzyme–substrate adduct occurs by operation of the enzyme's mechanism. Also included will be compounds which generate the pyridoxamine form of the coenzyme in enzymes for which only the pyridoxal form is a cofactor. All these types of inactivation are potentially valuable and readily explained in terms of the enzyme's normal mechanism.

A majority of the enzyme-activated inhibitors so far used have been directed at pyridoxal phosphate-dependent enzymes, partly because it is realtively easy to base an inactivating mechanism on the normal catalytic mechanism but also because so many different types of enzyme use this cofactor. This review will, for the most part, be confined to pyridoxal enzymes.

PYRIDOXAL PHOSPHATE IN CATALYSIS

Pyridoxal phosphate is a cofactor for enzymes catalysing several clearly distinguishable types of reaction on amino acid substrates. The involvement of the cofactor in all of these reactions can be explained in a satisfying way in terms of variations of a single, plausible, underlying mechanism. This mechanism will be presented here briefly and without evidence, simply to show how its availability has allowed the design of appropriate inhibitors. Several excellent presentations of the evidence for this mechanism, by authors closely involved in its elucidation, are available (Fasella, 1967; Dunathan, 1971; Boeker and Snell, 1972; Braunstein, 1973; Metzler, 1977).

Formation of external aldimine

In all the enzymes so far examined, the coenzyme is bound to the enzyme via a Schiff base with a lysine residue of the protein (internal aldimine, **I**). In a step which will be presumed to be common to all of these enzymes, the substrate amino group replaces the enzyme lysine amino group and an external aldimine (**II**) is formed (pyridoxal phosphate structures will be shown throughout without the 3 and 5 substituents).

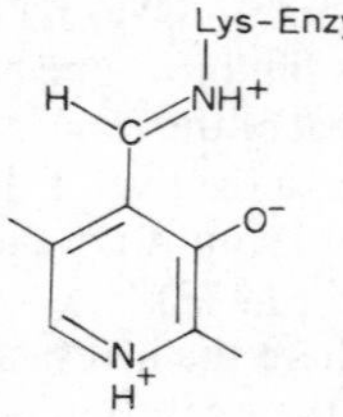

I

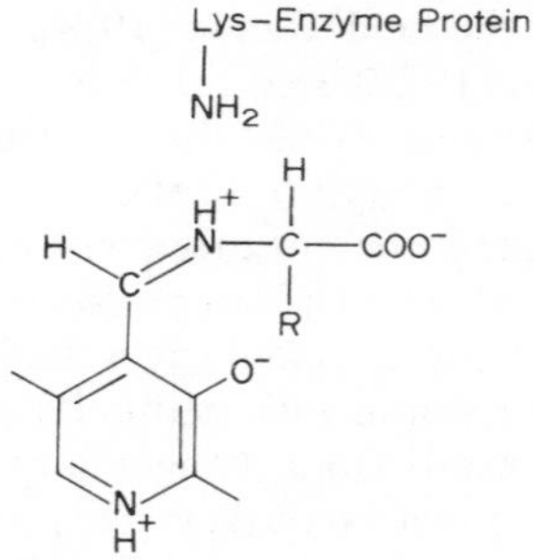

II

This reaction is transimination.

Breaking of bond to α-carbon to give quinonoid intermediate
Appropriate non-covalent binding of the substrate to the enzyme prevents rotation about the substrate C_α-N bond, thereby fixing the geometry of the three remaining bonds to C_α. The bond perpendicular to the plane formed by the coenzyme ring and the double bond of the aldimine is weakened because of delocalisation of its electrons by interaction with the π electrons of the resonant system. Since one of three bonds may occupy this susceptible position, the reactions of pyridoxal-dependent enzymes may initially be divided into three categories, namely those in which the hydrogen, the carboxyl group or the substituent R is eliminated.

Loss of any one of these groups leads to an intermediate which may be represented by the carbanionic and quinonoid canonical forms (III). For convenience this is referred to simply as the quinonoid intermediate but its carbanionic nature is essential to an explanation of the diverse reactions catalysed.

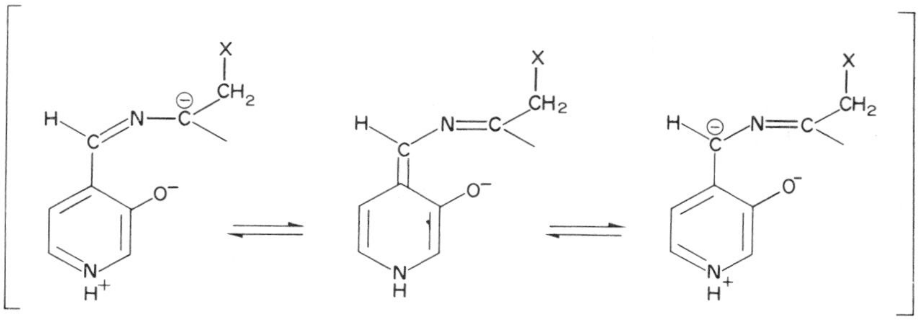

III

Fate of quinonoid intermediate
The quinonoid intermediate may be considered to have excess electrons which may, in effect, be lost in one of three ways, namely by protonation at one of the electron-rich carbons (4′-C of coenzyme or α-C of substrate) or by elimination of a β-substituent. Protonation at either the 4′-C or the α-C determines the position of the double bond of the coenzyme product imine and thus which form the coenzyme is left in after hydrolysis or transimination (as shown on the following page (A)):

Two possibilities arise when a β-substituent is eliminated.

(1) The resulting imine may hydrolyse to give an unsaturated product and the pyridoxal form of the coenzyme.
(2) The nucleophilic second substrate may add to the β-carbon and a proton to the α-carbon. Hydrolysis of the resulting imine again gives the pyridoxal form of the enzyme (as shown on the following page (B)).

A number of enzymes also catalyse elimination of a γ-substituent after abstraction of a β-proton (as shown on the following page (C)).

A. Protonation at 4′ as α-C

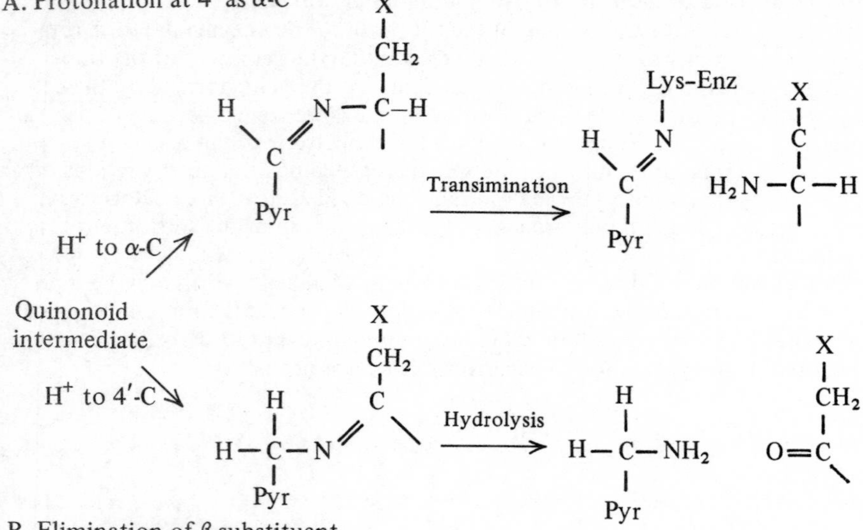

X
CH_2
H N — C–H
C
Pyr

Transimination

Lys-Enz
H N
C
Pyr

X
C
H_2N — C — H

H^+ to α-C

Quinonoid intermediate

H^+ to 4′-C

X
CH_2
H C
H — C — N
Pyr

Hydrolysis

H
H — C — NH_2
Pyr

X
CH_2
O=C

B. Elimination of β-substituent

X^-

Quinonoid intermediate

CH_2
H^+
H N—C
C
O^-
N
H^+

CH_2 (⊕)
H^+
H N=C
C
O^-
N
H

Lys-Enz
H N
C
O^-
N
H^+

\+

CH_2
$H_3\overset{+}{N}$—C

(1)

CH_2
H N—C
C
O^-
N
H^+

Lys-Enz
H NH^+
C
O^-
N
H^+

\+

B
CH_2
H_2N C
H

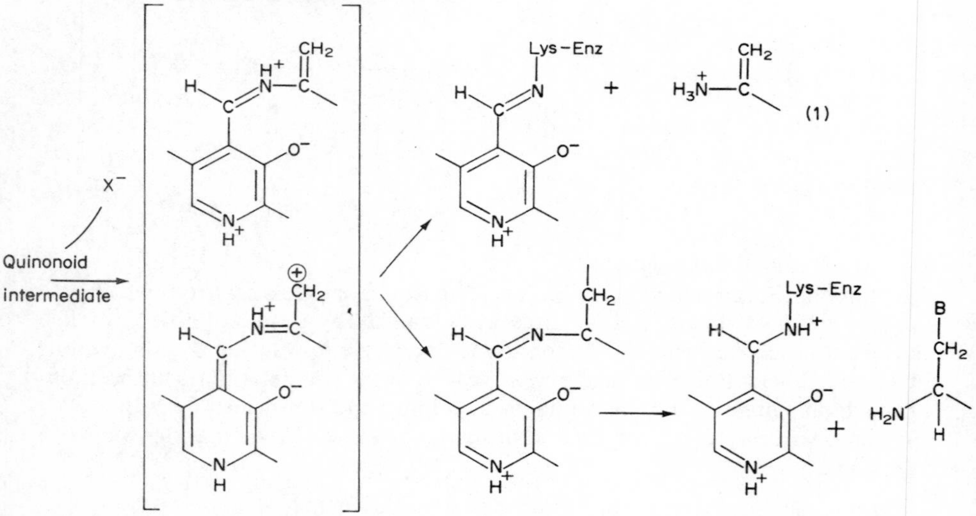

C. Elimination of γ-substituent

X
CH_2
H—C—H
H N=C
C
O^-
N
H

$-H^+$
$-X^-$

CH_2
H—C
H N=C
C
O^-
N
H

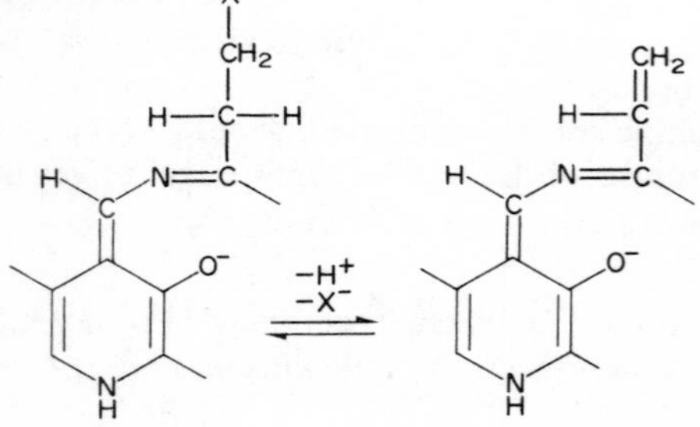

MECHANISMS OF ENZYME-INDUCED INACTIVATION

The many examples of enzyme-induced inactivation of pyridoxal phosphate-dependent enzymes can be explained in terms of alterations in the fate of the quinonoid intermediate brought about by the inclusion of appropriate groups on the substrate. The first three types result in formation of a covalent bond between substrate and enzyme. Two further mechanisms are also described in which irreversible inhibition depends on a distortion of the enzyme's normal mechanism but in which no covalent bond is formed with the enzyme protein.

Covalent modification of enzyme protein

Electronegative substituent on β-carbon
Probably the earliest example of enzyme-induced inactivation was observed when it was noted that threonine dehydratase, when presented with serine as a substrate, catalysed the formation of pyruvate and simultaneously its own destruction (Wood and Gunsalus, 1949). Evidence was later presented showing that the incorporation of radiodactive serine into enzyme protein was probably at a cysteine residue and that the modified enzyme was not completely inactive (Phillips and Wood, 1965; Phillips, 1968).

The principle underlying the design of synthetic inhibitors of this type is to incorporate into the β position of the substrate a group X of sufficient electronegativity that the extra electrons of the quinonoid intermediate are lost with the leaving group X instead of being neutralised by protonation. An unsaturated compound is formed and the electron withdrawing properties of the coenzyme imine leave a partial positive charge on the β-carbon. This new intermediate may be represented by the canonical forms **IV**.

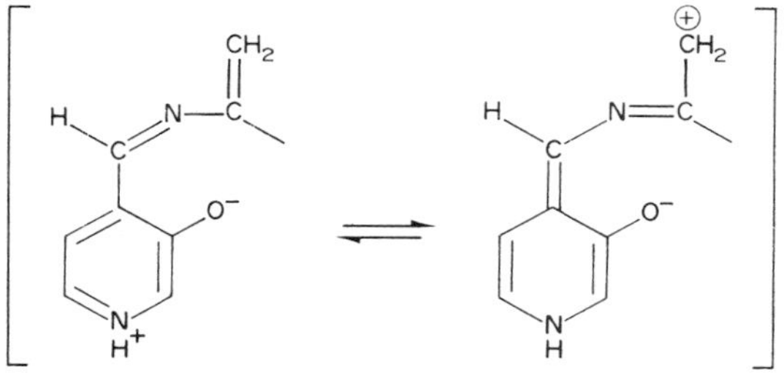

IV

Perault *et al.* (1961) calculated the positive charge in the β carbon to be $0.117e$ and pointed out that it must therefore represent a particularly reactive centre for attacks by nucleophilic agents. If an appropriately placed enzyme nucleophile is present, addition may take place giving first a new resonance-stabilised quinonoid intermediate which, as before, may be protonated at α-C of substrate or 4′-C of coenzyme. Two structures are thus likely and it is possible that either or both of these imines will hydrolyse to give free coenzyme. The nucleophile should remain covalently modified however (B represents enzyme nucleophile).

An example of this type of inactivation is to be found in the inactivation of aspartate transaminase by the glutamate analogue, serine-*O*-sulphate. After elimination of the sulphate, the lysine-258 is modified. This is the residue normally bound to the coenzyme in the native state (John and Fasella, 1969; John *et al.*, 1973). Evidence from absorption spectra of the modified coenzyme indicates that both structures **V** and **VI** are present in the final inactive product.

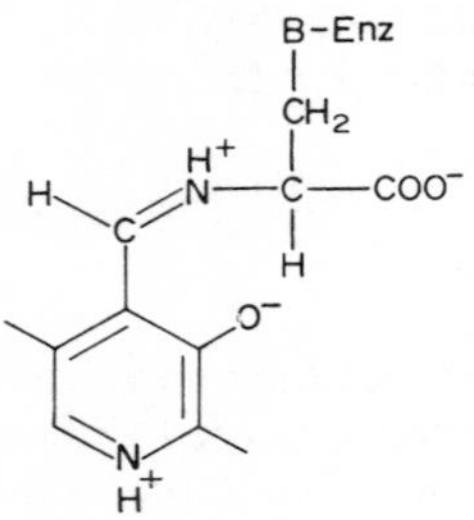

V 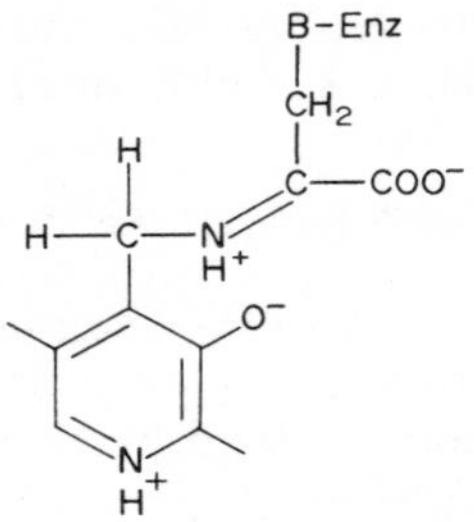 **VI**

Vinyl substituents (alkenes)
Loss of an α-substituent from a substrate that is already β-γ unsaturated leads to a quinonoid intermediate with three possibilities for protonation. Protonation at

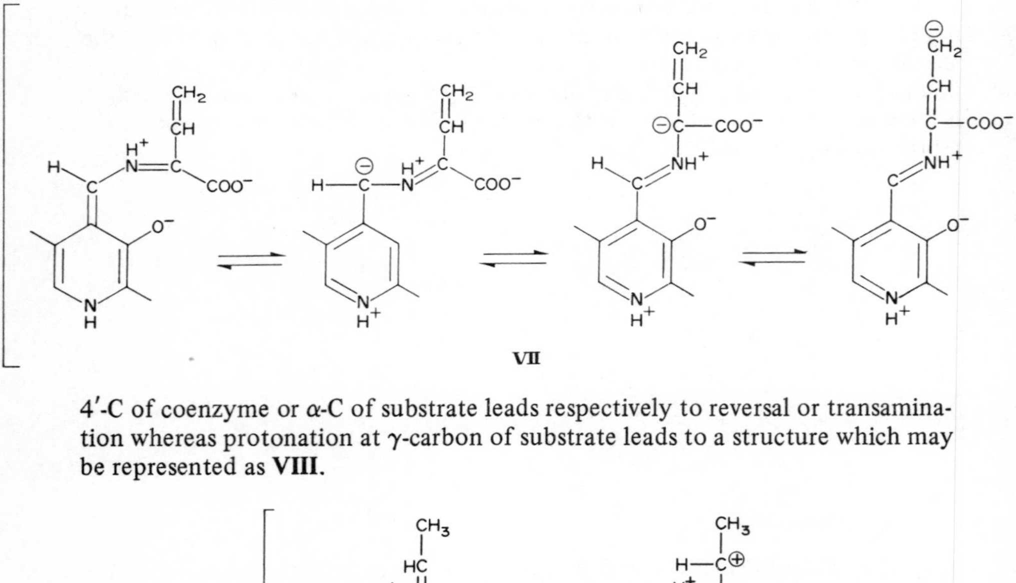

VII

4′-C of coenzyme or α-C of substrate leads respectively to reversal or transamination whereas protonation at γ-carbon of substrate leads to a structure which may be represented as **VIII**.

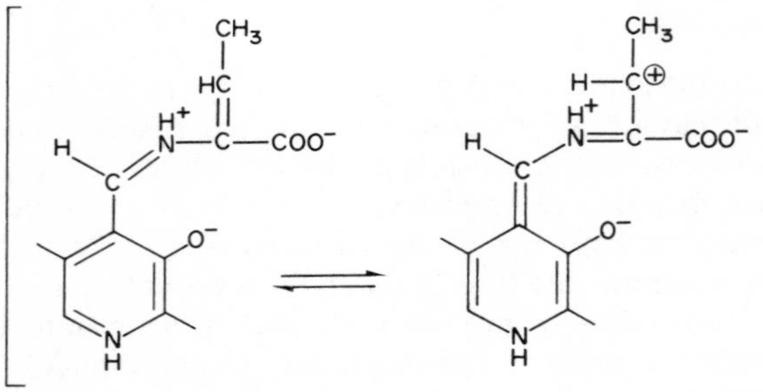

VIII

Addition of an enzyme nucleophile would thus be expected at the β-carbon of the substrate.

A well-characterised example is the use of vinyl glycine to inactivate aspartate transaminase (Rando, 1974*b*; Gehring *et al.*, 1977). Although the same lysine is modified as when serine-*O*-sulphate is used, there are significant differences kinetically. With vinyl glycine, the major event is inactivation and only 10 per cent of turnover is transamination. β-elimination cannot, of course, occur. With serine-*O*-sulphate, β-elimination is the major event, with transamination and inactivation being respectively about 100 and 1000 times slower.

Acetylenic substituents (alkynes)

Abstraction of an α substituent from a substrate with a β–γ triple bond followed by protonation at the γ-carbon would lead to a reactive conjugated allene. Alternatively, protonation at 4′-C would still leave a conjugated acetylene although conjugation does not extend to the coenzyme. Thus inactivation might occur by nucleophilic addition at the β or γ carbons.

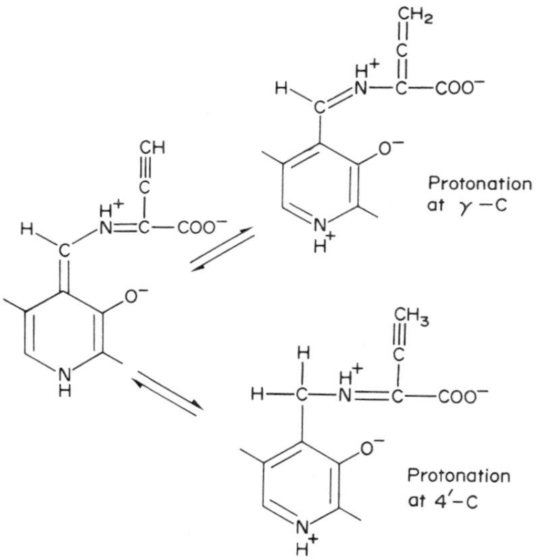

Protonation at γ–C

Protonation at 4′–C

An interesting example is propargylglycine which inactivates several enzymes, γ-cystathionase, cystathionine-γ-synthase, alanine transaminase and aspartate transaminase (Abeles and Walsh, 1973; Marcotte and Walsh, 1975; Tanase and Morino, 1976; Washtien and Abeles, 1977).

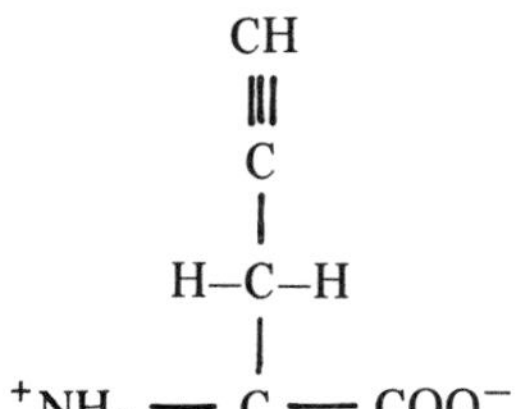

$$HC{\equiv}C{-}CH_2{-}CH(NH_3^+){-}COO^-$$

Propargylglycine

With γ-cystathionase and cystathionine-γ-synthase, the normal enzyme mechanism involves abstraction of a β-proton as well as an α-proton and the mechanism is proposed to proceed via the formation of a conjugated allene (Abeles and Walsh, 1973; Washtien and Abeles, 1977). Note that the abstraction of the β-proton is crucial to this mechanism because a saturated carbon separates the α-carbon from the triple bond. The conjugation supposed necessary for reactivity does not occur unless the β-proton is abstracted.

Proton abstracted from β − C and added to γ − C

Washtien and Abeles (1977), working with γ-cystathionase, provided excellent evidence for the enzyme-induced mechanism when they demonstrated almost identical isotope effects (2.2) for the normal reaction with α-deuterated substrate and inactivation with α-deuterated propargylglycine.

Abstraction of β-protons is not, however, a part of the mechanism of alanine and aspartate transaminases and it is difficult to see how these enzymes produce a very reactive species by abstraction of an α-proton alone. Marcotte and Walsh (1975) suggested that the inactivation of alanine transaminase by propargylglycine is due to enzyme-catalysed labilisation of protons on the β-carbon, supporting their view with the evidence of Babu and Johnston (1974), who showed that this enzyme catalyses exchange of the hydrogens on the β-carbon as well as those on the α-carbon. This cannot, however, be the explanation for the inactivation of aspartate transaminase by propargylglycine (Tanase and Morino, 1976) since this enzyme does not labilise the β hydrogens of glutamate significantly. Inclusion of formate increases the rates of alanine transamination by this enzyme (Saier and Jenkins, 1967), the rate of inactivation by chloroalanine (Morino and Okamoto, 1972; John and Tudball, 1972; Morino *et al.*, 1974) and the rate of inactivation by propargylglycine. The fact that all of these observations would be explained by increased labilisation of the α-proton when formate is bound to the enzyme supports the view that the α-proton must be abstracted before inactivation takes place.

An alternative view is that bound formate enhances the nucleophilicity of the base which abstracts the proton (presumably Lys-258) and that propargylglycine reacts more readily with this via something more akin to an affinity-labelling mechanism. In this context, it should be remembered that acetylenes themselves react slowly with primary amines (Kruse and Kleinschmidt, 1961) and that very great increases in rate can be obtained by converting a bimolecular to a unimolecular reaction (Bruice, 1962, 1976; Page and Jencks, 1971).

Nitrile substituents

Tryptophan synthetase is inactivated by cyanoglycine. The mechanism proposed is similar to that for acetylenic derivatives (Miles, 1975*a*).

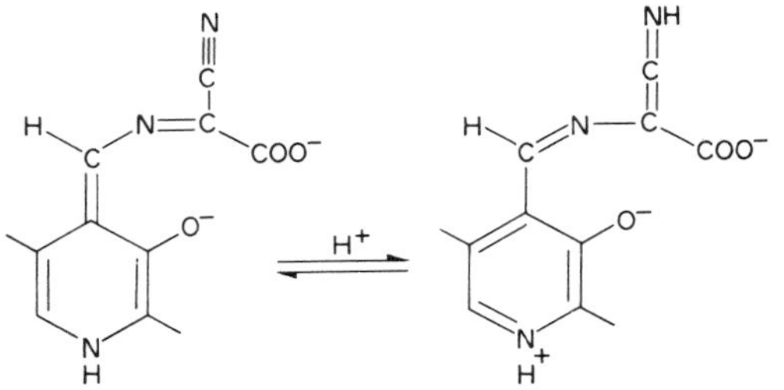

Increasing reactivity of the Michael acceptor

In all the foregoing examples, the irreversible step of the inactivation process is probably a Michael type reaction in which an enzyme nucleophile adds to an unsaturated carbon made electrophilic by conjugation with the coenzyme. Reactivity of the Michael type acceptor can be enhanced by substitution of electron withdrawing groups at the appropriate carbon. Silverman and Abeles (1976) showed that although monohaloalanines are β-elimination substrates for cystathionase, tryptophanase, tryptophan synthase and threonine dehydrase, they do not inactivate these enzymes; the di- and trihaloalanines, on the other hand, are effective inhibitors.

Comparison of GABA transaminase inhibitors

Inhibitors based on the β-elimination principle and on the incorporation of vinyl and acetylenic substituents have been designed as inhibitors of GABA transaminase. Their inactivating abilities and specificities have been fairly extensively examined *in vitro* and *in vivo*.

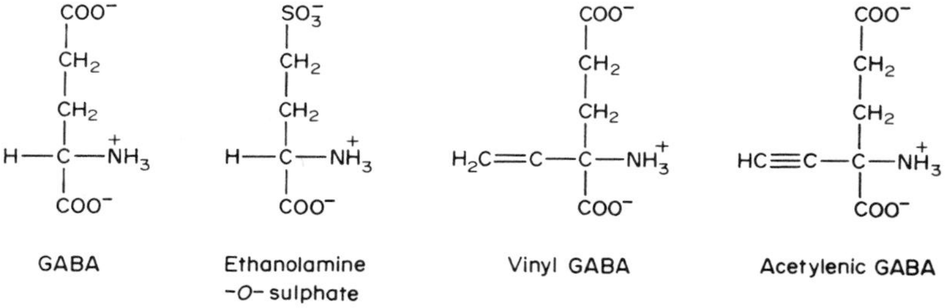

GABA | Ethanolamine -O- sulphate | Vinyl GABA | Acetylenic GABA

The acetylenic compound is considerably less specific than the other two. It inactivates ornithine transaminase more effectively than it does GABA transaminase and is also a fairly effective inactivator of aspartate transaminase (Jung and Seiler, 1978; John *et al.*, 1979).

Both the other compounds are reasonably specific. Ethanolamine-*O*-sulphate has no effect on aspartate transaminase or alanine transaminase even at 100 mM

(Fowler and John, 1972). It does, however, inactivate ornithine transaminase but about 40 times less effectively than it does GABA transaminase (John *et al.*, 1978). Vinyl GABA is about four times more effective as an inhibitor of GABA transaminase than is ethanolamine-*O*-sulphate. It inactivates ornithine transaminase about 80 times less effectively than it does GABA transaminase and inactivates aspartate transaminase only very slowly. The properties of enzyme-induced inactivators of GABA transaminase have been reviewed by Metcalf *et al.* (1978) and by Jung (1978). Of these compounds only ethanolamine sulphate is free from antagonist effects (Robin *et al.*, 1979) and this compound has recently been shown to elevate brain GABA levels when given orally (Fletcher and Fowler, 1980).

Inactivation without modification of enzyme protein

Irreversible inhibition can be achieved through the operation of an enzyme's mechanism without covalent linkage between the enzyme protein and the inactivating species. When an inactivating modification is introduced into the coenzyme which, nevertheless, still remains tightly bound, the enzyme is just as effectively inactivated as if the protein were covalently altered.

Enzymic generation of coenzyme–substrate analogue

Some coenzyme–substrate analogues bind tightly to the appropriate apoenzyme (for example, Borri-Voltattorni *et al.*, 1975). However, most of the binding interaction is due to the coenzyme moiety and, since most pyridoxal enzymes do not lose their coenzyme readily, such coenzyme–substrate analogues are unlikely to be very potent inhibitors of the majority of holoenzymes. This difficulty may be very neatly overcome by generating the coenzyme–substrate analogue *in situ*. Furthermore, as pointed out by Metcalf *et al.* (1978), such a mechanism does not require a conveniently placed enzyme nucleophile.

An example of this type of inactivation is to be found in gabaculine (Rando and Bangerter, 1976; Rando, 1977). This compound is an analogue of GABA

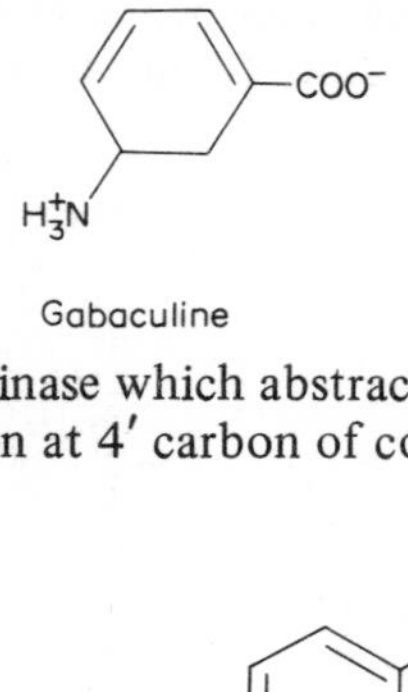

Gabaculine

and binds to GABA transaminase which abstracts the proton on the carbon α to the amino group. Protonation at 4′ carbon of coenzyme and loss of the β-proton

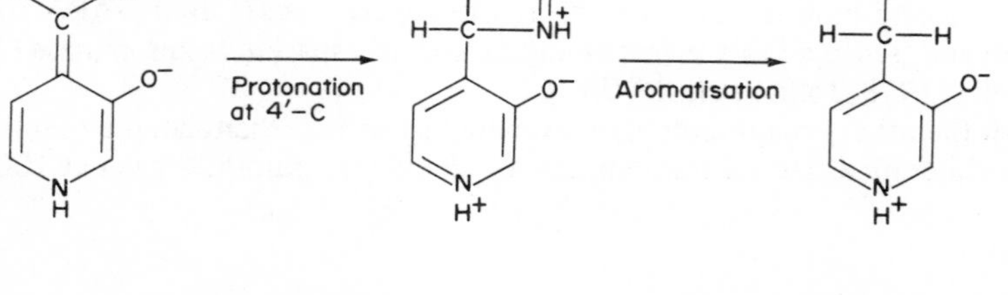

Protonation at 4′–C

Aromatisation

results in an aromatic ring. Gabaculine is also a potent inhibitor of ornithine transaminase. An isomer of gabaculine behaves similarly (Jung and Seiler, 1978).

Decarboxylation-dependent transamination
With the exception of the transaminases, the cofactor in pyridoxal phosphate-dependent enzymes cannot be pyridoxamine phosphate. Conversion of the normal coenzyme to the pyridoxamine form by a 'mistake' in the normal mechanism therefore results in inactivation. Such inhibition is only irreversible in the sense that it is not reversed unless the coenzyme is returned to its original form either by replacement or by reversal with a suitable 'substrate' aldehyde or ketone.

The mistake is due simply to protonation of the 4′ carbon of the coenzyme rather than the α carbon of the substrate in the quinonoid intermediate (Huntley and Metzler, 1968). This leads, after hydrolysis of the resulting imine, to the pyridoxamine form of the enzyme and a product which is an aldehyde rather than an amine.

The reaction occurs with the amino acid decarboxylases to an extent which is significant even with the normal substrate. However, with the α-methyl amino acids, the ratio of transamination to decarboxylation is greatly increased and these substrate analogues are effectively irreversible inhibitors. For example, dopa decarboxylase is progressively inactivated during its normal reaction due to a transamination which occurs in a small proportion of decarboxylations (O'Leary and Baughn, 1977).

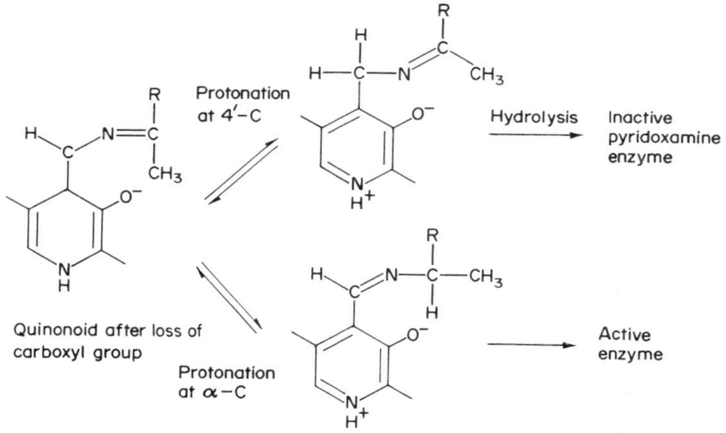

Protonation at 4′–C

Hydrolysis

Inactive pyridoxamine enzyme

Quinonoid after loss of carboxyl group

Protonation at α–C

Active enzyme

KINETICS OF ENZYME-INDUCED INACTIVATION

An examination of the kinetics of inactivation can assist in interpreting the underlying mechanism. The kinetics are often complex because of the occurrence of multiple reactions.

First-order behaviour and saturation kinetics
It has been stated (Walsh, 1977) that the observation of first order kinetics provides evidence that inactivation procedes via a two-step process, namely initial binding followed by an irreversible step (equation (5.1)) rather than by a simple bimolecular process (equation (5.2)). This is not so.

$$\mathrm{E + I} \underset{k_{-1}}{\overset{k_{+1}}{\rightleftharpoons}} \mathrm{EI} \xrightarrow{k_i} \mathrm{E_i} \qquad (5.1)$$

$$\mathrm{E + I} \xrightarrow{k_i} \mathrm{E_i} \qquad (5.2)$$

The observation of first-order kinetics does not distinguish between these mechanisms. Pseudo-first-order kinetics will arise in both cases when inhibitor concentration is sufficiently higher than enzyme concentration that it is not significantly depleted during the reaction.

The pseudo-first-order constants are

$$k_{obs} = k_i/(1 + K_i/[\mathrm{I}])$$

for equation (5.1) and

$$k_{obs} = k_i[\mathrm{I}]$$

for equation (5.2). K_i is an apparent dissociation constant and the derivation assumes that the first step is rapidly reversible.

Mechanism (5.1) is that proposed by Kitz and Wilson (1962) for irreversible inhibition of acetylcholinesterase by analogues of acetylcholine. It predicts saturation kinetics, that is k_{obs} approaches a limiting value, k_i, as [I] is increased. Thus a plot of $1/k_{obs}$ against [I] will give a straight line cutting the $1/k_{obs}$ axis at $1/k_i$ and the $1/[\mathrm{I}]$ axis at $1/K_i$. When this is demonstrable, it provides good evidence that a reversible complex is formed. However, passage of such a plot through the origin is consistent with both mechanisms (5.1) and (5.2). Such behaviour would be observed for the two-step mechanism (5.1) when the complex does not accumulate sufficiently to deplete the free enzyme sufficiently, that is when the values of [I] used are low relative to K_i.

An important case where saturation will be difficult to demonstrate arises when k_i is itself large. Under these circumstances, although enzyme-induced inactivation will tend to be very efficient, first-order behaviour need not necessarily be seen.

Non-first-order inactivation

In very many cases of enzyme-induced inactivation, the loss of enzyme activity does not follow first-order kinetics because multiple reactions occur.

Significant conversion to product

Consider a case in which the inactivating reaction, characterised by k_i, is accompanied by another reaction, characterised by k_e, in which product and free enzyme are released:

$$\mathrm{E + I} \rightleftharpoons \mathrm{EI} \begin{array}{l} \xrightarrow{k_i} \mathrm{E_i} \\ \xrightarrow{k_e} \mathrm{E + P} \end{array} \qquad (5.3)$$

In this case, the extent by which the initial inhibitor concentration $[I_0]$ must exceed the initial enzyme concentration $[E_0]$ in order to see pseudo-first-order kinetics depends directly on the ratio k_e/k_i. Put another way, if the inhibitor is significantly depleted by conversion to P then its concentration will fall throughout the reaction. The necessary condition for pseudo-first-order kinetics, that [I] remains essentially constant, will not be met. Figure 5.1 shows some theoretical semi-logarithmic plots for various conditions.

The examples given at the end of this article include many where k_e/k_i has been estimated.

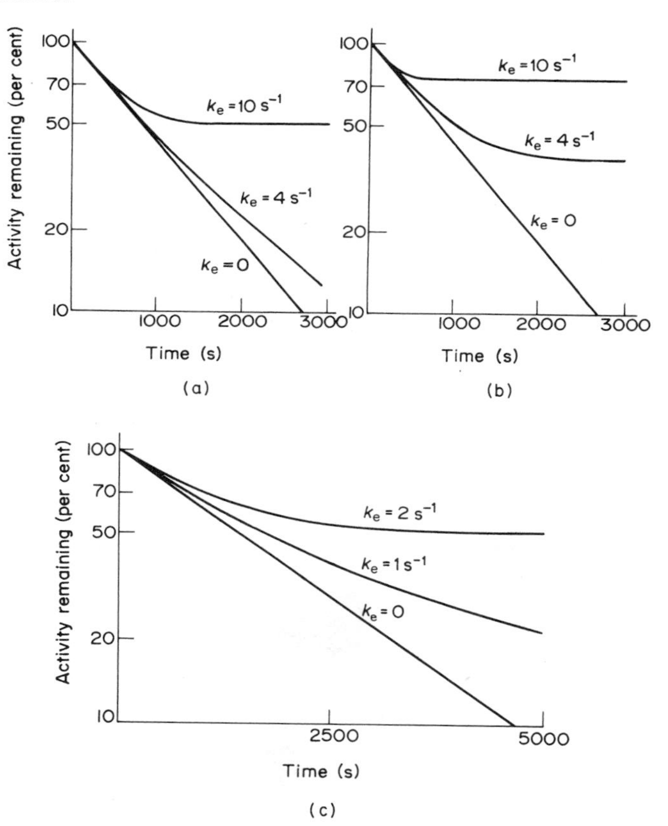

Figure 5.1 Semi-logarithmic plots showing inactivation for a theoretical reaction following mechanism (5.3). The inactivation was determined according to an equation derived from mechanism (5.3):

$$t = \frac{1}{k_i} \ln \frac{[E_0]}{[E]} + \frac{K_i}{k_i [I_0] - k_e[E_0]} \ln \left[\frac{[E_0]}{[E]} \left\{ 1 - \frac{k_e}{k_i [I_0]} ([E_0] - [E]) \right\} \right]$$

[E] and $[E_0]$ are the concentrations of active enzyme initially and at time, t; k_i and k_e are rate constants as indicated in mechanism (5.3). K_i is the apparent dissociation constant for initial binding.

In all the plots $K_i = 1$ mM and $k_i = 10^{-3}$ s^{-1}

(a) $\{I_0\} = 5$ mM; $\{E_0\} = 1$ μM. (b) $\{I_0\} = 5$ mM; $\{E_0\} = 2$ μM. (c) $\{I_0\} = 1$ mM; $\{E_0\} = 1$ μM.

Protection by transamination

The occurrence of transamination is a further complication since it generates the pyridoxamine form of the enzyme (E_m) and this cannot combine with an amino acid inhibitor. The enzyme thus protects itself. Protection by this mechanism might also be afforded against affinity-labelling inhibitors.

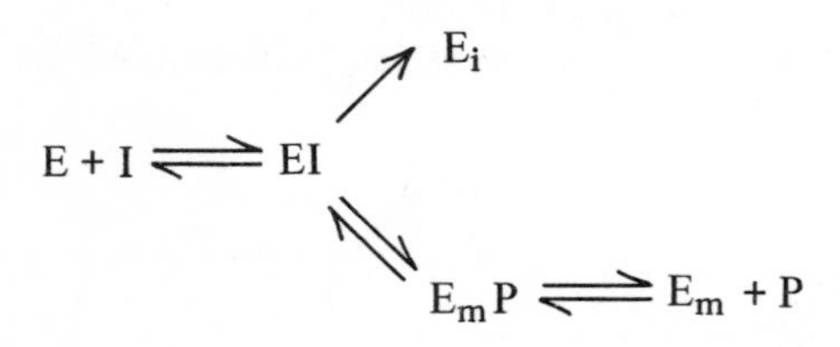

This mechanism has been discussed by John *et al.* (1978). When transamination is fast relative to inactivation, and the transamination equilibrium in favour of E_m, it might be inferred that the compound under investigation is not an inhibitor. Tanase and Morino (1976) observed it when treating both isoenzymes of aspartate aminotransferase with propargyl glycine. Only in the presence of keto acid, which reverses the transamination, is inactivation observed. This self-protection mechanism probably accounts for the conclusion of other workers that propargyl glycine does not inactivate aspartate transaminase (Marcotte and Walsh, 1975; Walsh, 1977).

The relative magnitudes of the rate and equilibrium constants determine what is observed. When serine-*O*-sulphate is used to inactivate the cytoplasmic and mitochondrial isoenzymes of aspartate transaminase, the course of inactivation is quite different although the underlying mechanism is the same.

In these cases, however, three reactions occur simultaneously (John and Fasella, 1969; John *et al.*, 1978):

$$E + I \rightleftharpoons EI \rightleftharpoons E_mP \rightleftharpoons E_m + P$$
$$EI \rightarrow E_i$$
$$EI \rightarrow E + P'$$

The production of pyruvate, a keto acid capable of reversing the transamination, via the β-elimination mechanism, reverses the protection afforded by the initial transamination:

$$E_m + P' \rightleftharpoons E + \text{Ala}$$

When the reaction between the mitochondrial form is followed over a long period, inactivation proceeds after a lag during which pyruvate slowly accumulates by β-elimination.

EXAMPLES OF ENZYME-INDUCED INACTIVATION

Some of the kinetic data presented in the following list are not taken directly from the papers quoted but are derived approximately from results quoted therein.

Aspartate transaminase

The pig heart cytosol enzyme is described except when otherwise stated.

Serine-O-sulphate
β-Elimination, transamination and inactivation occur simultaneously; k_{elim} = 12 s^{-1}, transamination is about 100 times slower and when [E] = 10^{-6} M, k_i is about 1000 times slower. Lys-258 covalently modified, possibly Cys-390 also (John and Fasella, 1969; John *et al.*, 1973). Also inhibits other transaminases.

β-Chloroalanine
β-Elimination, transamination and inactivation all occur. Formate increases k_i and k_{elim}, probably by binding to the enzyme cation which binds the distal carboxyl of normal substrates. $k_{elim}/k_i \sim 1700$; also inhibits alanine transaminase, aspartate β-decarboxylase, and amino acid racemase (see list) but not tryptophan synthase, tryptophanase, threonine dehydrase or β- and γ-cystathionase (Silverman and Abeles, 1976).

β-Bromoalanine
See Morino and Okamoto, 1970.

Bromopyruvate and pyridoxamine enzyme
Transaminates, eliminates and inactivates. $k_{elim:trans}/k_i \sim 40$, Cys-390 modified (Okamoto and Morino, 1973). Inactivation may be due to enhanced reactivity of Cys-390 (Birchmeier and Christen, 1974).

Vinyl glycine
Transamination occurs but inactivation is 90 per cent of total turnover. Lys-258 modified. Pyridoxyl-bearing lysine of chicken mitochondrial enzyme also modified (Rando, 1974*b*; Gehring *et al.*, 1977). Does not inactivate tryptophan synthase (Miles, 1975*b*) or alanine transaminase (Soper *et al.*, 1977*a*).

α-Amino-γ-methoxybutenoate
See Rando *et al.*, 1976.

Propargyl glycine
Transaminates and inactivates (Tanase and Morino, 1976). Enzyme-induced mechanism requires that enzyme abstracts hydrogen from β-carbon (see Marcotte and Walsh, 1975).

GABA-transaminase

Ethanolamine-O-sulphate
β-Eliminates and inactivates. Does not transaminate. K_i = 15 mM, k_i = 0.01 s^{-1};

$k_{elim}/k_i \sim 6$ (Fowler and John, 1972). Does not inactivate glutamate decarboxylase, alanine transaminase or aspartate transaminase. Very slow inactivation of ornithine transaminase (John *et al.*, 1978).

Vinyl-GABA
Does not transaminate. $k_{cat} = k_i = 0.02\ s^{-1}$. $K_i = 5$ mM. Does not inactivate glutamate decarboxylase. Very slow inactivation of ornithine transaminase, aspartate transaminase and alanine transaminase (Lippert *et al.*, 1977; John *et al.*, 1979).

Acetylenic-GABA
Does not transaminate. $k_{cat} = k_i = 4 \times 10^{-3}\ s^{-1}$. $K_i = 1.3$ mM. Also inactivates glutamate decarboxylase, ornithine transaminase and aspartate transaminase (Jung and Metcalf, 1975; John *et al.*, 1979).

Gabaculine
Probably the most potent GABA transaminase inhibitor. $k_i = 1.15 \times 10^{-2}\ s^{-1}$, $K_i = 3\ \mu M$ (Rando and Bangerter, 1976; Rando, 1977; Rando and Bangerter, 1977). Also a potent inhibitor of ornithine transaminase.

D-Amino acid transaminase

DL-Vinyl glycine
Transaminates and inactivates. $k_{trans}/k_i \sim 450$; $k_i = 6 \times 10^{-3}\ s^{-1}$; $K_i = 2$ mM (Soper *et al.*, 1977*a*).

D-Chloroalanine
Eliminates and inactivates. $k_{elim}/k_i \sim 1500$; $k_i = 10^{-2}\ s^{-1}$; K_i (measured by competition with D-alanine) = 10 μM (Soper *et al.*, 1977*b*).

Ornithine transaminase

Acetylenic-GABA
Inactivates and transaminates. Ketoglutarate permits inactivation to go to completion. $k_i = 0.01\ s^{-1}$, $K_i = 3$ mM. Works *in vivo* (Jung and Seiler, 1978; John *et al.*, 1979). Also inactivates GABA transaminase.

Gabaculine
Half activity lost in a few minutes in a non-exponential process at 10^{-5} M inhibitor (Jung and Seiler, 1978). Also inactivates GABA transaminase.

Alanine transaminase

Chloroalanine
Elimination and inactivation occur; k_{elim}/k_i increases with pH from 500 to 1000; $k_i \sim 0.01\ s^{-1}$, varies with buffer anion; $K_i = K_{elim} = 0.15$ mM (pH 7) (Golichowski and Jenkins, 1978; Morino *et al.*, 1979).

Alanine racemase

D-*Chloroalanine*
Eliminates and inactivates; $k_{elim}/k_i = 920$; $k_i = 0.003\ s^{-1}$ when [I] = 0.5 mM. Inhibits growth of bacteria and successfully treats mice with pneumococcal and streptococcal infections (Manning *et al.*, 1974; Wang and Walsh, 1978). Also inactivates D-amino acid transaminase (Soper *et al.*, 1977*b*).

D-*Fluoroalanine*
Eliminates and inactivates; $k_{elim}/k_i = 790$; $k_i = 0.023\ s^{-1}$ when [I] = 0.1 mM (Wang and Walsh, 1978).

O-*Carbamoyl*-D-*serine*
Eliminates and inactivates; $k_{elim}/k_i = 800$; $k_i = 0.003\ s^{-1}$ when [I] = 1 mM (Wang and Walsh, 1978).

O-*Acetyl* D-*serine*
Eliminates and inactivates; $k_{elim}/k_i = 800$; $k_i = 0.0015\ s^{-1}$ when [I] = 1 mM.

Threonine dehydrase

Serine
Eliminates and inhibits. Inhibition is irreversible but leaves the enzyme 7 per cent active. *N*-Ethylmaleimide enzyme not susceptible (Wood and Gunsalus, 1949; Phillips and Wood, 1965; Phillips, 1968).

Trifluoroalanine
Does not β-eliminate but inactivates; $k_i = 8 \times 10^{-4}\ s^{-1}$ when [I] = 40 mM (Silverman and Abeles, 1976).

γ-Cystathionase

Propargylglycine
Inactivates with $k_i = 5 \times 10^{-2}\ s^{-1}$; $K_i = 0.7$ mM. Inactivation and normal reaction show same isotope effect when α-deuterated compounds used (Abeles and Walsh, 1973; Washtien and Abeles, 1977). Also inactivates aspartate transaminase (Tanase and Morino, 1976), alanine transaminase and cystathionine-γ-synthase (Marcotte and Walsh, 1975).

Trifluoroalanine
Inactivates, whereas the monohaloalanines β-eliminate without inactivating. Activity lost with $t_{1/2} < 1$ min when [I] = 20 mM. Also inactivates β-cystathionase, tryptophanase, tryptophan synthase, and threonine dehydrase but not cystathionine synthase, aspartate transaminase or alanine transaminase (Silverman and Abeles, 1976).

Dichloroalanine
Inactivates; k_i = 0.006 s^{-1} at [I] = 20 mM. Does not inactivate β-cystathionase, serine dehydrase or cystathionine synthase but does inactivate tryptophan synthase (Silverman and Abeles, 1976).

β-Cystathionase

Trifluoroalanine
Inactivates with k_i = 0.014 s^{-1} when [I] = 1 mM.

Tryptophan synthase

Cyanoglycine
Inactivates with $k_i = 9 \times 10^{-3}$ s^{-1} when [I] = 1 mM (Miles, 1975*a*).

α-Amino-γ-methoxy butenoate
Inactivation is accompanied by another reaction that forms a keto product; k_i = 1.5×10^{-3} s^{-1} at 4 mM inhibitor (Miles, 1975*b*; Miles, 1978).

Trifluoroalanine
No significant β-elimination above that required for inactivation; $k_i = 10^{-3}$ s^{-1} when [I] = 40 mM (Silverman and Abeles, 1976).

Dichloroalanine
As for trifluoroalanine.

Tryptophanase

Trifluoroalanine
$k_i = 10^{-3}$ s^{-1} when [I] = 80 mM.

Dichloroalanine
$k_i = 2 \times 10^{-4}$ s^{-1} when [I] = 80 mM (Silverman and Abeles, 1976).

Aspartate decarboxylase
Chloroalanine
Eliminates and inactivates; $k_{elim} = 9$ s^{-1}, $k_{elim}/k_i \sim 1000$, K_{elim} = 63 μM. Modification identified as enzyme glutamate esterified with β-hydroxypyruvate (Tate *et al.*, 1969; Relyea *et al.*, 1974).

Dopa decarboxylase

α-Methyl dopa
Inactivates by decarboxylation-dependent transamination; $t_{1/2}$ = 2 min when [I] = 0.01 mM (Borri-Voltattorni *et al.*, 1971). A slow 'normal' decarboxylation also occurs.

α-Chloromethyl dopa
Inactivates with $t_{1/2}$ = 2 min at [I] = 0.04 mM. Loss of activity is not exponential (Bey, 1978).

α-Fluoromethyl dopa
Non-exponential loss of activity with 90 per cent inactivation in 12 min at [I] = 5 μM.

α-Difluoromethyl dopa
Non-exponential loss of activity with 90 per cent inactivation after 25 min when [I] = 0.01 mM (Bey, 1978; Palfreyman *et al.*, 1978).

α-Vinyl dopa
Inactivation first-order but only 70% activity lost. Slow return of activity and loss of bound inhibitor on dialysis (Taub and Patchett, 1977; Maycock *et al.*, 1978).

α-Acetylenic dopa
As for α-vinyl dopa.

Glutamate decarboxylase

α-Methyl glutamate
Inactivates by decarboxylation-dependent transamination (Huntley and Metzler, 1968).

Acetylenic-GABA
Inactivates with $k_i = 2 \times 10^{-3}$ s^{-1}; K_i = 1 mM. Radioactive inhibitor incorporated irreversibly. Also inhibits ornithine, GABA and aspartate transaminases.

Ornithine decarboxylase

α-Methyl ornithine
Inactivates by decarboxylation-dependent transamination; also decarboxylates; $k_{decarbox}/k_i \sim 25$. Pyridoxal phosphate restores activity. Decarboxylation of α-methyl compound is 6000 times slower than that of ornithine (O'Leary and Herreid, 1978).

α-Cyanomethyl ornithine
Inactivates with $k_i = 4 \times 10^{-3}$ s^{-1}; K_i = 8.7 mM (Bey, 1978).

α-Fluoromethyl ornithine
Inactivates with $k_i = 7 \times 10^{-3}$ s^{-1}; K_i = 0.075 mM (Bey, 1978).

α-Difluoromethyl ornithine
Inactivates with $k_i = 4 \times 10^{-3}$ s^{-1}; K_i = 0.039 with DL compound and K_i = 0.018 with single enantiomer.

REFERENCES

Abeles, R. and Maycock, A. K. (1976). *Accounts chem. Res.*, **9**, 313
Abeles, R. and Walsh, C. (1973). *J. Am. chem. Soc.*, **95**, 6124
Babu, U. M. and Johnston, R. B. (1974). *Biochem. biophys. Res. Commun.*, **58**, 460
Bey, P. (1978). In *Enzyme-Activated Irreversible Inhibitors* (ed. N. Seiler, M. J. Jung and J. Koch-Weser), Elsevier/North-Holland, Amsterdam, pp. 27–41
Birchmeier, W. and Christen, P. (1974). *J. biol. Chem.*, **249**, 6311
Boeker, E. A. and Snell, E. E. (1972). In *The Enzymes*, vol. 6 (ed. P. D. Boyer), Academic Press, New York and London, pp. 217–48
Borri-Voltattorni, C., Minelli, A. and Turano, C. (1971). *Fedn. Eur. biochem. Socs Lett.*, **17**, 231
Borri-Voltattorni, C., Orlacchio, A., Giartosio, A., Conti, F. and Turano, C. (1975). *Biochem. biophys. Res. Commun.*, **66**, 863
Braunstein, A. E. (1973). In *The Enzymes*, vol. 9 (ed. P. D. Boyer), Academic Press, New York and London, pp. 379–480
Bruice, T. C. (1962). *Brookhaven Symp. Biol.* **15**, 52
Bruice, T. C. (1976). *A. Rev. Biochem.*, **45**, 331
Dunathan, H. C. (1971). *Adv. Enzymol.*, **35**, 79
Fasella, P. (1967). *A. Rev. Biochem.*, **36**, 185
Fasella, P. and John, R. A. (1969). *Proc. int. Congr. Pharmacol.*, **5**, 184
Fletcher, A. and Fowler, L. J. (1980). *Br. J. Pharmacol.* in press
Fowler, L. J. and John, R. A. (1972). *Biochem. J.*, **130**, 569
Gehring, H., Rando, R. and Christen, P. (1977). *Biochemistry*, **16**, 4832
Golichowski, A. and Jenkins, W. T. (1978). *Arch. Biochem. Biophys.*, **189**, 109
Huntley, T. and Metzler, D. E. (1968). In *Symposium on Pyridoxal Enzymes* (ed. K. Yamada, N. Kebenumia and H. Wada), Maruzen, Tokyo, p. 81
John, R. A. and Fasella, P. (1969). *Biochemistry*, **8**, 4477
John, R. A. and Tudball, N. (1972). *Eur. J. Biochem.*, **31**, 135
John, R. A., Bossa, F., Barra, D. and Fasella, P. (1973). *Biochem. Soc. Trans.*, **4**, 862
John, R. A., Charteris, A. T. and Fowler, L. J. (1978). In *Enzyme-Activated Irreversible Inhibitors* (ed. N. Seiler, M. J. Jung and J. Koch-Weser), Elsevier/North-Holland, Amsterdam, pp. 109–22
John, R. A., Jones, E. D. and Fowler, L. J. (1979). *Biochem. J.*, **177**, 721
Jung, M. J. (1978). In *Enzyme-Activated Irreversible Inhibitors* (ed. N. Seiler, M. J. Jung and J. Kock-Weser), Elsevier/North-Holland, Amsterdam, pp. 123–34
Jung, M. J. and Metcalf, B. W. (1975). *Biochem. biophys. Res. Commun.*, **67**, 301
Jung, M. J. and Seiler, N. (1978). *J. biol. Chem.*, **253**, 7431
Kitz, R. and Wilson, I. B. (1962). *J. biol. Chem.*, **237**, 3245
Kruse, C. W. and Kleinschmidt, R. F. (1961). *J. Am. chem. Soc.*, **83**, 213
Lippert, B., Metcalf, B. W., Jung, M. J. and Casara, P. (1977). *Eur. J. Biochem.*, **74**, 441
Manning, J. M., Merrifield, N. E., Jones, W. M. and Gotschlich, E. C. (1974). *Proc. natn. Acad. Sci. U.S.A.*, **71**, 417
Marcotte, P. and Walsh, C. (1975). *Biochem. biophys. Res. Commun.* **62**, 677
Maycock, A. L., Aster, S. D. and Patchett, A. A. (1978). In *Enzyme-Activated Irreversible Inhibitors* (ed. N. Seiler, M. J. Jung and J. Koch-Weser), Elsevier/North-Holland, Amsterdam, pp. 211–20
Metcalf, B. W., Lippert, B. and Casara, P. (1978). In *Enzyme-Activated Irreversible Inhibitors* (ed. N. Seiler, M. J. Jung and J. Koch-Weser), Elsevier/North-Holland, Amsterdam, pp. 123–4
Metzler, D. E. (1977). *Biochemistry, the Chemical Reactions of Living Cells*, Academic Press, New York, San Francisco and London, pp. 444–61
Miles, E. W. (1975a). *Biochem. biophys. Res. Commun.*, **64**, 248
Miles, E. W. (1975b). *Biochem. biophys. Res. Commun.*, **66**, 94
Miles, E. W. (1978). In *Enzyme-Activated Irreversible Inhibitors* (ed. N. Seiler, M. J. Jung and J. Koch-Weser), Elsevier/North-Holland, Amsterdam, pp. 73–85
Morino, Y. and Okamoto, M. (1970). *Biochem. biophys. Res. Commun.*, **40**, 600
Morino, Y. and Okamoto, M. (1972). *Biochem. biophys. Res. Commun.*, **47**, 498

Morino, Y., Abdalla, M. O. and Okamoto, M. (1974). *J. biol. Chem.*, **249**, 6684
Morino, Y., Kojima, H. and Tanase, S. (1979). *J. biol. Chem.*, **254**, 279
Okamoto, M. and Morino, Y. (1973). *J. biol. Chem.*, **248**, 82
O'Leary, M. and Baughn. R. L. (1977). *J. biol. Chem.*, **252**, 7168
O'Leary, M. and Herreid, R. M. (1978). *Biochemistry*, **17**, 1010
Page, M. I. and Jencks, W. T. (1971). *Proc. natn. Acad. Sci. U.S.A.*, **68**, 1678
Palfreyman, M. G., Danzin, C., Jung, M. J., Fozard, J. R., Wagner, J., Woodward, J. K., Aubry, M., Dage, R. C. and Koch-Weser, J. (1978). In *Enzyme-Activated Irreversible Inhibitors* (ed. N. Seiler, M. J. Jung and J. Koch-Weser), Elsevier/North-Holland, Amsterdam, pp. 221–33
Perault, A.-M., Pullman, B. and Valdemoro, C. (1961). *Biochim. biophys. Acta*, **46**, 555
Phillips, A. T. (1968). *Biochim. biophys. Acta*, **151**, 526
Phillips, A. T. and Wood, W. A. (1965). *J. biol. Chem.*, **240**, 4703
Rando, R. R. (1973). *J. Am. chem. Soc.*, **95**, 4438
Rando, R. R. (1974*a*). *Science, N.Y.* **185**, 320
Rando, R. R. (1974*b*). *Biochemistry*, **13**, 3859
Rando, R. R. (1977). *Biochemistry*, **16**, 4606
Rando, R. R. and Bangerter, F. W. (1976). *J. Am. chem. Soc.*, **98**, 6762
Rando, R. R. and Bangerter, F. W. (1977). *Biochem. biophys. Res. Commun.*, **76**, 1276
Rando, R. R., Relyea, N. and Cheng, L. (1976). *J. biol. Chem.*, **251**, 3306
Relyea, N. M., Tate, S. S. and Meister. A. (1974). *J. biol. Chem.*, **249**, 1519
Robin, M. M., Palfreyman, M. G. and Schechter, P. J. (1979). *Life Sci.*, **25**, 1103
Saier, M. H. and Jenkins, W. T. (1967). *J. biol. Chem.*, **242**, 101
Seiler, N., Jung, M. J. and Koch.-Weser, J. (eds) (1978). *Enzyme-Activated Irreversible Inhibitors*, Elsevier/North-Holland, Amsterdam
Silverman, R. B. and Abeles, R. H. (1976). *Biochemistry*, **15**, 4718
Soper, T. S., Manning, J. M., Marcotte, P. A. and Walsh, C. T. (1977*a*). *J. biol. Chem.*, **252**, 1571
Soper, T. S., Jones, W. M., Lerner, B., Trop, M. and Manning, J. M. (1977*b*). *J. biol. Chem.*, **252**, 3170
Tanase, S. and Morino, Y. (1976). *Biochem. biophys. Res. Commun.*, **68**, 1301
Tate, S. S., Relyea, N. M. and Meister, A. (1969). *Biochemistry*, **9**, 5016
Taub, D. and Patchett, A. A. (1977). *Tetrahedron Lett.*, No. 32, 2745
Walsh, C. (1977). *Horizons Biochem. Biophys.*, **3**, 36
Wang, E. and Walsh, C. (1978). *Biochemistry*, **17**, 1313
Washtien, W. and Abeles, R. H. (1977). *Biochemistry*, **16**, 2485
Wood, W. A. and Gunsalus, I. C. (1949). *J. biol. Chem.*, **181**, 171

6

Biochemistry and pharmacology of enzyme-activated irreversible inhibitors of some pyridoxal phosphate-dependent enzymes

M. J. Jung, J. Koch-Weser and A. Sjoerdsma
(Centre de Recherche Merrell International,
16 rue d'Ankara, 67084-Strasbourg Cedex, France)

INTRODUCTION

In recent years a new approach to the design of inhibitors of certain enzymes has evolved which is based on the principle of substrate-induced irreversible inhibition (see Seiler *et al.*, 1978*a*). Such enzyme-activated inhibitors have also been called suicide enzyme inactivators (Abeles and Maycock, 1976), k_{cat} inhibitors (Rando, 1974*a*) or mechanism-based irreversible inhibitors (Rando, 1977*a*). Inhibitors of this type may offer advantages over classical inhibitors as new drugs. The specificity of the inhibition may be expected to decrease unwanted side effects. Since the inhibition is not dependent on sustained tissue concentration of inhibitor and since new enzyme has to be synthesised to overcome the inhibition, high potency and long duration of action may also be achieved.

Pyridoxal-5′-phosphate-dependent enzymes catalyse many reactions. Among the most important are transamination of α- and ω-amino acids, decarboxylation of α-amino acids, amine oxidation and racemisation of L- or D-amino acids (Dunathan, 1971; Davis and Metzler, 1972). Some of these reactions, such as aspartate and alanine transamination, are essential to the living organism and not known to function excessively in disease states, so that it would be of little therapeutic interest to inhibit them in mammals. Aspartate and alanine aminotransferase are, however, the two pyridoxal-dependent enzymes for which enzyme-activated inhibitors have been most widely studied. Such compounds are serine-*O*-sulphate (John and Fasella, 1969), fluoro- and chloroalanine (Morino and Okamoto, 1973; Silverman and Abeles, 1976), L-2-amino-4-methoxy-*trans*-2-butenoic acid (Rando *et al.*, 1976), vinylglycine (Rando, 1974*b*), and propargylglycine (Marcotte and Walsh, 1975). The experience accumulated in these studies was very useful for designing inhibitors of other enzymes as exemplified by the inhibition of 4-aminobutyric acid aminotransferase with ethanolamine-*O*-sulphate (Fowler and John, 1972). It should be mentioned that the D-enantiomer of most of these glycine or alanine analogues are inhibitors of bacterial alanine racemase (Wang and Walsh, 1978) and hence should possess antibacterial activity, as has been shown for D-fluoroalanine (Kollonitsch *et al.*, 1973).

The pyridoxal-dependent enzymes we chose as target for inhibition were 4-aminobutyric acid:2-oxoglutarate aminotransferase (EC 2.6.1.19; GABA-T), L-glutamate-1-carboxylyase (EC 4.1.1.15; GAD), L-ornithine-1-carboxylyase (EC 4.1.1.17; Orn-DC) and L-aromatic amino acid 1-carboxylyase (EC 4.1.1.28; AADC). This choice was governed by the importance of the physiological roles of the products or substrates of these enzymes. Inhibitors were known for all four enzymes. However, they were either of a competitive nature, such as α-methylornithine for Orn-DC (Abdel-Monem *et al.*, 1974) and D-glutamate or α-methylglutamate for GAD (Wu, 1976), or pyridoxal scavengers, such as aminooxyacetic acid for GABA-T (Wallach, 1961), α-hydrazino-δ-aminovaleric acid for Orn-DC (Kato *et al.*, 1976) and α-methylhydrazinodopa for AADC (Sletzinger *et al.*, 1963). With the exception of α-methylhydrazinodopa in combination with dopa (Bianchine, 1976), none of these compounds has found therapeutic application, despite their interesting pharmacologic activities. Inhibitors of the four enzymes of concern to us, GABA-T, GAD, Orn-DC and AADC, were developed and studied in parallel. Unexpected results with one enzyme were often useful in designing irreversible inhibitors for another. With the exception of the GABA-T inhibitors ethanolamine-*O*-sulphate and gabaculine, all the compounds discussed here were synthesised and studied by scientists at our Centre.

INHIBITORS OF 4-AMINOBUTYRIC ACID:2-OXOGLUTARATE AMINOTRANSFERASE

γ-Aminobutyric acid (GABA) is widely accepted as the main inhibitory neurotransmitter in the mammalian central nervous system (Krnjevic, 1976). It is synthesised through decarboxylation of L-glutamic acid catalysed by GAD and metabolised mainly by transamination with 2-oxoglutarate catalysed by GABA-T (scheme 6.1). Three types of enzyme-activated inhibitors are now known for GABA-T. For each type the mechanism of the inhibition is based on redirecting or taking advantage of the electron flow following abstraction of the α-hydrogen by the enzyme. The normal mechanism of GABA transamination and the postulated or demonstrated mechanisms of inhibition are detailed in scheme 6.2.

The first enzyme-activated irreversible inhibitor of GABA-T was ethanolamine-*O*-sulphate (Fowler and John, 1972; scheme 6.2(a)). The mechanism of the inhibition requires elimination of the sulphate group, thus generating an electrophilic species in the active site of the enzyme, which can then alkylate the enzyme. Alternatively, the electrophilic species can be generated by the normal prototropy as depicted in scheme 6.2(b) for γ-vinyl GABA (4-aminohex-5-enoic acid; Lippert *et al.*, 1977) or by allenic rearrangement (scheme 6.2(c)) as has been postulated for γ-acetylenic GABA (4-aminohex-5-ynoic acid; Jung and Metcalf, 1975). Finally, gabaculine (5-amino-1, 3-cyclohexadienyl carboxylic acid; Rando, 1977*b*)

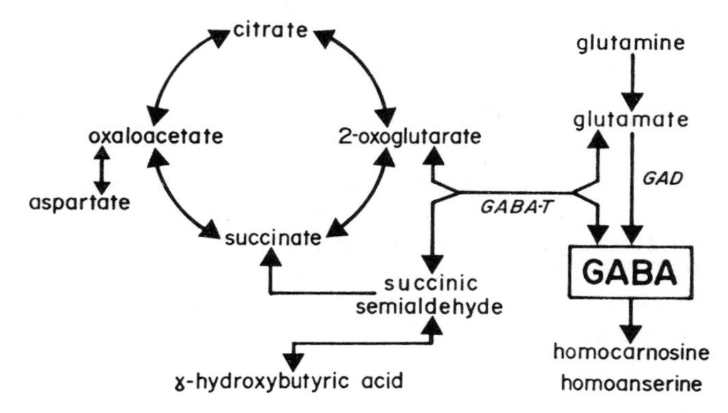

Scheme 6.1 Metabolic pathways of GABA

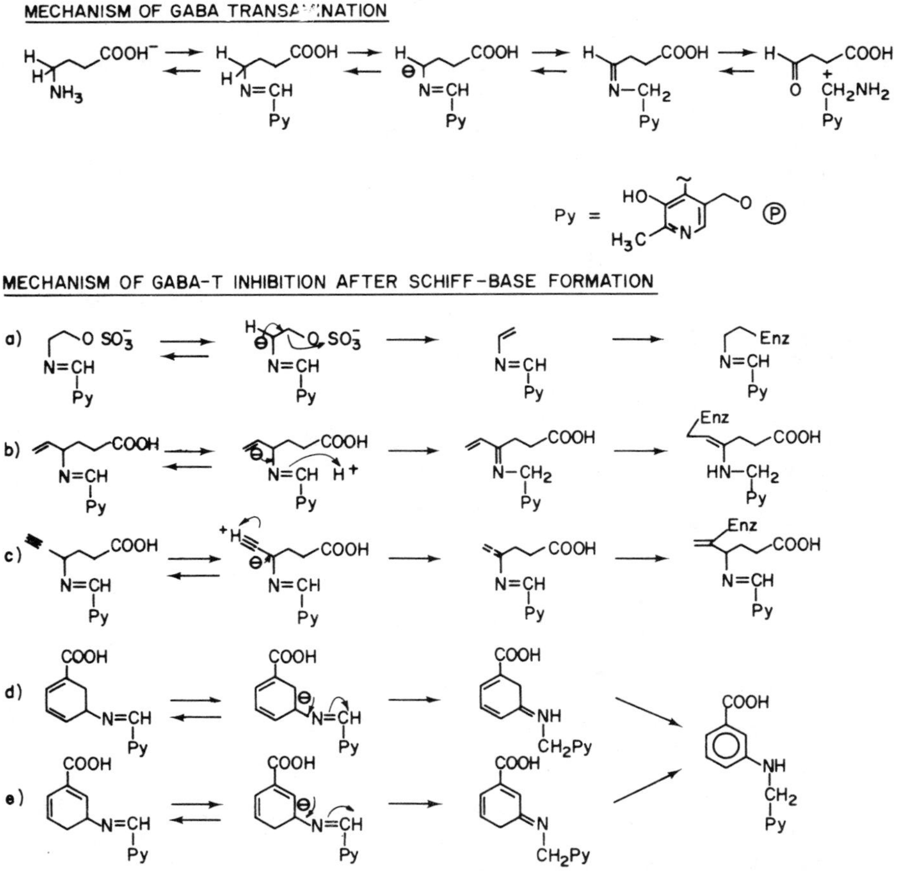

MECHANISM OF GABA TRANSAMINATION

Py =

MECHANISM OF GABA-T INHIBITION AFTER SCHIFF-BASE FORMATION

a)

b)

c)

d)

e)

Scheme 6.2

and isogabaculine (3-amino-1, 5-cyclohexadienyl carboxylic acid; Metcalf *et al.*, 1978*a*) inhibit GABA-T by the mechanisms shown in schemes 6.2(d) and 6.2(e), respectively. In these cases, no covalent bond is formed between the enzyme and the inhibitor. Instead, a tight-binding, stable inhibitor is generated from the GABA analogue and the coenzyme.

In vitro, all five compounds inhibit GABA-T in a time-dependent manner. The rate of inhibition follows pseudo-first-order kinetics at least for several half-lives and is concentration-dependent as shown in figure 6.1 for isogabaculine (Metcalf and Jung, 1979). The rates of inhibition of pig brain GABA-T by 0.1 mM ethanolamine-*O*-sulphate, γ-acetylenic and γ-vinyl GABA are comparable. Isogabaculine is about 10 times and gabaculine 100 times more effective.

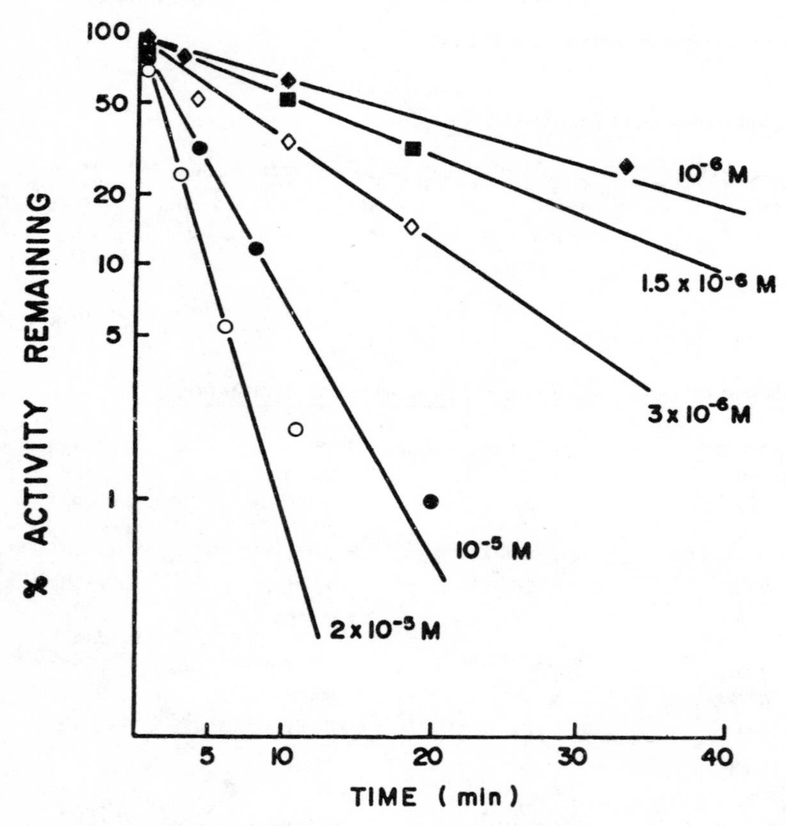

Figure 6.1 Inhibition of pig brain GABA-T by isogabaculine

When given systemically to rats or mice, these compounds produce a time-dependent decrease of brain GABA-T activity, accompanied by a significant elevation of brain GABA concentrations. The potencies and the rapidity of onset and duration of effects are compared in table 6.1. The differences in potencies are

Table 6.1 Comparison between the effects of five enzyme-activated inhibitors of GABA-T (Jung, 1978)

Compound	GABA-T inhibition			GABA			References
	Dose (mmol kg^{-1})	Time needed for 50% inhibition (min)	Inhibition at 24 h (%)	Rate of GABA accumulation (μmol g^{-1} h^{-1})	Maximal GABA levels (time in parenthesis (μg (g brain)$^{-1}$)	GABA levels at 24 h (μg g^{-1})	
Ethanolamine-*O*-sulphate	15.8	120	85	0.58	–	–	Leach and Walker (1977)
γ-Acetylenic GABA	0.78	45	85	4.0	950 (8 h)	350	Jung *et al.* (1977*b*)
γ-Vinyl GABA	11.6	50	85	6.5	1350 (6–24 h)	1350	Jung *et al.* (1977*a*)
Gabaculine	0.72	30	87	5.3	2400 (24 h)	2400	Rando and Bangerter (1977)
Isogabaculine	0.24	30	80	2.5	1840 (12–24 h)	1200	Schechter *et al.* (1979)

most probably due to the accessibility of the brain to the product. Ethanolamine-*O*-sulphate was initially thought not to act centrally after peripheral administration (Fowler, 1973), but it does so when large enough doses are administered (Leach and Walker, 1977). γ-Vinyl GABA, while as potent as γ-acetylenic GABA in inhibiting purified GABA-T, has to be administered in 10–15 times larger doses in order to obtain comparable concentrations of free drug in the brains of mice (Jung *et al*., 1977*a,b*). Isogabaculine and gabaculine are 10–100 times more potent than γ-acetylenic GABA as inhibitors of GABA-T *in vitro*; nevertheless similar doses have to be used *in vivo* (Schechter *et al*., 1979; Rando and Bangerter, 1977). The inhibition of brain GABA-T by ethanolamine-*O*-sulphate is much slower than in the case of the other compounds, and this is reflected by the rate of GABA accumulation. Although the duration of GABA-T inhibition is similar for all compounds (80–90 per cent inhibition at 24 h), the patterns of GABA changes are not. For instance, peak GABA concentrations are reached 8 h after γ-acetylenic GABA (Jung *et al*., 1977*b*), while GABA continues to accumulate for 24 h after gabaculine (Rando and Bangerter, 1977). Long-lasting plateau values are reached 4 h after γ-vinyl GABA (Jung *et al*., 1977*a*) or 12 h after isogabaculine (Schechter *et al*., 1979). The reasons for these differences are not understood.

The compounds also differ in their enzyme selectivity. None of them has any appreciable action on aspartate or alanine aminotransferase. However, L-ornithine: 2-oxoacid aminotransferase is inhibited *in vitro* and *in vivo* by γ-acetylenic GABA and gabaculine (Jung and Seiler, 1978) and by isogabaculine (Metcalf and Jung, 1979). This inhibitory activity is not surprising, since this enzyme seems to accept GABA as a weak substrate (Jung and Seiler, 1978). γ-Acetylenic GABA also irre-

Table 6.2 Summary of pharmacological effects of four enzyme-activated GABA-T inhibitors (Schechter and Tranier, 1978)

Test	γ-Acetylenic GABA	γ-Vinyl GABA	Gabaculine	Isogabaculine
Spontaneous motor activity	Depresses	Depresses	Depresses	Depresses
Body temperature	Depresses	Depresses	Depresses	Depresses
Preconvulsive behaviour at high doses	Yes	Never	Yes	Yes
Protection against:				
Audiogenic seizures	Active	Active	Active	Active
Isoniazid seizures	Active	Active	Slightly active	Slightly active
Thiosemicarbazide seizures	Active		Slightly active	Slightly active
Strychnine seizures	Active	Active	Inactive	Inactive
Pentylenetetrazol seizures	Inactive	Inactive	Slightly active	Slightly active
Electroshock threshold	Active	Active	Inactive	Inactive
Picrotoxin seizures	Inactive	Inactive	Inactive	Inactive
Bicuculline seizures	Inactive	Inactive	Inactive	Inactive
LD_{50} (CL_{95}) mice IP	180 mg kg^{-1} (155–209)	3000 mg kg^{-1} (1950–4830)	48 mg kg^{-1} (39–58)	51 mg kg^{-1} (31–81)

versibly inhibits GAD from mammalian brain or bacteria (Jung *et al.*, 1978*b*), as will be discussed in the next section.

The various GABA-T inhibitors produce similar effects in animals, chiefly sedation, hypothermia and, at high doses, catatonia (Schechter and Tranier, 1978). They protect genetically-sensitive mice against audiogenic seizures in a dose-dependent manner. The amount of protection has been tentatively correlated with the elevation of total brain GABA concentration (Schechter *et al.*, 1977). The protection provided by γ-acetylenic GABA, γ-vinyl GABA, gabaculine and isogabaculine differ to some extent (table 6.2). To explain these differences, we have invoked regional differences in GABA elevation (Jung, 1978) or differential effects on distribution of GABA at the cellular level (Jung, 1978; Sarhan and Seiler, 1979). It is apparent that a variety of compounds is now available which can markedly and selectively modify GABA metabolism in the brain. Whether these compounds will be of benefit in diseases possibly associated with disturbances in GABAergic mechanisms (for example epilepsy, Huntington's chorea, tardive dyskinesia and schizophrenia; Schechter and Tranier, 1978) remains to be seen.

INHIBITORS OF L-GLUTAMATE-1-CARBOXYLASE

While investigating the effects of γ-acetylenic GABA on brain GABA metabolism at our Centre, Lippert discovered that this compound also irreversibly inhibits mouse brain GAD (Jung *et al.*, 1977*b*). This finding was unwelcome, since our objective was to increase brain GABA levels. We soon realised that the inhibition was due to the (+)-enantiomer, which is unfortunately the one which inhibits GABA-T (B. Lippert, unpublished results). The absolute configuration of the two optical isomers could be determined by oxidation of γ-acetylenic GABA to gluta-

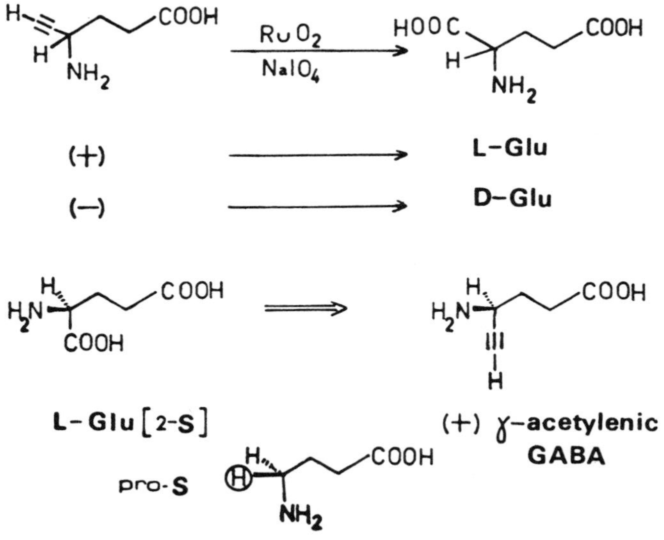

Figure 6.2 Absolute configuration of the enantiomers of (±)-γ-acetylenic GABA

MECHANISM OF L-GLUTAMATE DECARBOXYLATION

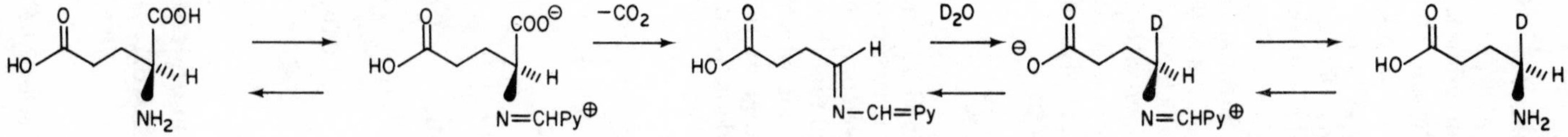

$-CO_2$

D_2O

MECHANISM OF BACTERIAL GAD INHIBITION BY (−)−γ−ACETYLENIC GABA.

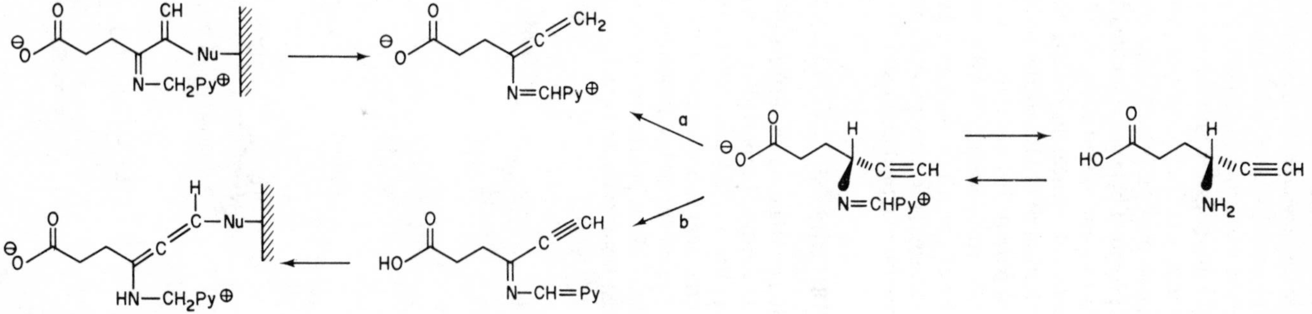

a

b

Scheme 6.3

mic acid (figure 6.2): the (+)-enantiomer has the (**S**) configuration, the (−)-enantiomer the (**R**). Using commerical bacterial enzymes, we found that (4**S**)-4-amino-hex-5-ynoic acid inhibits GABA-T, while the **R**-enantiomer inhibits GAD. From the fact that decarboxylation of L-glutamic acid by bacterial GAD occurs with retention of configuration (Yamada and O'Leary, 1978), the inhibition of bacterial GAD by (4**R**)-4-amino-hex-5-ynoic acid can be rationalised (scheme 6.3). Assuming that γ-acetylenic GABA can replace GABA in the active site of GAD and form a Schiff base with pyridoxal phosphate, hydrogen abstraction from carbon 4 (step 3 → 2, scheme 6.3), the reverse of the protonation step in the normal decarboxylation reaction, would lead to the same reaction intermediate postulated for the mechanism of GABA-T inhibition by γ-acetylenic GABA (scheme 6.2(c)).

The analogous inhibition of mammalian GAD by (4**S**)-(+)-4-amino-hex-5-ynoic acid via reversion of the protonation step invoked above for the inhibition of bacterial GAD would imply that decarboxylation of L-glutamate by this enzyme occurs with inversion of configuration. This is not the case, however (M. Bouclier *et al.*, 1979). A possible mechanism is catalysis by mammalian GAD of α-hydrogen abstraction from analogues of L-glutamate which cannot be decarboxylated, that is GAD could function as a glutamate transaminase.

The following compounds are rather inefficient irreversible inhibitors of purified glutamic acid decarboxylase: α-vinylglutamate (B. W. Metcalf, unpublished results); α-chloromethyl glutamate and α-fluoromethyl glutamate (P. Bey, unpublished results; Kollonitsch *et al.*, 1978). None is potent enough to reduce the synthesis of GABA in the brain. A more active compound would be an interesting investigative tool but probably of little therapeutic interest, since compounds like mercaptopropionic acid (Adcock and Taberner, 1978) and allylglycine (Fisher and Davies, 1976), which inhibit the biosynthesis of GABA as well as other reactions, are potent convulsant agents. However, the concept of inhibition of an amino acid decarboxylase by the acetylenic analogue of the product amine, which was proposed and studied at our Centre, has been applied with success in the inhibition of Orn-DC (Metcalf *et al.*, 1978*b*).

INHIBITORS OF L-ORNITHINE-1-CARBOXYLASE

Orn-DC catalyses the decarboxylation of L-ornithine to putrescine, which serves as a precursor of spermidine and spermine (scheme 6.4). Evidence accumulated over the last 10 years concerning the role of these di- and polyamines in cellular growth and division (Russell, 1973; Jänne *et al.*, 1978), suggests that a potent Orn-DC inhibitor might be useful for controlling diseases characterised by excessive cell proliferation. In fact, α-hydrazino-δ-aminovaleric acid, a competitive inhibitor of Orn-DC, slows the growth of a mouse sarcoma (Kato *et al.*, 1976). However, competitive inhibitors of Orn-DC cause an accumulation of the enzyme, presumably by stabilising it against degradation (Harik *et al.*, 1974). This may result in an overproduction of putrescine, once the inhibitor has been cleared from the organism. It was, therefore, decided to search for irreversible inhibitors.

We noted in the preceding section that γ-acetylenic GABA inhibits mammalian brain GAD irreversibly but that the mechanism of inhibition has not been totally clarified. Metcalf suggested that (±)-α-acetylenic putrescine (5-hexyne-1, 4-diamine) should, by analogy, inhibit Orn-DC. In fact, when Orn-DC prepared

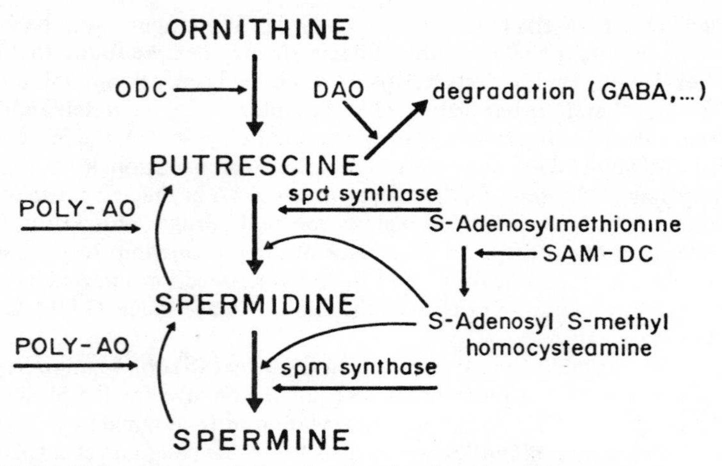

Scheme 6.4 Metabolic pathways of di- and polyamines in eukaryotic systems

either from rat liver or rat prostate was incubated with α-acetylenic putrescine at concentrations of 0.1–10 μM, there was a time-dependent irreversible decrease of enzyme activity. For instance, at a concentration of 2 μM the enzyme preparation was 50 per cent inhibited within 20 min (Metcalf *et al.*, 1978*b*). Rat liver Orn-DC is inhibited by (–)-5-hexyne-1, 4-diamine; however, the absolute configuration of this enantiomer could not be determined. Therefore, it remains unknown whether α-acetylenic putrescine inhibits Orn-DC by a mechanism involving the reversion of the protonation step as was discussed for bacterial GAD and (–)-γ-acetylenic GABA or by the abnormal transamination invoked for the inhibition of mammalian GAD by (+)-γ-acetylenic GABA. This problem is not a purely academic one, since it is possible that bacterial Orn-DC is inhibited by one isomer and mammalian Orn-DC by the other.

When injected into rats at doses of 10–200 mg kg^{-1}, α-acetylenic putrescine reduced Orn-DC activity in prostate, testis and thymus in a dose-dependent, long-lasting manner (Danzin *et al.*, 1979*c*). After a dose of 100 mg kg^{-1}, the prostatic enzyme was totally inhibited within 1 h. This level of inhibition was maintained for about 6 h, and even 24 h after drug administration prostatic Orn-DC activity had returned only to 25–30 per cent of control. The time-course of enzyme inhibition in thymus and testis was similar, but the maximal effects were less pronounced (80 per cent inhibition in testis and 50 per cent in thymus).

Further investigations with this compound were complicated by the finding that doses above 100 mg kg^{-1} resulted in similar behavioural changes as comparable doses of γ-acetylenic GABA, the irreversible GABA-T inhibitor discussed previously. In fact, brain GABA-T activity was markedly decreased and brain GABA concentrations elevated in animals treated with 200 mg acetylenic putrescine kg^{-1} (Danzin *et al.*, 1979*b*). These effects can be blocked by injection of pargyline, a monoamine oxidase inhibitor, prior to administration of α-acetylenic putrescine. Injection of aminoguanidine, a diamine oxidase inhibitor, did not prevent the changes in GABA metabolism. We demonstrated the presence of γ-acetylenic GABA in liver and brain of mice injected with α-acetylenic putrescine. Based on the inhibitory action of pargyline, we propose that γ-acetylenic GABA is formed by the dual action of mitochondrial monoamine oxidase and of a non-specific aldehyde oxidase. Modifications of the 5-hexyne-1, 4-diamine skeleton may prevent this undesirable metabolic transformation while retaining the Orn-DC inhibitory action.

Meanwhile, we had found that α-difluoromethylornithine is also an enzyme-activated inhibitor of Orn-DC (Metcalf *et al.*, 1978*b*). The proposed mechanism of action is shown in scheme 6.5. At 20 μM, α-difluoromethylornithine and α-

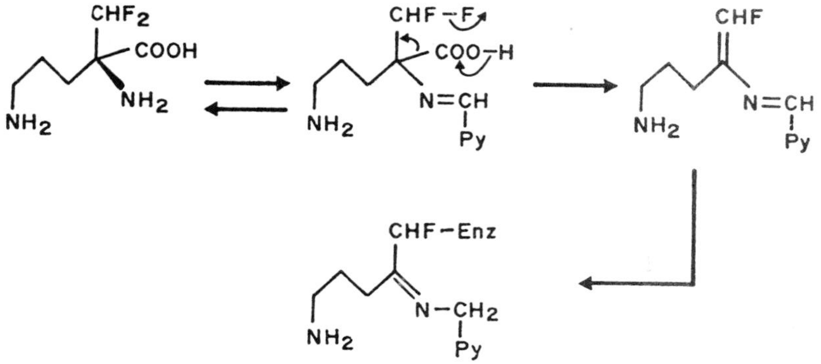

Scheme 6.5 Mechanism of Orn-DC inhibition by α-difluoromethylornithine (Bey, 1978)

acetylenic putrescine inactivate rat liver Orn-DC with a half-life of 10 min. This similar potency is also apparent *in vivo*. A dose of 100 mg α-difluoromethylornithine kg^{-1} decreases Orn-DC activity in prostate, thymus and testis of rats to a similar extent and duration as an equivalent dose of α-acetylenic putrescine (Danzin *et al.*, 1979*a*). α-Difluoromethylornithine has minimal toxicity, doses as high as 1 g kg^{-1} having been administered twice daily for 10 days without any obvious effect. Such treatment considerably reduces the concentrations of putrescine and spermidine in prostate, thymus and testis (table 6.3). That of spermidine is significantly reduced only in the prostate. The reduction of polyamine concentration is accompanied by a reduced weight gain of prostate and thymus compared with control animals of the same age.

Since Orn-DC activity begins to recover 6 h after difluoromethylornithine administration, injection of the drug every 6 h or, even better, oral administration in drinking water have been used to test its effects in different animal models of rapid tissue proliferation. The first model was testosterone-induced regrowth of the ventral prostate of castrated rats. During involution of the prostate following castration, there is a marked decrease of Orn-DC activity and a dramatic reduction of putrescine, spermidine and spermine concentrations in this gland (Takyi *et al.*, 1977, and references cited therein). The effects of castration are reversed by daily injections of testosterone. Administration of difluoromethylornithine intraperitoneally at 6 h intervals blocks the testosterone-induced increase of putrescine and

Table 6.3 Polyamine levels in rat tissues after repeated administration of -difluoromethylornithine

Tissue		Putrescine† (nmol g^{-1})	Spermidine† (nmol g^{-1})	Spermine† (nmol g^{-1})	Organ weight† (g)
Prostate	C	202 ± 37	7334 ± 193	2968 ± 293	0.38 ± 0.02
	100 mg kg^{-1}	4 ± 4	1188 ± 289	2741 ± 346	0.35 ± 0.03
	1 g kg^{-1}	2 ± 2	305 ± 64	2207 ± 197	0.24 ± 0.02
Thymus	C	126 ± 8	1536 ± 12	604 ± 24	0.71 ± 0.05
	100 mg kg^{-1}	58 ± 3	1154 ± 63	609 ± 19	0.62 ± 0.05
	1 g kg^{-1}	31 ± 3	779 ± 62	900 ± 43	0.45 ± 0.03
Testis	C	19 ± 3	203 ± 34	330 ± 34	1.38 ± 0.05
	100 mg kg^{-1}	1.6 ± 1	200 ± 22	347 ± 46	1.38 ± 0.02
	1 g kg^{-1}	1.2 ± 0.2	118 ± 10	395 ± 17	1.29 ± 0.04
Spleen	C	37 ± 2	796 ± 12	504 ± 18	0.66 ± 0.05
	100 mg kg^{-1}	30 ± 3	792 ± 28	524 ± 14	0.77 ± 0.06
	1 g kg^{-1}	–	–	–	0.69 ± 0.04
Liver	C	4.7 ± 0.2	1028 ± 120	660 ± 68	9.62 ± 0.80
	100 mg kg^{-1}	4.9 ± 0.7	860 ± 198	774 ± 146	10.29 ± 0.42
	1 g kg^{-1}	–	–	–	8.30 ± 0.09

†Mean ± SEM, *n* = 5.

spermidine and slows the accumulation of spermine, as well as the recovery of weight (Danzin *et al.*, 1979*d*). Cytological studies indicate that α-difluoromethylornithine inhibits the restoration of prostatic secretions induced by testosterone. This finding may point to a regulatory role of putrescine or spermidine in the secretory activity of the prostate.

Difluoromethylornithine prolongs the survival time of mice inoculated with L1210 leukemia cells. The time between intraperitoneal injection of 10^6 cells and death of 50 per cent of the animals was increased from 7.5 days for control animals to 9.5 days for treated animals (Prakash *et al.*, 1978). The Orn-DC inhibitor markedly reduces the growth rate of a solid tumour in mice. For instance, 25 days after subcutaneous inoculation of 10^5 EMT_6 cells into mice, the tumours weighed 5 g in control animals but only 2 g in the difluoromethylornithine-treated group (Seiler *et al.*, 1978*b*).

These few examples demonstrate the usefulness of Orn-DC inhibitors but also show two limitations. Because Orn-DC is synthesised rapidly, adequate levels of irreversible Orn-DC inhibitors have to be constantly maintained in the target tissues in order to keep the synthetic capacity for putrescine as low as possible. Furthermore, even when putrescine concentrations are low, the organism can maintain or partly restore the spermine pool. Combining an Orn-DC inhibitor with an inhibitor of one of the other enzymes in the biosynthetic pathway of polyamines (that is, for example, *S*-adenosyl methionine decarboxylase, spermidine synthetase or spermine synthetase) may permit reduction of the spermine pool as well and may enhance the promising effects observed with the Orn-DC inhibitors.

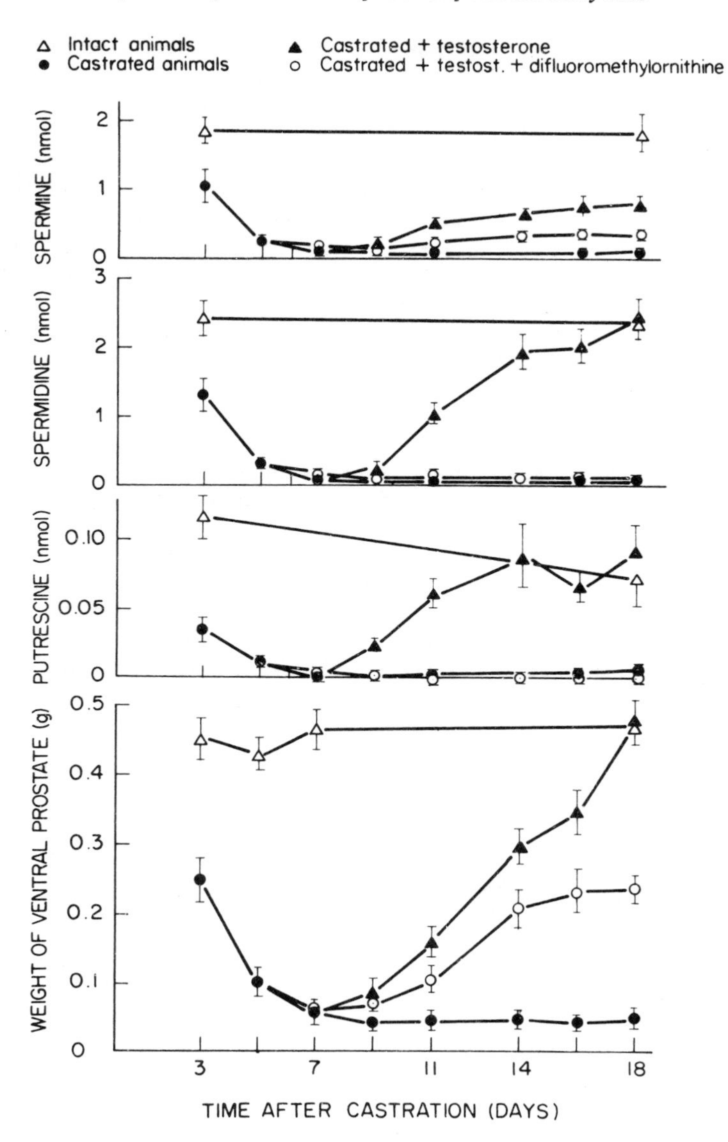

Figure 6.3 Effect of difluoromethylornithine on testosterone-induced regrowth of prostate of castrated rats and the replenishment of polyamines (Danzin *et al.*, 1979*d*)

INHIBITORS OF AROMATIC AMINO ACID DECARBOXYLASE

AADC is an essential enzyme for the 'biosynthesis of dopamine, noradrenaline, adrenaline and serotonin in the peripheral and central nervous systems. Since amino acid decarboxylation is normally not the rate-limiting step in the biosynthetic pathway, an inhibitor of this enzyme may have a dual utility depending on the extent and site of the inhibition.

Partial peripheral inhibition

Inhibitors of the enzyme in peripheral tissues slow the decarboxylation of exogenously supplied L-dopa or L-5-hydroxytryptophan to their corresponding monoamines, even if the inhibition is not marked enough to reduce the synthesis of the endogenous amines. As a result, more of the amino acid reaches the brain to be decarboxylated there. Examples of such inhibitors are carbidopa (α-methylhydrazinodopa; Sletzinger *et al.*, 1963) and benserazide (Burkard *et al.*, 1963). Both compounds are used in combination with dopa for the treatment of parkinsonism. Difluoromethyldopa is also an inhibitor of this kind (Palfreyman *et al.*, 1978*a*).

Purified hog kidney AADC loses its activity in a time- and concentration-dependent, irreversible manner when incubated with 10–100 μM difluoromethyldopa. The kinetics of inactivation are more complex than in the cases discussed above; 80 per cent of enzyme activity is lost within 5–6 min (10 μM difluoromethyldopa), but after this initial rapid phase, the rate of inhibition slows. A mechanism identical to that of Orn-DC inhibition by difluoromethylornithine is suggested (scheme 6.5). When given intraperitoneally or orally to mice, difluoromethyldopa produces a rapid, long-lasting and dose-dependent inhibition of AADC in kidney and heart, but has only a marginal effect on the brain enzyme at the highest doses (figure 6.4 (a)). The inhibitory effect is most pronounced in the kidney.

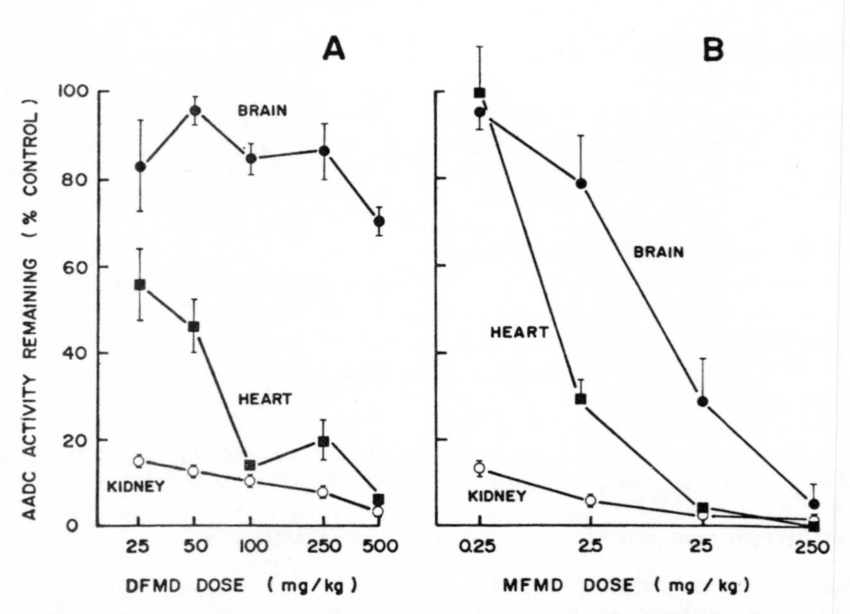

Figure 6.4 Inhibition of AADC by difluoromethyldopa (A) and monofluoromethyldopa (B) in brain, heart and kidney of mice (Palfreyman *et al.*, 1978*a*; Jung *et al.*, 1979)

Daily administration of 1 g difluoromethyldopa kg^{-1} for 8 days does not significantly reduce the catecholamine content of brain and heart or the serotonin content of heart and kidney (Palfreyman *et al.*, 1978*b*). When administered orally together with L-dopa or 5-hydroxytryptophan, it reduces the peripheral decarboxylation of these aminoacids and increases their entry into the brain where they can be decarboxylated (table 6.4). In agreement with the biochemical results,

Table 6.4 Effect of α-difluoromethyldopa on metabolism of exogenous dopa and 5-hydroxytryptophan (Palfreyman *et al.*, 1978*b*)

	Dopa ($ng\ g^{-1}$)	DA ($ng\ g^{-1}$)	NA ($ng\ g^{-1}$)	Brain 5-HT ($ng\ g^{-1}$)	Kidney 5-HT ($ng\ g^{-1}$)
Control	8 ± 1	501 ± 13	217 ± 5	1.00 ± 0.14	0.39 ± 0.03
DFMD (100 mg kg^{-1})	14 ± 2	717 ± 91	240 ± 3	1.00 ± 0.04	0.37 ± 0.04
L-Dopa (250 mg kg^{-1}, IP)	28 ± 5	448 ± 8	224 ± 6	–	–
DFMD + dopa	2810 ± 960	1886 ± 450	194 ± 8	–	–
5-HTP (50 mg kg^{-1})	–	–	–	0.99 ± 0.05	28.0 ± 2.4
DFMD + 5-HTP	–	–	–	2.50 ± 0.1	16.3 ± 3.7

DA = dopamine; NA = noradrenaline; 5-HT = 5-hydroxytryptamine; DFMD = difluoromethyl dopa; 5-HTP = 5-hydroxytryptophan

difluoromethyldopa potentiates in a dose-related manner the 'wet-dog' shaking behaviour of rats injected with 5-hydroxytryptophan. This response is correlated with the elevation of brain serotonin (Fozard and Palfreyman, 1978). Difluoromethyldopa also prevents the peripheral pressor effect of small intravenous doses of dopa in dogs and potentiates the central depressor effects (Palfreyman *et al.*, 1978*b*).

Total peripheral and/or central inhibition

As already mentioned, AADC is not the rate-limiting enzyme in the endogenous synthesis of serotonin and the catecholamines. From the ratio of tyrosine hydroxylase to AADC activity in brain, one can estimate that AADC activity must be inhibited by over 99 per cent to affect catecholamine synthesis. Monofluoromethyldopa, another enzyme-activated inhibitor of AADC (Kollonitsch *et al.*, 1978) has such an intense effect (Jung *et al.*, 1979).

At a concentration of 10 μM, monofluoromethyldopa completely blocks activity of purified hog kidney AADC within a few minutes. The mechanism of the inhibition has been unequivocally demonstrated (Kollonitsch *et al.*, 1978) and supports in retrospect those suggested for Orn-DC and AADC inhibition by the corresponding difluoromethyl analogues. The potency of monofluoromethyldopa is equally dramatic *in vivo*, but organ selectivity is maintained at the lower doses (figure 6.4(b)). Doses of 0.25–2.5 mg kg^{-1} markedly reduce AADC activity in kidney and heart of mice. Inhibition of the brain enzyme becomes significant at doses higher than 5 mg kg^{-1} and is complete at doses of 100-250 mg kg^{-1}. The onset of inhibition is as rapid as with difluoromethyldopa, but the inhibition is of longer duration. For instance, 24 h after a dose of 100 mg kg^{-1} of monofluoromethyldopa, AADC activity in kidney, heart and brain remains less than 8-10 per cent of control.

The degree of AADC inhibition after a single dose of monofluoromethyldopa is sufficient to decrease catecholamine concentrations significantly in all three organs. To prove that this decrease is due solely to the inhibition of dopa decarboxylation, changes in concentrations of dopa, dopamine, noradrenaline and dihydroxyphenylacetic acid (dopac) were measured in brains of mice after treatment with monofluoromethyldopa, reserpine, monofluoromethyldopa plus reserpine, pargyline and monofluoromethyldopa plus pargyline (figure 6.5).

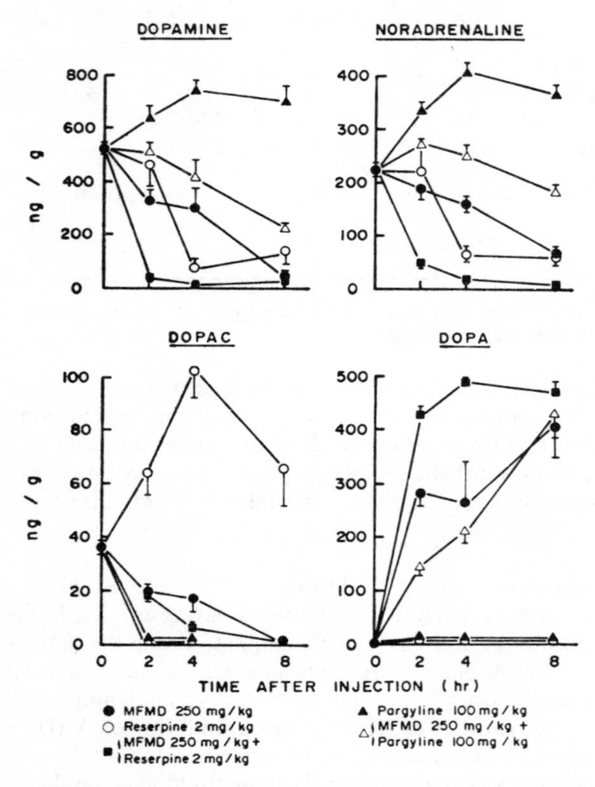

Figure 6.5 Effect of monofluoromethyldopa (MFMD) on brain catecholamine metabolism

Monofluoromethyldopa potentiates the decrease of dopamine and noradrenaline caused by reserpine. It converts the increase in concentration of these amines by pargyline into a decrease. The concentration of dopac, the main metabolite of dopamine, decreases progressively after monofluoromethyldopa. In brains of mice treated with reserpine, the concentration of dopac is increased three- to four-fold, but no increase is seen when the AADC inhibitor is given concomitantly. Dramatic increases of dopa concentration are found in the brains of animals treated either with inhibitor alone or in combination with reserpine or pargyline. From the accumulation of dopa after monofluoromethyldopa administration, it can be concluded that there is little or no inhibition of tyrosine hydroxylase. The decrease of dopac suggests that monofluoromethyldopa does not cause a release of catecholamines as does reserpine. The recovery of AADC activity and amine concentrations within 3-5 days indicates that, unlike 6-hydroxydopamine, the dopa analogue does not destroy nerve terminals. Finally, monofluoromethyldopa does not inhibit uptake or release of catecholamines in isolated synaptosomes (C. Gardner, unpublished results).

After five doses of 100 mg monofluoromethyldopa kg^{-1} at 12 h intervals, mice show a marked ptosis, not accompanied by diarrhoea, and extreme hypothermia (Jung *et al.*, 1979). In these animals, brain, heart and kidney are almost totally depleted of catecholamines. The concentration of serotonin is decreased by 80 per cent in brain, 60 per cent in heart and 30 per cent in kidney. AADC activity is undetectable in all three organs (table 6.5).

Table 6.5 Effect of repeated administration of monofluoromethyldopa on the biogenic amine content of mouse tissues (Jung *et al.*, 1979)

Tissue	Treatment	DA ($ng\ g^{-1}$)	NA ($ng\ g^{-1}$)	5-HT ($ng\ g^{-1}$)	AADC ($nmol\ g^{-1}\ min^{-1}$)
Brain	Saline	722 ± 47	310 ± 15	657 ± 11	64.0 ± 1.3
		(5)	(5)	(5)	
	MFMD	18.6 ± 2	19.6 ± 3	110 ± 10	~0
		(4)	(4)	(4)	
		−97%	−94%	−83%	
Heart	Saline	<14	499 ± 37	753 ± 113	27.8 ± 1.3
			(5)	(5)	
	MFMD	<14	46 ± 17	280 ± 9	~0
			(4)	(4)	
			−91%	−63%	
Kidney	Saline	<7	160 ± 4	293 ± 34	100 ± 15
			(3)	(5)	
	MFMD	<7	19 ± 3	199 ± 12	~0
			(4)	(4)	
			−88%	−32%	

MFMD = monofluoromethyldopa; AADC = L-aromatic amino acid decarboxylase. For other abbreviations see footnote to table 6.4.

At doses lower than 5 mg kg^{-1}, some effects of monofluoromethyldopa are similar to those of the difluoromethyl analogue. Upon systemic administration of dopa or 5-hydroxytryptophan, it increases the concentrations of these amino acids in the brain and enhances the rises in brain dopamine, noradrenaline and serotonin (Palfreyman *et al.*, 1979).

Thus, structural modifications of α-methyldopa yield two enzyme-activated irreversible inhibitors of AADC with distinct actions. Replacement of the methyl by a difluoromethyl group results in a peripherally-selective decarboxylase inhibitor which should be useful for combined administration with dopa or 5-hydroxytryptophan. Substitution of the methyl by a monofluoromethyl group yields an extremely potent inhibitor which only acts peripherally in low doses but centrally as well in higher doses. This is the first compound to decrease the concentration of endogenous biogenic amines by inhibiting AADC. Both compounds are valuable investigative tools and might well find therapeutic applications.

CONCLUSIONS

α-Acetylenic amines and α-vinyl amines can irreversibly inhibit the transamination of ω-amino acids. α-Acetylenic amines and α-monofluoromethyl or α-difluoromethyl analogues of α-amino acids irreversibly inhibit the parent amino acid decarboxylases. These few examples serve to illustrate the importance of understanding the molecular mechanism of enzyme catalysis. The design of enzyme-activated inhibitors of pyridoxal-dependent enzymes is greatly facilitated by the covalent nature of the substrate–coenzyme complex which allows the reactive intermediate to alkylate the enzyme before diffusion out of the active site can occur. However, even for this well-known class of enzymes, the design of enzyme-activated inhibitors is not without surprises. For instance, α-acetylenic dopamine is but a poor inhibitor of AADC (G. Gayon, unpublished results) while acetylenic putrescine is a potent inhibitor of Orn-DC. The potencies of di- and monofluoromethylornithine as inhibitors of rat liver Orn-DC are comparable (Bey, 1978), while difluoromethyldopa must be used at 100 times the concentration of the monofluoromethyl analogue to obtain similar degrees of inactivation of AADC. This lack of predictability requires studies on purified enzymes employing the whole array of methods and techniques available to the enzymologist. The pharmacologist must adapt to a situation in which pharmacological action is not dependent on the continuous presence of unchanged drug. Because of the specificity of their action, some of the compounds we have discussed here hold promise as therapeutic agents.

ACKNOWLEDGEMENTS

The authors acknowledge the important contributions of many colleagues to this work: P. Bey and B. W. Metcalf in chemistry; M. Bouclier, C. Danzin and B. Lippert in biochemistry; M. G. Palfreyman, N. Prakash, P. J. Schechter and J. K. Woodward in pharmacology; J. Grove and J. Wagner in analytical biochemistry.

REFERENCES

Abeles, R. H. and Maycock, A. L. (1976). *Accounts Chem. Res.*, 9, 313
Abdel-Monem, M. M., Newton, N. E. and Weeks, C. E. (1974). *J. med. Chem.*, 17, 447
Adcock, T. and Taberner, P. V. (1978). *Biochem. Pharmac.*, 27, 246
Aures, D., Hakanson, R. and Clark, W. G. (1970). In *Handbook of Neurochemistry* vol. 5 (ed. A. Lajtha), Plenum Press, New York, p. 165
Bey, P. (1978). In *Enzyme Activated Irreversible Inhibitors* (ed. N. Seiler, M. J. Jung, and J. Koch-Weser), Elsevier/North-Holland, Amsterdam, p. 27
Bianchine, J. R. (1976). *New Engl. J. Med.*, **295**, 814

Bouclier, M., Jung, M. J. and Lippert, B. (1979). *Eur. J. Biochem.*, **98**, 363
Burkard, W., Gey, K. F. and Pletscher, A. (1963). *Experientia*, **18**, 411
Danzin, C., Jung, M. J., Grove, J. and Bey, P. (1979*a*). *Life Sci.*, **24**, 519
Danzin, C., Jung, M. J., Seiler, N. and Metcalf, B. W. (1979*b*). *Biochem. Pharmac.*, **28**, 633
Danzin, C., Jung, M. J., Metcalf, B. W., Grove, J. and Casara, P. (1979*c*). *Biochem. Pharmac.*, **28**, 627
Danzin, C., Jung, M. J., Claverie, N., Grove, J., Sjoerdsma, A. and Koch-Weser, J. (1979*d*). *Biochem. J.*, **180**, 507
Davis, L. and Metzler, D. E. (1972). In *The Enzymes* vol. 7 (ed. P. D. Boyer), Academic Press, New York and London, p. 33
Dunathan, H. C. (1971). In *Adv. Enzymol.*, **35**, 79
Fisher, S. K. and Davies, W. E. (1976). *Biochem. Pharmac.*, **25**, 1881
Fowler, L. J. (1973). *J. Neurochem.*, **21**, 437
Fowler, L. J. and John, R. A. (1972). *Biochem. J.*, **130**, 569
Fozard, J. R. and Palfreyman, M. G. (1978). *Br. J. Pharmac.*, **63**, 387P
Harik, S. I., Hollenberg, M. D. and Snyder, S. H. (1974). *Molec. Pharmac.*, **10**, 41
Jänne, J., Pösö, H. and Raina, A. (1978). *Biochim. biophys. Acta*, **473**, 241
John, R. A. and Fasella, P. (1969). *Biochemistry*, **8**, 4477
Jung, M. J. (1978). In *Enzyme Activated Irreversible Inhibitors* (ed. N. Seiler, M. J. Jung, and J. Koch-Weser), Elsevier/North Holland, Amsterdam, p. 135
Jung, M. J. and Metcalf, B. W. (1975). *Biochem. biophys. Res. Commun.*, **67**, 301
Jung, M. J. and Seiler, N. (1978). *J. biol. Chem.*, **253**, 7431
Jung, M. J., Metcalf, B. W., Lippert, B. and Casara, P. (1978). *Biochemistry*, **17**, 262
Jung, M. J., Lippert, B., Metcalf, B. W., Böhlen, P. and Schechter, P. J. (1977*a*). *J. Neurochem.*, **29**, 797
Jung, M. J., Lippert, B., Metcalf, B. W., Schechter, P. J., Böhlen, P. and Sjoerdsma, A. (1977*b*). *J. Neurochem.*, **28**, 717
Jung, M. J., Palfreyman, M. G., Wagner, J., Bey, P., Ribéreau-Gayon, G., Zraïka, M. and Koch-Weser, J. (1979). *Life Sci.*, **24**, 1037
Kato, Y., Inoue, H., Gohda, E., Tamada, F. and Takeda, Y. (1976). *Gann*, **67**, 569
Kollonitsch, J., Barash, L., Kahan, F. M. and Knopp, H. (1973). *Nature, Lond.*, **243**, 346
Kollonitsch, J., Patchett, A. A., Marbrug, S., Maycock, A. L., Perkins, L. M., Doldouras, G. A., Duggan, D. E. and Aster, S. D. (1978). *Nature, Lond.*, **274**, 906
Krnjevic, K. (1976). In *GABA in Nervous System Function* (ed. E. Roberts, T. N. Chase and D. B. Tower), Raven Press, New York, p. 269
Leach, M. J. and Walker, J. M. G. (1977). *Biochem. Pharmac.*, **26**, 1569
Lippert, B., Metcalf, B. W., Jung, M. J. and Casara, P. (1977). *Eur. J. Biochem.*, **74**, 441
Marcotte, M. and Walsh, C. (1975). *Biochem. biophys. Res. Commun.*, **62**, 677
Metcalf, B. W. and Jung, M. J. (1979). *Molec. Pharmac.*, **16**, 539
Metcalf, B. W., Lippert, B. and Casara, P. (1978*a*). In *Enzyme Activated Irreversible Inhibitors* (ed. N. Seiler, M. J. Jung and J. Koch-Weser), Elsevier/North Holland, Amsterdam, p. 123
Metcalf, B. W., Bey, P., Danzin, C., Jung, M. J., Casara, P. and Vevert, J. P. (1978*b*). *J. Am. chem. Soc.*, **100**, 2551
Morino, Y. and Okamoto, M. (1973). *Biochem. biophys. Res. Commun.*, **50**, 1061
Palfreyman, M. G., Jung, M. J., Danzin, C., Ribéreau-Gayon, G., Bey, P., Zraïka, M. and Sjoerdsma, A. (1979). In *Catecholamines: Basic and Clinical Frontiers* (ed. E. Usdin), Pergamon Press, Oxford, p. 132
Palfreyman, M. G., Danzin, C., Bey, P., Jung, M. J., Ribéreau-Gayon, G., Aubry, M., Vevert, J. P. and Sjoerdsma, A. (1978*a*). *J. Neurochem.*, **31**, 927
Palfreyman, M. G., Danzin, C., Jung, M. J., Fozard, J. R., Wagner, J., Woodward, J. K., Aubry, M., Dage, R. C. and Koch-Weser, J. (1978*b*). In *Enzyme Activated Irreversible Inhibitors* (ed. N. Seiler, M. J. Jung and J. Koch-Weser), Elsevier/North Holland, Amsterdam, p. 221
Prakash, N. J., Schechter, P. J., Grove, J. and Koch-Weser, J. (1978). *Cancer Res.*, **38**, 3059
Rando, R. R. (1974*a*). *Science, N.Y.*, **185**, 320
Rando, R. R. (1974*b*). *Biochemistry*, **13**, 3859
Rando, R. R. (1977*a*). In *Methods in Enzymology*, vol. 46 (ed. W. B. Jakoby and M. Wilchek), Academic Press, New York, p. 28

Rando, R. R. (1977*b*). *Biochemistry,* **16**, 4604
Rando, R. R. and Bangerter, F. W. (1977). *Biochem. biophys. Res. Commun.*, **76**, 1276
Rando, R. R., Relyea, N. and Cheng, L. (1976). *J. biol. Chem.*, **251**, 3306
Russell, D. H. (ed.) (1973). *Polyamines in Normal and Neoplastic Growth*, Raven Press, New York
Sarhan, S. and Seiler, N. (1979). *Neurochem. Res.*, in press
Schechter, P. J. and Tranier, Y. (1978). In *Enzyme Activated Irreversible Inhibitors* (ed. N. Seiler, M. J. Jung and J. Koch-Weser), Elsevier/North-Holland, Amsterdam, p. 149
Schechter, P. J., Tranier, Y. and Grove, J. (1979). *Life Sci.*, **24**, 1173
Schechter, P. J., Tranier, Y., Jung, M. J. and Böhlen, P. (1977). *Eur. J. Pharmac.*, **45**, 319
Seiler, N., Jung, M. J. and Koch-Weser, J. (eds) (1978*a*). *Enzyme Activated Irreversible Inhibitors*, Elsevier/North-Holland, Amsterdam
Seiler, N., Danzin, C., Prakash, J. and Koch-Weser, J. (1978*b*). In *Enzyme Activated Irreversible Inhibitors* (ed. N. Seiler, M. J. Jung and J. Koch-Weser), Elsevier/North-Holland, Amsterdam, p. 55
Silverman, R. B. and Abeles, R. H. (1976). *Biochemistry,* **15**, 4718
Sletzinger, M., Chemerda, J. M. and Bollinger, F. W. (1963). *J. med. Chem.* **6**, 101
Takyi, E. E. K., Fuller, D. J. M., Donaldson, L. J. and Thomas, G. H. (1977). *Biochem. J.*, **162**, 87
Wallach, D. P. (1961). *Biochem. Pharmac.*, **5**, 323
Wang, E. and Walsh, C. (1978). *Biochemistry,* **17**, 1313
Wu, J. Y. (1976). In *GABA in Nervous System Function* (ed. E. Roberts, T. N. Chase and D. B. Tower), Raven Press, New York, p. 7
Yamada, H. and O'Leary, M. (1978). *Biochemistry,* **17**, 669

7
Acetylcholinesterase and other esterase inhibitors

W. N. Aldridge, (Toxicology Unit, Medical Research Council, Woodmansterne Road, Carshalton, Surrey SM5 4EF, UK)

INTRODUCTION (BASIC MECHANISMS)

The mode of action of substances with inhibitory action against the cholinesterases and esterases is rather well understood and in some instances the detailed structure of the enzyme has been worked out. As an introduction to this paper, I shall present a summary of the processes involved and indicate the kind of drugs to be considered and where they fit into the mechanisms of inhibition. It is too well documented to require elaboration here that good inhibitors require to be a good fit to the enzyme catalytic centre or need to possess a molecular complimentariness. However, it is as well to bear in mind that the similarity may well not be very clear in any predictive sense. For example, it would, in the absence of current knowledge, be difficult to predict the similarity between acetylcholine and diisopropylphosphorofluoridate which leads to the latter being a good inhibitor of acetyl or butyryl cholinesterase. In fact, of course, the requirements are quite simple—both are esters or anhydride-like and possess the comparable C=O and P=O necessary for attachment to the enzyme. The organophosphorus compound must not contain a negative charge and since, unlike a substrate, one cycle of the reaction leads to inhibition of the enzyme, diisopropylphosphorofluoridate is consequently an effective inhibitor.

A diagrammatic representation of the reaction of esterases with organophosphorus compounds is shown in figure 7.1; only esterases which utilise a serine in their catalytic centre are inhibited. The whole reaction is formally the same as the reaction of esterases with substrates (Aldridge and Reiner, 1972). Stage 1 (k_{+1}, k_{-1}) is the formation of the reversible Michaelis-type complex. A compound such as 'Tensilon' or edrophonium chloride (3-hydroxyphenyldimethylethylammonium chloride) would only partake in this part of the reaction and not proceed to the next stage. Stage 2 (k_{+2}) is the acylation step which occurs in phosphorylation, carbamylation and sulphonylation by organophosphorus compounds, carbamates and sulphonates. In some cases this form of the enzyme spontaneously reactivates to yield the original catalytic esterase (stage 3, k_{+3}) but this is by a reaction quite

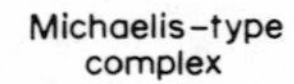

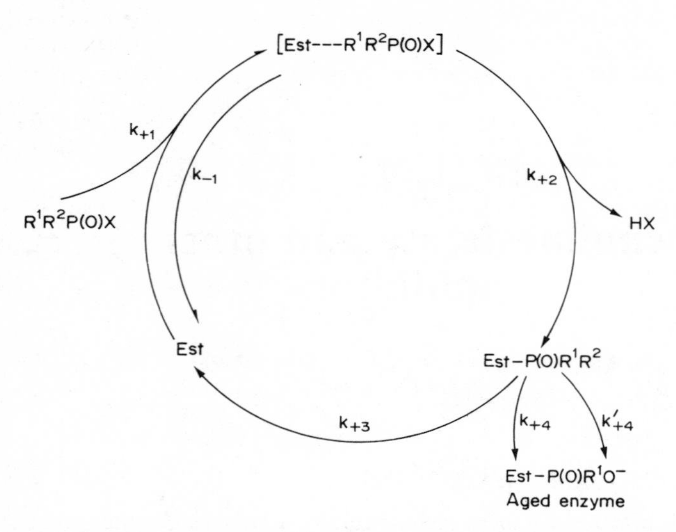

Figure 7.1 Diagrammatic representation of the reaction of esterases with organophosphorus compounds. Est = serine or B-esterase; $R^1R^2P(O)X$ = organophosphorus compound with X the leaving group. When the concentration of the Michaelis complex is small the reaction rate, Est → Est – $P(O)R^1R^2$ is given by a bimolecular rate constant $k_a = k_{+1}k_{+2}/(k_{-1} + k_{+2})$.

different from the reversible step of stage 1 (k_{-1}), and involves the breakage of the covalent attachment of phosphorus moeity to the serine of the catalytic centre (Aldridge, 1953; Reiner and Aldridge, 1967). Stage 4 (k_{+4}, k'_{+4}) involves the loss of a group attached to the phosphorus yielding a monosubstituted phosphorylated esterase and is often called the aging reaction. This reaction, for many years considered to be restricted to the cholinesterases, is now known to occur with other esterases (table 7.1). The reaction occurs by two distinct mechanisms (Aldridge, 1975), one involving carbonium ion formation (Michel *et al.*, 1967) and the other more like the enzyme-catalysed deacylation which occurs in substrate reactions as well as with some phosphorylated and carbamylated esterases (k_{+3}; Reiner and Aldridge, 1967). Whatever is the mechanism of production of the monosubstituted phosphorylated esterase, it is of profound importance for the

Table 7.1 Phosphorylated enzymes which age

Acetylcholinesterase (Hobbiger, 1956)
Cholinesterase (Hobbiger, 1955)
Chymotrypsin (Lee and Turnbull, 1958, 1961; Bender and Wedler, 1972)
Carboxylesterase; sheep, chicken, pig, horse (Hamilton *et al.*, 1975)
Esterase on CNS membrane, 'Neurotoxic esterase'; phenyl-*n*-valerate (Clothier and Johnson, 1979)

treatment of poisoning by some organophosphorus compounds and for the understanding of the mechanism of delayed neuropathy (see later).

For esterases inhibited by carbamates, sulphonates and phosphinates the aging reaction cannot occur. For some phosphorylated acetylcholinesterases, both aging and spontaneous reactivation occur at similar rates (Aldridge, 1975); an example is given in table 7.2.

Table 7.2 Reactions taking place after inhibition of human acetylcholinesterase by dichlorvos (O, O-dimethyl-2, 2-dichlorovinyl phosphate) (Škrinjarić-Špoljar *et al.*, 1973)

	Rate constant
k_a (stage 1 and 2)	1.17×10^5 l mol^{-1} min^{-1}
k_{+3} (stage 3)	1.36×10^{-2} min^{-1} (half-life 0.85 h)
k_{+4} (stage 4)	0.30×10^{-2} min^{-1} (half-life 3.84 h)

The stages refer to those shown in figure 7.1 and all measurements were made at pH 7.4 and 37 °C

Both organophosphorus compounds and carbamates react with esterases in an essentially similar manner (Aldridge and Reiner, 1972). All the esters of carbamic acids in current use as insecticides or in medical treatment are mono- or dimethyl carbamates. Monomethyl or dimethyl carbamyl acetylcholinesterases are more unstable than many phosphorylated enzymes such as those produced from dimethyl- or diethylphosphates. However, 2-chloroethylphosphorylated acetylcholinesterase is as unstable as the carbamylated enzymes. For a summary of published results see Reiner (1971).

THERAPY OF ORGANOPHOSPHORUS POISONING

Knowledge of the mechanisms of reaction of organophosphorus compounds with esterases led to the discovery of rational therapeutic drugs for the treatment of poisoning (Hobbiger, 1963, 1968). In the original work on the effectiveness of hydroxamic acids and oximes as reactivators of several inhibited esterases, the quaternised pyridine-2-aldoximes were found to be effective agents both *in vivo* and *in vitro* in reactivating acetylcholinesterase. The most effective agents are the *bis*-pyridinium aldoximes (Hobbiger *et al.*, 1960; Erdmann, 1976; Simeon *et al.*, 1979). The aged enzyme cannot be reactivated by hydroxamic acids or oximes and therapy is limited to treatment with atropine, etc., that is prevention of the actions of the accumulated acetylcholine resulting from the inhibition of acetylcholinesterase. Indeed the discovery of the aging process followed the inability to reactivate inhibited cholinesterases after they had been stored.

The reactivation of phosphorylated esterases, in the case of chymotrypsin and trypsin, shows a direct relationship between pK of the reactivator and reactivating potency (Cohen and Erlanger, 1960; Cohen *et al.*, 1962). In other words the dissociation constant of the nucleophile is an expression of its affinity for protons which is directly related to its affinity for the positively charged phosphorus. For reactivation of inhibited acetylcholinesterase, the above relationship does not hold (Wilson *et al.*, 1958; Hobbiger *et al.*, 1960; Aldridge and Reiner, 1972) and it is assumed that other structural requirements play a dominating role—for example the positively charged nitrogen atom. It is now known that the rate of reactivation of inhibited cholinesterase is dependent upon the structure and p*K* of the nucleophilic reagent, the structure of the phosphorus-containing group attached to the esterase, to asymmetric centres in either this group attached to esterases or in the nucleophilic reagent and to the stability and inhibitory potency of the phosphorylated oximes generated during the reaction. Many phosphorylated oximes are very effective inhibitors of acetylcholinesterase, some having rate constants as high as 3×10^7 l mol^{-1} min^{-1} (Schoene, 1976).

The mechanism of reactivation by nucleophilic reagents is summarised in figure 7.2 in a way analogous to the inhibitory process (figure 7.1).

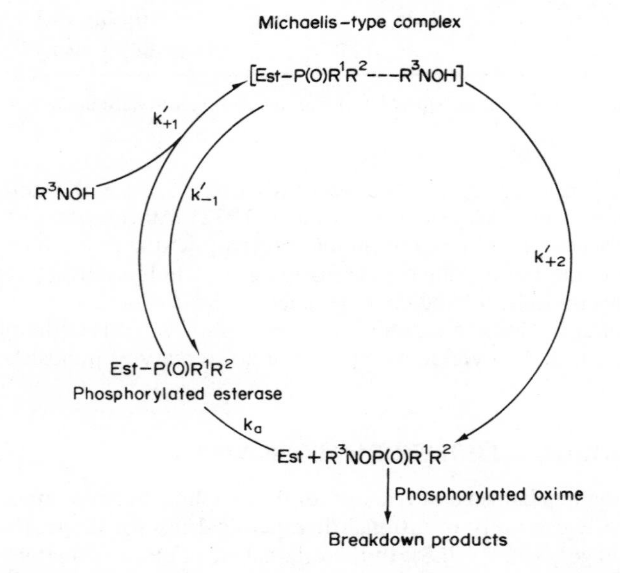

Figure 7.2 Diagrammatic representation of the reactivation of phosphylated esterases by nucleophilic oximes or hydroxamic acids.

Est = serine or B-esterase; R^3NOH = substituted oxime or hydroxamic acid. For convenience, the reinhibition of Est by the phosphorylated oxime defined by the rate constant k_a has not been given as the separate steps shown in figure 7.1.

ANTAGONISM OF TOXICITY OF ORGANOPHOSPHORUS COMPOUNDS BY CARBAMATES

It was shown many years ago that after pretreatment of cats with a small dose of eserine, which produces mild signs of poisoning, the animals could then withstand several times the LD_{50} of diisopropylphosphorofluoridate (DFP) given a few minutes later (Koster, 1946). Other experiments are described by Barnes (1972) which led to the same conclusion. The toxicity of carbamate administered after the DFP appears to be additive. Other experiments have been published showing antagonism between carbamates and organophosphorus compounds in rats (Barnes, 1972), in hens (Johnson, 1969*b*; Johnson and Barnes, 1970), in rabbits after subconjunctival injection (Krishna and Leopold, 1960) and in human eyes (Leopold and McDonald, 1948). Clinically, in patients with myasthenia gravis, the blocking action of physostigmine (Comroe *et al.*, 1946) or neostigmine (Harvey *et al.*, 1947) against diisopropylphosphorofluoridate has been demonstrated. In contrast, the toxicity of carbamates when given after the diisopropylphosphorofluoridate appears to be additive. This seemingly paradoxical situation, of two compounds with the same mechanism of toxicity one of which protects against the toxicity of the other, is difficult to explain. It seems probable that the explanation lies in a complex readjustment of reaction rates at sites directly related to the toxicity (figure 7.3). Although this may be a correct general statement, we are far from understanding the details of this effect, for example the comparative distribution of the two inhibitors, the rates of their destruction and the importance of certain biological sites of acetylcholinesterase location for the development of the acute toxic response.

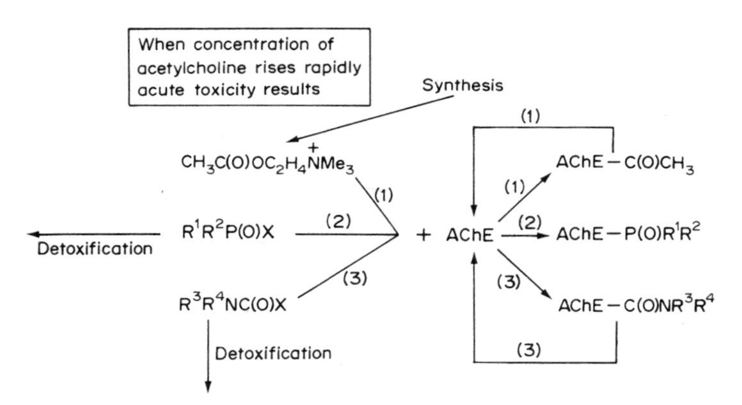

Figure 7.3 Diagrammatic representation of the reactions involved in the antagonism between carbamates and organophosphorus compounds.

AChE = acetylcholinesterase. If reaction (1) is prevented by reaction (2), acute toxicity results. For most direct-acting organophosphorus compounds, the mortality-dose relationship is very steep. If reaction (1) is prevented by reaction (3) acute toxicity results, but the mortality-dose relationship is flat due to the recycling of AChE in reaction (3). With the correct doses, when reaction (3) takes place before reaction (2), the animals are protected. If reaction (2) takes place before reaction (3), their toxicities are additive.

DELAYED NEUROPATHY PRODUCED BY ORGANOPHOSPHORUS COMPOUNDS

Due to the work of Johnson, at Carshalton, we now understand the nature of the primary chemical interactions necessary for the production of the clinical signs of ataxia in hens 10-14 days after dosing which often leads to permanent paralysis of the legs. A protein in the nervous system must be phosphorylated or phosphonylated (Johnson, 1969*a*). This protein hydrolyses esters such as phenyl phenylacetate and phenyl-*n*-valerate (Johnson, 1969*b*, 1975*c*) and there is an excellent correlation between inhibition of this esterase by phosphates and phosphonates and the development of neuropathy (Johnson, 1970).

Inhibition of the esterase is not by itself sufficient for the genesis of this disease for, in addition, a group attached to the phosphorus atom must be lost from the inhibited esterase (Clothier and Johnson, 1979; figure 7.4). This is the aging process described earlier (see figure 7.1, k_{+4}). Although some carbamates and phosphinates can inhibit the delayed neuropathy esterase, because they cannot age they do not cause the neuropathy (Johnson, 1970, 1974).

The ineffectiveness of the carbamates and phosphinates is very interesting because it implies that this particular esterase has no function in the sense that its inhibition does not lead to the clinical condition or the morphological lesions in the central and/or peripheral nervous system. Nevertheless, the conversion to a monosubstituted phosphorus group (Est–P(O)R′O$^-$) on the esterase catalytic centre, begins the biological processes necessary for the damage to the nerves (figure 7.4).

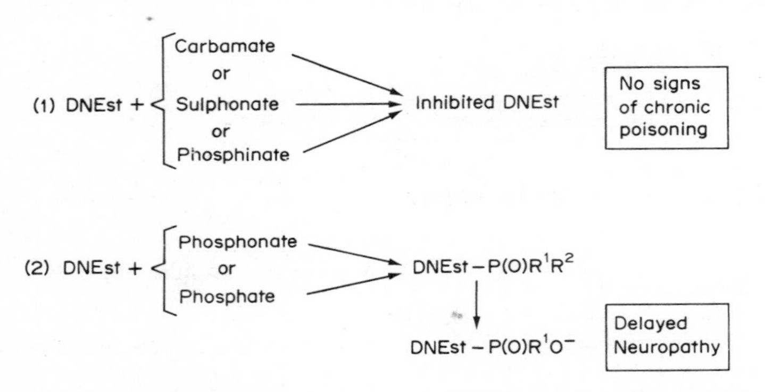

Figure 7.4 Diagrammatic representation of primary chemical reaction necessary for the initiation of delayed neuropathy (2) and for protection by other compounds (1).
DNEst = delayed neuropathy esterase. If animals are predosed with compounds and reaction (1) takes place, reaction (2) cannot take place and the animals are protected. For reviews see Johnson (1975*a*, *b*).

After inhibition *in vitro*, delayed neuropathy esterase may be reactivated by fluoride (Clothier and Johnson, 1979), and by some oximes and hydroxamic acids (Clothier, 1979). Nevertheless the aging reaction is so fast that these substances are of no value *in vivo*. The aged enzyme, with a negatively charged phosphorus-containing group attached to the esterase, cannot be reactivated.

The biological/biochemical changes which result from the primary chemical attack (which may be complete with an hour) are unknown and therefore therapy analogous to treatment of acute toxicity (linked to inhibition of acetylcholinesterase) by atropine, scopolamine, etc., is not available. The only possible and unsatisfactory approach would be prophylaxis by carbamates and phosphinates which will inhibit the esterase involved but the inhibited enzyme cannot age.

THERAPY OF *SCHISTOSOMA HAEMOTOBIUM* INFECTION WITH METRIPHONATE

Metriphonate has been used for the treatment of schistosomiasis in man (for a review see Holmstedt *et al.*, 1978) and two WHO trials have been carried out (Davis and Bailey, 1969; Plestina *et al.*, 1972). It is of interest that treatment is effective in infections by *S. haemotobium* and ineffective for *S. mansoni* (Abdalla *et al.*, 1965; Katz *et al.*, 1968). This year a very large number of people will be treated under the auspices of WHO.

Metriphonate is a phosphonate which rearranges to the phosphate, dichlorvos, a well-known insecticide (figure 7.5). The evidence, that the inhibition of the

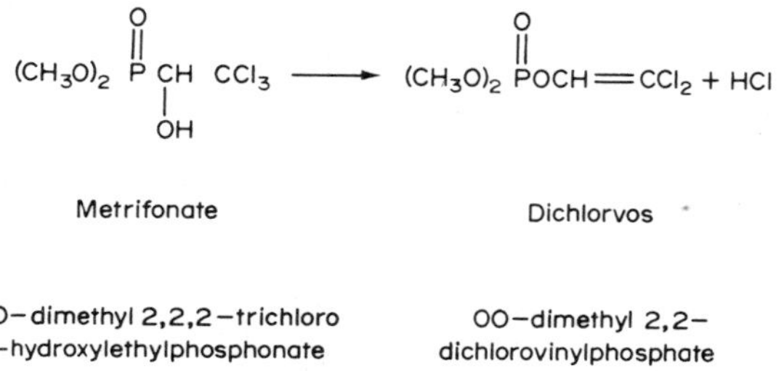

$$(CH_3O)_2\overset{O}{\overset{\|}{P}}\underset{OH}{\underset{|}{C}H}\ CCl_3 \longrightarrow (CH_3O)_2\overset{O}{\overset{\|}{P}}OCH{=}CCl_2 + HCl$$

Metrifonate — Dichlorvos

OO–dimethyl 2,2,2–trichloro –1–hydroxylethylphosphonate — OO–dimethyl 2,2–dichlorovinylphosphate

Figure 7.5 Structural formulae and chemical names of metriphonate and dichlorvos. The half-life of metriphonate at pH 7 in water at 37.5 °C is 6.4 h (Metcalf *et al.*, 1959).

cholinesterases is entirely due to its transformation to dichlorvos, has been indirect (Van Die, 1957; Metcalf *et al.*, 1959; Reiner *et al.*, 1975). It has rested on the difficulty of envisaging a leaving group in metriphonate, the low inhibitory power of the pure compound and chemical studies on the rearrangement. The best evidence is from the work of Reiner *et al.* (1975), who provided sound experimental results to support the view that metriphonate is not a cholinesterase inhibitor and that the whole of the inhibition of cholinesterase is due to a compound produced from it. Recent work (Nordgren *et al.*, 1978) has provided direct chemical proof using gas chromatographic and mass fragmentographic methods that both dichlorvos and metriphonate are present in the mouse brain. Taking published values for the intraperitoneal toxicity to male mice as 28 and 650 mg kg^{-1} for dichlorvos and metriphonate (Holmstedt *et al.*, 1978) and using the peak concentrations found in brain, it seems that equitoxic doses would be expected to produce similar peak concentrations of dichlorvos. Thus it appears certain that the cholinesterase-inhibiting activity of metriphonate is due to its rearrangement and transformation into dichlorvos. The production of dichlorvos is a chemical reaction influenced by the composition of the medium (for example pH) but not catalysed by an enzyme. Metriphonate acts as a depot from which dichlorvos is released at a constant rate in the rather constant *in vivo* conditions. It has recently been shown that dichlorvos can be detected and determined in the plasma of a patient 24 h after treatment with 7.5 mg metriphonate kg^{-1} (B. Holmstedt, personal communication).

The mechanism whereby metriphonate is an effective therapeutic agent in infections with *S. haematobium* is not known. Schistosomes contain acetylcholinesterase and acetylcholine is present in them (Gear and Fripp, 1974; Bueding, 1952; Barker *et al.*, 1966; E. Reiner *et al.*, personal communication). There is disagreement whether the cholinesterase activity is due only to acetylcholinesterase (E. Reiner *et al.*, personal communication) or whether in addition another cholinesterase is present (Bueding, 1952). There is no doubt that the acetylcholinesterase of *S. mansoni* and *S. haematobium* in infected hamsters is inhibited after treatment with metriphonate or dichlorvos (Bueding *et al.*, 1972). Although acetylcholinesterase is present in the sucker organ (Bueding *et al.*, 1972) and it has been suggested that the disappearance of eggs from the urine might be associated with a shift of the functionally damaged parasites to an 'ectopic' site such as the lung (Abdalla *et al.*, 1965; Forsyth and Rashid, 1967*a, b*; Davis and Bailey, 1969; James *et al.*, 1972), the involvement of acetylcholinesterase is not proved. One puzzling feature is the very high dose (200 mg kg^{-1} each day for up to 7 days) required in infected hamsters (James *et al.*, 1972; Bueding *et al.*, 1972; Katz *et al.*, 1968), when compared with those effective in man (7.5-12.5 mg kg^{-1}, three doses at intervals of 14 days; Plestina *et al.*, 1972). It will also be important to determine if the dimethylphosphorylated acetylcholinesterase from the schistosome is unstable and/or ages rapidly (see table 7.2). After several doses of metriphonate, it seems certain that the inhibited erythrocyte cholinesterase of the host does age (Plestina *et al.*, 1972) because the rate of return of enzyme activity approximates to the rate of synthesis of erythrocytes, that is about 60 days. After one dose of metriphonate to infected hamsters, the slow recovery of activity suggests that the aging reaction also occurs in the schistosome (Bueding *et al.*, 1972). Although it seems plausible that the therapeutic effect of metriphonate is due to the dichlorvos produced from it, the circulating concentration of metriphonate is 160 times higher than the concentration of that drug (Nordgren *et al.*, 1978). Metriphonate is not chemically inert in all respects, for example its low alkylating potential is 17 times less than dichlorvos (Bedford and Robinson, 1972).

There are many reports that organophosphorus compounds have some activity in the treatment of other parasitic infections (Plestina *et al.*, 1972; Haas, 1970; Cerf *et al.*, 1962; Cervoni *et al.*, 1969; Chavarria *et al.*, 1969). In view of our extensive knowledge of the mechanisms of action of organophosphorus compounds and the availability of effective therapy, more research is justified into the mechanisms and structure-activity relationships of organophosphorus compounds in the treatment of parasitic diseases.

MYASTHENIA GRAVIS

In recent years, there have been considerable advances in understanding of the pathogenesis of myasthenia gravis (Heilbronn and Stålberg, 1978). Anticholinesterases have been used for many years for the treatment and diagnosis of the disease (Grob, 1963).

Myasthenia gravis is a chronic disease characterised by weakness and abnormal fatiguability of skeletal muscle. A diagnosis can usually be made on the basis of the history, distribution and usual fluctuating nature of the weakness. The diagnosis is confirmed by the improvement in strength which occurs following the administration of an anticholinesterase such as neostigmine or edrophonium (Harvey and Whitehill, 1937; Osserman and Kaplan, 1952). There is a neuromuscular block in myasthenia gravis with a progressively reduced action potential after repetitive nerve stimulation which also improves following acetylcholinesterase inhibitors (Harvey and Masland, 1941). These results on the efficacy of anticholinesterase drugs indicate a deficient action of acetylcholine on the end plates but does not differentiate between various possibilities of how this may be brought about. It could be due to decreased release of acetylcholine, its more rapid removal or to interference with its action at the end-plate. Recent research has indicated the major importance of the latter.

It has been known for some time that improvement in the condition of patients has often been claimed after thymectomy (Grob, 1963). The involvement of an immunological process is now on a firmer basis following the demonstration of anti-receptor antibodies in the sera from myasthenics together with other indications of immunological involvement (for a review, see Heilbronn and Stålberg, 1978). Key experimental research stimulating and supporting this advance has followed the purification of the acetylcholine receptor and its use for the production of antibodies and the discovery of an experimental animal model for myasthenia gravis. The disease produced in receptor-immunised animals has now been characterised by immunological, biochemical, electrophysiological and electron microscopic methods (see Grob, 1976). However, in the human condition, the anti-acetylcholine receptor antibody concentration is not reduced after thymectomy (Seybold *et al.*, 1978).

The current concept therefore is that the progressive muscular weakness is caused by a destruction of the postsynaptic area of the motor end-plate resulting in a reduction of the number of functioning receptor sites. The effect of anticholinesterases can now be explained by an increase in concentration of acetylcholine in the synapse due to inhibition of its destruction. This increased amount of acetylcholine presumably causes a greater proportion of the remaining functional receptors to respond. Such a conclusion requires that, under the conditions at the end-plate, the affinity constant of acetylcholine for the receptor is such that there are always free receptors. Whether the affinity constant is identical for receptors from normal and myasthenic animals or patients suffering from the disease has not been established. Experiments with cultured muscle cells may help to resolve these questions (Appel *et al.*, 1977; Heinemann *et al.*, 1977).

OTHER THERAPEUTIC USES OF ANTICHOLINESTERASE

A wide variety of anticholinesterase carbamates and organophosphorus compounds have been used for the control of intraocular pressure in glaucoma (Leopold and Krishna, 1963).

The parasympathomimetic drug, pilocarpine, is still used but other candidate drugs such as some of the β-blockers are now being considered (Editorial, 1978).

Physostigmine (eserine) has also been used after acute poisoning with a variety of drugs, to ensure the restoration of vital functions until the drugs are eliminated from the body (Bernards, 1973; Rumack, 1973; Slovis *et al.*, 1971; Di Liberti, 1975).

REFERENCES

Abdalla, A., Saif, N., Taha, A., Aschmawy, H., Tawfik, J., Abdel-Fattah, F., Sabat, S. and Abdel-Meguid, M. (1965). *J. Egypt. med. Assoc.*, **48**, 262

Aldridge, W. N. (1953). *Biochem. J.*, **54**, 442

Aldridge, W. N. (1975). *Croat. chem Acta*, **47**, 215

Aldridge, W. N. and Reiner, E. (1972). *Enzyme Inhibitors as Substrates. Interactions of Esterases with Esters of Organophosphorus and Carbamic Acids*, North-Holland, Amsterdam, pp. 1–328

Appel, S. H., Anwyl, R., McAdams, M. W. and Elias, S. (1977). *Proc. natn. Acad. Sci. U.S.A.*, **74**, 2130

Barker, L. R., Bueding, E. and Timms, A. R. (1966). *Br. J. Pharmac.*, **26**, 656

Barnes, J. M. (1972). In *Forensic Toxicology* (ed. B. Ballantyne), John Wright and Son, Bristol, p. 79

Bedford, C. T. and Robinson, J. (1972). *Xenobiotica*, **2**, 307

Bender, M. L. and Wedler, F. C. (1972). *J. Am. chem. Soc.*, **94**, 2101

Bernards, W. (1973). *Anaesthesia Analgesia*, **52**, 938

Bueding, E. (1952). *Br. J. Pharmac.*, **7**, 563

Bueding, E., Lin, C. L. and Rogers, S. H. (1972). *Br. J. Pharmac.*, **46**, 480

Cerf, J., Lebrun, A. and Dierichtx, J. (1962). *Am. J. trop. Med. Hyg.*, **11**, 514

Cervoni, W. A., Oliver-Gonzalez, J., Kaye, S. and Slomka, M. B. (1969). *Am. J. trop. Med. Hyg.*, **18**, 912

Chavarria, A. P., Schwertzwelder-Villarejos, V. M., Kotcher, E. and Argnedas, J. (1969). *Am. J. trop. Med. Hyg.*, **18**, 907

Clothier, B. (1979). Thesis for M.Phil., Council for National Academic Awards, pp. 1–159

Clothier, B. and Johnson, M. K. (1979). *Biochem. J.*, **177**, 549

Cohen, W. and Erlanger, B. F. (1960). *J. Am. chem. Soc.*, **82**, 3928

Cohen, W., Lache, M. and Erlanger, B. F. (1962). *Biochemistry*, **1**, 686

Comroe, J. H., Todd, J., Leopold, J. H., Koelle, G. B., Bodensky, O. and Gilman, A. (1946). *Am. J. med. Sci.*, **212**, 641

Davis, A. and Bailey, D. R. (1969). *Bull. Wld Hlth Org.*, **41**, 209

Di Liberti, J., O'Brien, M. L. and Turner, T. (1975). *J. Pediatr.*, **86**, 106

Editorial (1978). *Br. med. J.*, **1**, 460

Erdmann, W. D. (1976). In *Medical Protection against Chemical Warfare Agents* (ed. Stockholm International Peace Research Institute), Almqvist and Wiksell, Stockholm, p. 46

Forsyth, D. M. and Raschid, C. (1967*a*). *Lancet*, **1**, 130

Forsyth, D. M. and Raschid, C. (1967*b*). *Lancet*, **2**, 909

Gear, N. R. and Fripp, P. J. (1974). *Comp. Biochem. Physiol.*, **47B**, 743

Grob, D. (1963). In *Cholinesterases and Anticholinesterase Agents* (ed. G. B. Koelle), Springer-Verlag, Berlin, p. 1028

Grob, D. (1976). *Ann. N.Y. Acad. Sci.*, **274**, 1

Haas, D. K. (1970). In *Topics in Medicinal Chemistry*, vol. 3 (ed. J. L. Rabinowsky and R. M. Myerson), John Wiley, New York, p. 171

Hamilton, S. E., Dudman, N. P. B., De Jersey, J., Stoops, J. K. and Zenner, B. (1975). *Biochim. biophys. Acta*, **377**, 282

Harvey, A. M. and Masland, R. L. (1941). *Bull. Johns Hopkins Hosp.*, **69**, 1

Harvey, A. M. and Whitehill, M. R. (1937). *J. Am. med. Assoc.*, **108**, 1329

Harvey, A. M., Lilienthal, J. L., Grob, D., Jones, B. F. and Talbot, S. A. (1947). *Bull. Johns Hopkins Hosp.*, **81**, 267

Heilbronn, E. and Stålberg, E. (1978). *J. Neurochem.*, **31**, 5

Heinemann, S., Bevan, S., Kullberg, R., Lindstrom, Y. and Rice, J. (1977). *Proc. natn. Acad. Sci. U.S.A.*, **74**, 3090

Hobbiger, F. (1955). *Br. J. Pharmac.*, **10**, 356

Hobbiger, F. (1956). *Br. J. Pharmac.*, **11**, 295

Hobbiger, F. (1963). In *Cholinesterases and Anticholinesterase Agents* (ed. G. B. Koelle), Springer-Verlag, Berlin, p. 921
Hobbiger, F. (1968). In *Recent Advances in Pharmacology* (ed. J. M. Robson and R. S. Stacey), Churchill, London
Hobbiger, F., Pitman, M. and Sadler, P. W. (1960). *Biochem. J.*, **75**, 363
Holmstedt, B., Nordgren, I., Sandoz, M. and Sundwall, A. (1978). *Arch. Toxicol.*, **41**, 3
James, C., Webbe, G. and Preston, J. M. (1972). *J. trop. Med. Parasitol.*, **66**, 467
Johnson, M. K. (1969*a*). *Biochem. J.*, **111**, 487
Johnson, M. K. (1969*b*). *Biochem. J.*, **114**, 711
Johnson, M. K. (1970). *Biochem. J.*, **120**, 523
Johnson, M. K. (1974). *J. Neurochem.*, **23**, 785
Johnson, M. K. (1975*a*). *Arch. Toxicol.*, **34**, 259
Johnson, M. K. (1975*b*). *Crit. Rev. Toxicol.*, **3**, 289
Johnson, M. K. (1975*c*). *Biochem. Pharmac.*, **24**, 797
Johnson, M. K. and Barnes, J. M. (1970). *Biochem. Pharmac.*, **19**, 3045
Katz, N., Pellegrino, J. and Pereira, J. P. (1968). *Rev. Soc. bras. Med. trop. II*, **5**, 237
Koster, R. (1946). *J. Pharmac. exp. Therap.*, **88**, 39
Krishna, N. and Leopold, I. N. (1960). *Am. J. Ophthalmol.*, **49**, 270
Lee, W. and Turnbull, J. H. (1958). *Biochim. biophys. Acta*, **30**, 655
Lee, W. and Turnbull, J. H. (1961). *Experientia*, **17**, 360
Leopold, I. H. and Krishna, N. (1963). In *Cholinesterase and Anticholinesterase Agents* (ed. G. B. Koelle), Springer-Verlag, Berlin, p. 1051
Leopold, I. N. and McDonald, P. R. (1948). *Arch. Ophthalmol., Chicago*, **40**, 176
Metcalf, R. L., Fukuto, R. B. and March, R. B. (1959). *J. econ. Entomol.*, **52**, 44
Michel, H. O., Hackley, B. E., Berkovitz, L., List, G., Hackley, E. B., Gillilan, W. and Parkau, M. (1967). *Arch. Biochem. Biophys.*, **121**, 29
Nordgren, I., Bergstrom, M., Holmstedt, B. and Sandoz, M. (1978). *Arch. Toxicol.*, **41**, 31
Osserman, K. E. and Kaplan, L. I. (1952). *J. Am. med. Assoc.*, **150**, 265
Plestina, R., Davis, A. and Bailey, D. R. (1972). *Bull. Wld Hlth Org.*, **46**, 747
Reiner, E. (1971). *Bull. Wld Hlth Org.*, **44**, 109
Reiner, E. and Aldridge, W. N. (1967). *Biochem. J.*, **105**, 171
Reiner, E., Krauthacker, B., Simeon, B. and Skrinjarić-Špoljar, M. (1975). *Biochem. Pharmac.*, **24**, 717
Rumack, B. H. (1973). *Pediatrics*, **52**, 449
Schoene, K. (1976). In *Medical Protection against Chemical Warfare Agents* (ed. Stockholm International Peace Research Institute), Almqvist and Wiksell, Stockholm, p. 88
Seybold, M. E., Baersgen, R. N., Nave, B. and Lindstrom, J. M. (1978). *Br. med. J.*, **2**, 1051
Simeon, V., Wilhelm, K., Granov, A., Besarovic–Lazarev, S., Buntic, A., Fajdetic, A. and Binefeld, Z. (1979). *Arch. Toxicol.*, **41**, 301
Skrinjarić-Špoljar, M., Simeon, V. and Reiner, E. (1973). *Biochim. biophys. Acta*, **315**, 363
Slovis, T. L., Ott, J. E., Teitalbaum, D. T. and Lipscomb, W. (1971). *Clin. Toxicol.*, **4**, 451
Van Die, J. (1957). *Koninke. Med. Akad., Wetenschap. Proc. B*, **60**, 227
Wilson, I. B., Ginsburg, S. and Quan, C. (1958). *Arch. Biochem. Biophys.*, **77**, 286

8

Inhibitors of adenylate cyclase and adenosine 3′, 5′-monophosphate phosphodiesterase

Don N. Harris, Marie B. Phillips, Harold J. Goldenberg and Magdi M. Asaad
(Department of Pharmacology, The Squibb Institute for Medical Research, Princeton, New Jersey 08540, USA)

INTRODUCTION

The role of adenosine-3′,5′-monophosphate (cyclic AMP) as 'second messenger' in a great many important physiological and pathological conditions has been described over the past 20 years (Greengard *et al.*, 1972; Schultz and Gratzner, 1973; Weiss, 1975; Greengard and Robison, 1975; Volicer, 1977). Briefly, most hormones, neurotransmitters, and some prostaglandins, interact with a specific receptor on the surface of the cell to activate an enzyme, adenylate cyclase, located on the plasma membrane. The adenylate cyclase, in turn, converts ATP to cyclic AMP and the cyclic nucleotide then activates a cyclic AMP-dependent protein kinase. The protein kinase phosphorylates cellular enzymes which catalyse the biological reactions typical for that cell. Excess cyclic AMP is broken down to 5′-adenosine monophosphate (5′-AMP) by cyclic AMP phosphodiesterase (cyclic AMP PDE), thus interrupting the action of the hormone.

Since cyclic AMP mediates so many important biological processes, it provides a potentially useful system for the development of new therapeutic agents (Amer and McKinney, 1973; Smith, 1974, 1978; Amer and Kreighbaum, 1975; Weiss and Hait, 1977). There are several different approaches to finding agents to manipulate the levels of cyclic AMP in tissues. First, compounds could be designed to compete with the natural agonist for the receptor site on adenylate cyclase. Such agents would therefore stimulate adenylate cyclase and increase cyclic AMP levels in those conditions characterised by low cyclic AMP levels due to a defective agonist. Another point of attack would be to design inhibitors of adenylate cyclase which would be useful in conditions where there are elevated levels of cyclic AMP due to abnormal stimulation of adenylate cyclase. Other agents could be designed to regulate the cyclic AMP PDE system. Inhibitors of cyclic AMP PDE could be used to elevate cyclic AMP levels in cells of tissues where cyclic AMP levels are subnormal. Development of new agents of this type have therapeutic potential. Our laboratories have shown that many existing drugs are inhibitors of cyclic AMP PDE (Weinryb *et al.*, 1972). Conversely, stimulators of cyclic AMP PDE could be used where elevated levels of cyclic AMP are a characteristic of the disease. Finally, exogenous cyclic AMP analogues could be designed to mimic endogenous cyclic AMP in activating specific protein kinases.

There is a great deal of evidence that cyclic AMP mediates the inhibition of platelet aggregation (Haslam, 1978; Haslam *et al.*, 1978; Salzman, 1977, 1978). Platelet aggregation *in vitro*, induced by such diverse agents as arachidonic acid, epinephrine, collagen and adenosine diphosphate (ADP) is inhibited by compounds able to elevate cyclic AMP levels in blood platelets. Many investigators have shown that prostaglandin I_2 (PGI_2) (Gorman *et al.*, 1978), prostaglandin D_2 (PGD_2) (Tateson *et al.*, 1977), prostaglandin E_1 (PGE_1) (Gorman *et al.*, 1977; Miller *et al.*, 1977), adenosine (Londos and Wolff, 1977; Haslam and Lynham, 1972) and isoprenaline (Abdulla, 1969) stimulate adenylate cyclase preparations of human platelet homogenates. Furthermore, these same agonists have been shown to elevate cyclic AMP levels in intact platelets (Miller *et al.*, 1977; Gorman *et al.*, 1977). Third, both inhibition of platelet aggregation and increase of cyclic AMP concentrations by these agents are potentiated by inhibitors of cyclic AMP PDE. Finally, several laboratories have shown that exogenous N^6-2′-*O*-dibutyryl cyclic AMP (dibutyryl cyclic AMP) inhibits platelet aggregation (Marquis *et al.*, 1969; Abdulla, 1969; Salzman and Levine, 1971). Inhibition of platelet aggregation thus conforms with Sutherland's four criteria for the mediation of a biological action by cyclic AMP (Robison *et al.*, 1971). More recently, it has been proposed that an additional criterion for establishing cyclic AMP as a mediator of biological actions would be provided by use of specific inhibitors of adenylate cyclase (Haslam *et al.*, 1978).

In the first part of this paper, we report some of our experimental results in discovering inhibitors of adenylate cyclase and cyclic AMP PDE, and how they effect cyclic AMP levels and ADP-induced platelet aggregation in the presence and absence of PGE_1. In the second part of the report, we discuss some of the diseases where cyclic AMP is strongly implicated and where inhibitors of adenylate cyclase and cyclic AMP PDE, such as those discussed below, may be used as potential therapeutic and pharmacological agents.

EFFECTS OF INHIBITORS OF ADENYLATE CYCLASE AND CYCLIC AMP PDE ON PLATELET AGGREGATION AND CYCLIC AMP LEVELS IN PLATELETS

In our laboratory we utilised the following biochemical approaches with human blood platelets in detecting potential pharmacological and therapeutic agents which influence cyclic AMP. As primary screening tests, we utilised partially purified subcellular preparations of cyclic AMP PDE and particulate adenylate cyclase preparations from blood platelets. Promising active compounds were tested for their effect on cyclic AMP levels and on platelet aggregation induced by ADP with and without PGE_1. These experimental approaches have allowed us to detect inhibitors of adenylate cyclase and cyclic AMP PDE which reverse and potentiate, respectively, the effects of PGE_1 on cyclic AMP levels and platelet aggregation. An additional benefit is that the experiments were carried out with an unaltered, human, whole cell, which plays a major role in haemostasis and thrombosis.

Methods

Cyclic AMP PDE

Cyclic AMP PDE was prepared at 4 °C from human blood platelets (Mills and Smith, 1971). Phosphodiesterase activity was assayed at low concentrations of substrate (0.06 μM) by a radioisotopic procedure (Brooker *et al.*, 1968). All assays were performed under conditions of 20–30 per cent hydrolysis of cyclic AMP to adenosine, where the hydrolysis of cyclic AMP was a linear function of enzyme protein concentration and time. Inhibitory potency was determined by comparing micromolar concentrations of the compounds which caused 50 per cent inhibition of enzymic activity.

Adenylate cyclase

Adenylate cyclase was prepared at 4 °C from lysed human blood platelets (Haslam and Lynham, 1972). Adenylate cyclase activity was determined by incubating the test compounds, PGE_1, $[^{32}P]$-ATP and enzyme protein for 5 min at 30 °C (Weinryb and Michel, 1974). Cyclic AMP was isolated by a procedure which included the use of $[^{3}H]$-cyclic AMP as a recovery standard, Dowex 50 chromatography and $BaSO_4$ precipitation treatment. The specific activity of the enzyme was expressed as picomoles cyclic AMP produced per milligram protein per minute. The data were normalised using percentage inhibition, because the stimulation of adenylate cyclase varied from day to day and with different enzyme preparations.

Platelet aggregation

Platelet aggregation was studied photometrically (Born, 1962), using a Born MK III aggregometer or a Chronolog aggregometer connected to a linear recorder. The rate of increase of optical transmission, a measure of the initial velocity of aggregation, was measured by determining the slope of the steepest part of the curve. Results are expressed as percentages of control values.

Cyclic AMP assay

Experiments to show the effects of compounds on concentration of cyclic AMP in intact platelets were carried out as described by McDonald and Stuart (1973), except that cyclic AMP concentrations were determined by radioimmunoassay (Weinryb, 1972). The concentration of cyclic AMP is expressed as picomoles of cyclic AMP formed per millilitre of platelet-rich plasma (PRP).

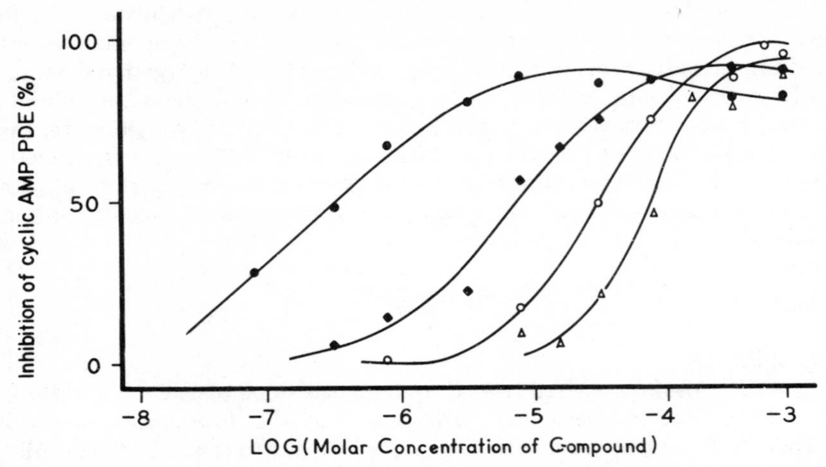

Figure 8.1 Inhibition of cyclic AMP phosphodiesterase by papaverine (●—●), bishydroxycoumarin (○—○), theophylline (△—△) and haloxon (♦—♦). Assays were performed as described in the text. Each point represents at least two determinations.

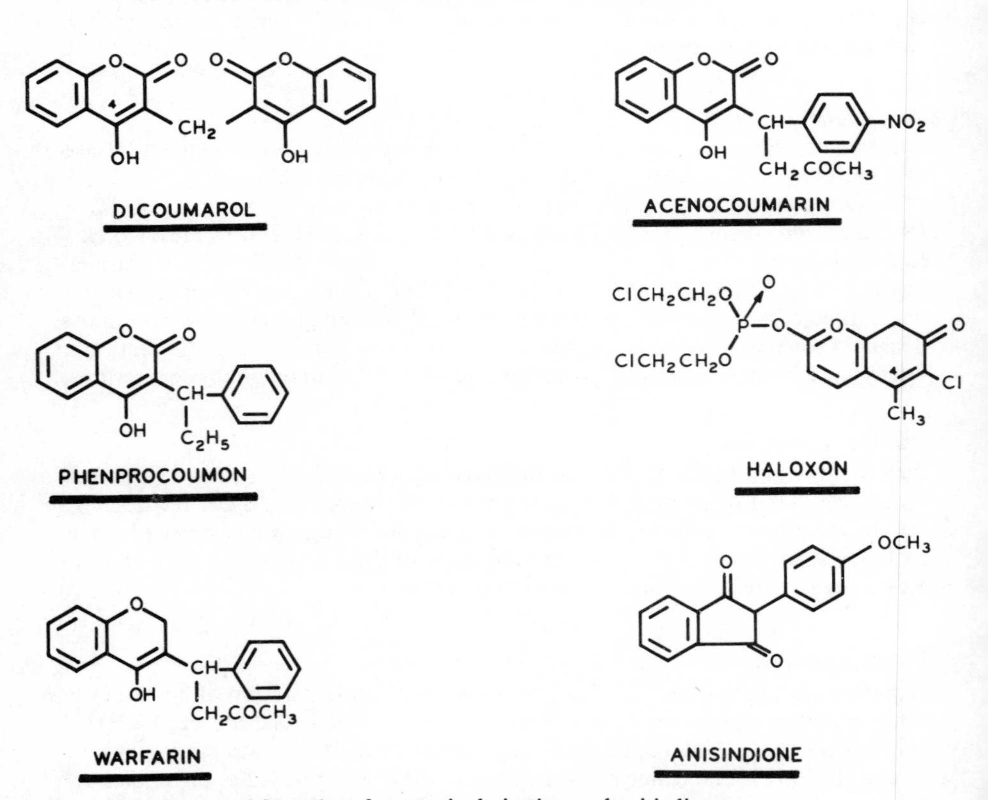

Figure. 8.2 Structural formulae of coumarin derivatives and anisindione.

Results and discussion

Inhibitors of cyclic AMP PDE

The activities of haloxon, bishydroxycoumarin (dicoumarol), theophylline and papaverine as inhibitors of cyclic AMP PDE prepared from human platelets are shown in figure 8.1. All four compounds inhibited the enzyme at low concentrations of substrate (0.06 μM). Papaverine was the most potent inhibitor; a concentration of 10^{-7} M caused 30 per cent inhibition, and at 10^{-5} M the inhibition curve plateaued at about 90 per cent inhibition. I_{50} values for these four compounds, as well as some other coumarin derivatives, anisindione and SQ 20 009 (etazolate hydrochloride) are shown in table 8.1 (see structures in figures 8.2 and 8.3).

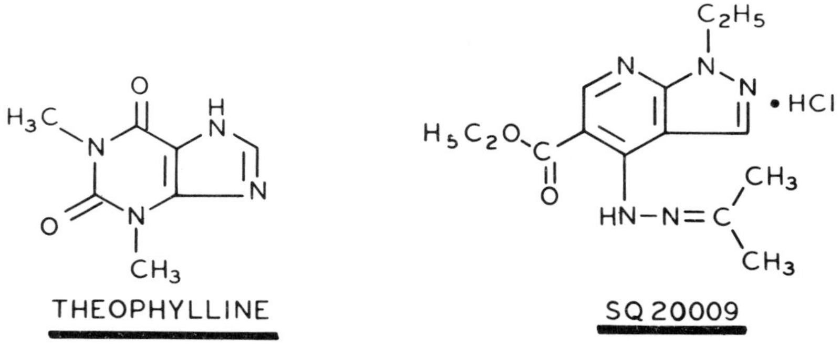

THEOPHYLLINE

SQ 20009

Figure 8.3 Structural formulae of theophylline and SQ 20 009 (etazolate).

Table 8.1 Inhibition of cyclic AMP phosphodiesterase of human blood platelets†

Compound	I_{50}‡ (μM)
Bishydroxycoumarin	40
Acenocoumarin	180
Warfarin	2000
Phenprocoumon	2600
Anisindione	170
Theophylline	100
Papaverine	0.4
SQ 20 009	50
Haloxon	10

†Assays were performed as described in the text.
‡Concentration causing a 50 per cent inhibition of enzymic activity.

The effect of PGE_1 on cyclic AMP levels and on the aggregation of human platelets treated with ADP was determined (table 8.2) both in the presence and absence of etazolate hydrochloride. PGE_1 inhibited platelet aggregation and raised the levels of cyclic AMP in a dose-related manner. The addition of etazolate alone had no effect on cyclic AMP levels or platelet aggregation (data not shown). The addition of etazolate to platelets containing PGE_1 potentiated the action of PGE_1. Inhibition of platelet aggregation was increased to a maximal level (100 per cent) and cyclic AMP levels were increased to a maximum of tenfold.

Table 8.2 Effect of etazolate on PGE_1-induced changes in ADP–platelet aggregation and cyclic AMP levels

Compounds additions	Concentration (μM)	Picomoles cyclic AMP per millilitre PRP	Percentage inhibition of platelet aggregation
PGE_1	0.05	ND†	70
+ Etazolate	100	ND†	58
+ Etazolate	1000	160	87
PGE_1	0.1	35	75
+ Etazolate	100	45	83
+ Etazolate	1000	350	94
PGE_1	0.5	220	84
+ Etazolate	100	275	93
+ Etazolate	1000	700	100

†ND = None detected (concentration too low to detect).

As shown in table 8.1, coumarin derivatives constitute another class of cyclic AMP PDE inhibitor. We therefore tested haloxon, dicoumarol and warfarin (coumarin derivatives with different potencies as inhibitors of cyclic AMP) for their effects on the PGE_1-induced elevation of platelet cyclic AMP and inhibition of ADP-induced platelet aggregation to see if their potencies as cyclic AMP PDE inhibitors would correlate with their abilities to potentiate the effect of PGE_1. All three compounds had relatively little effect on platelet aggregation in the absence of PGE_1. In the presence of haloxon, however, the effectiveness of PGE_1 as an inhibitor of ADP-induced platelet aggregation increased fivefold (table 8.3). Similar results were obtained with dicoumarol and PGE_1 where the effectiveness of PGE_1 was increased twofold. Somewhat different results were obtained with warfarin and PGE_1. Warfarin caused no change or slightly inhibited the effects of PGE_1 on ADP-induced platelet aggregation.

Haloxon, dicoumarol and warfarin were also tested for their effect on cyclic AMP levels in intact platelets in the presence and absence of PGE_1 (table 8.3). All three compounds were relatively ineffective in the absence of PGE_1. Haloxon had little effect by itself, but increased cyclic AMP levels twofold in the presence of PGE_1. A similar response was obtained with dicoumarol, where cyclic AMP levels were increased about 1.5-fold over the increase by PGE_1 alone. Warfarin, which had very weak cyclic AMP PDE inhibitory activity (I_{50} = 2000 μM), caused a slight decrease in cyclic AMP levels both in the presence and absence of PGE_1.

Thus, the two potent inhibitors of cyclic AMP PDE, haloxon and dicoumarol, had similar effects in potentiating the effects of PGE_1 on ADP-induced platelet aggregation and on cyclic AMP levels in platelets. The other coumarin derivative,

Table 8.3 Effect of haloxon, dicoumarin and warfarin on PGE_1-induced changes in ADP–platelet aggregation and cyclic AMP concentration

SQ no.	Concentration (μM)	Compound	Picomoles cyclic AMP per millilitre PRP	Percentage of control	Percentage inhibition of platelet aggregation
–	–	Control (95 per cent EtOH)	160	–	–
–	0.5	PGE_1	390	244	19
20 678	1000	Haloxon	170	106	15
–	1000	PGE_1 + haloxon	704	440	100
–	–	Control (1 N NaOH)	116	–	–
–	0.5	PGE_1	331	287	24
8225	1000	Dicoumarol	145	125	32
–	1000	PGE_1 + dicoumarol	500	433	46
–	–	Control (saline)	95	–	–
–	0.5	PGE_1	316	331	34
21 935	1000	Warfarin	70	73	14
–	1000	PGE_1 + warfarin	272	285	30

warfarin, which was a very weak inhibitor of cyclic AMP PDE, had no effect on platelet aggregation or on platelet cyclic AMP levels. These data provide additional evidence that inhibition of platelet aggregation is mediated by cyclic AMP; together with those obtained with etazolate, they satisfy Sutherland's third criterion for identification of cyclic AMP as the second messenger of a biological action, since both the elevation of cyclic AMP and inhibition of platelet aggregation were potentiated by these inhibitors of cyclic AMP PDE.

Inhibitors of adenylate cyclase

Adenylate cyclase activity increased three- to sixfold when PGE_1 (1 μM) was incubated with platelet lysates. A series of 9-substituted adenine derivatives (structures shown in figure 8.4) were tested for their ability to inhibit this PGE_1-

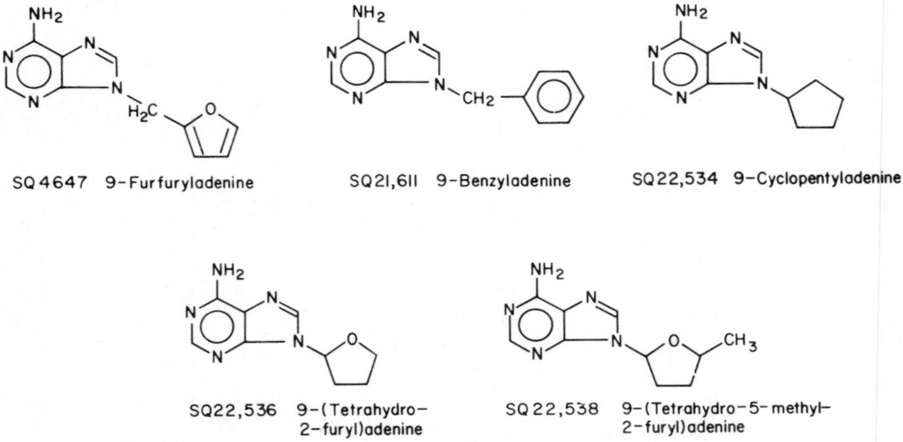

SQ 4647 9-Furfuryladenine

SQ21,611 9-Benzyladenine

SQ22,534 9-Cyclopentyladenine

SQ22,536 9-(Tetrahydro-2-furyl)adenine

SQ 22,538 9-(Tetrahydro-5-methyl-2-furyl)adenine

Figure 8.4 Structural formulae of 9-substituted adenine derivatives.

stimulated activity. Such increased activity was inhibited in a dose-related manner. The I_{50} values (in micromolarities) for the compounds are given in table 8.4. The adenine derivatives also inhibited basal adenylate cyclase activity, but the inhibition was not dose-related (data not shown). In addition, 9-(tetrahydro-2-furyl)-adenine, 9-furfuryl adenine and 9-benzyladenine were tested for their effect on cyclic AMP levels in intact platelets in combination with or without PGE_1 (table 8.5). PGE_1 caused a four- to fivefold increase in platelet cyclic AMP levels. The PGE_1-induced increase in cyclic AMP concentration was greatly inhibited by preincubating PRP with the 9-substituted adenine derivatives. Each compound completely inhibited the PGE_1-induced changes. Moreover, the compounds without PGE_1 addition had no effect on cyclic AMP levels. Additional studies showed that the adenine derivatives blocked the PGE_1-induced increases in cyclic AMP concentration in a dose-related manner (Harris *et al.*, 1979). I_{50} values for 9-(tetrahydro-2-furyl)adenine, 9-cyclopentyladenine and 9-benzyladenine were 1, 4 and 60 μM, respectively.

Table 8.4 Effect of 9-substituted adenine derivatives on PGE_1-stimulated adenylate cyclase activity of human blood platelets

Compound	I_{50} (μM)†
9-(Tetrahydro-2-methyl-2-furyl)adenine	21
9-Cyclopentyladenine	57
9-Furfuryladenine	68
9-(Tetrahydro-2-furyl)adenine	82
9-Benzyladenine	140

†Concentration of compound required to inhibit adenylate cyclase activity by 50 per cent.

Table 8.5 Effect of adenine derivatives alone, and in combination with PGE_1, on cyclic AMP levels in intact human blood platelets

Compound	Concentration (μM)	Picomoles of cyclic AMP per millilitre PRP
Control	–	30
PGE_1	0.5	130
9-Benzyladenine	1000	25
9-Benzyladenine + PGE_1		40
9-Furfuryladenine	1000	30
9-Furfuryladenine + PGE_1		35
9-(Tetrahydro-2-furyl)adenine	1000	25
9-(Tetrahydro-2-furyl)adenine + PGE_1		20

Finally, some of the 9-substituted adenine derivatives were analysed for their effectiveness on ADP-induced platelet aggregation with and without PGE_1 (table 8.6). 9-(Tetrahydro-2-furyl)adenine and 9-furfuryladenine were very effective at 100 μM, achieving reversal of inhibition of 80 per cent or better. The other two compounds were less effective, achieving reversal of aggregation of only 30 per cent. Additional studies also showed that these effects of the adenine derivatives were dose-related (Harris *et al.*, 1979). The I_{50} value for 9-(tetrahydro-2-furyl)-adenine and 9-furfuryladenine in each case was about 30 μM. The adenine compounds had no effect on ADP-induced platelet aggregation in the absence of PGE_1 (data not shown). When the data for the effect of the adenine derivatives on percentage ADP-induced aggregation were analysed as a function of their effect on the number of picomoles cyclic AMP formed per millilitre of PRP, by linear regression and Student's t tests, a significant negative correlation was found ($r = -0.63$, $P < 0.01, N = 18$).

We have shown in these experiments that a series of 9-substituted adenine derivatives inhibits PGE_1-activated adenylate cyclase activity, and reverses the effects of PGE_1 on ADP-induced platelet aggregation and cyclic AMP levels in intact platelets. In this respect, the adenine derivatives resemble thromboxane

Table 8.6 Effect of 9-substituted adenine derivatives on PGE_1-induced inhibition of ADP-induced platelet aggregation

Compound	Concentration (μM)	Percentage ADP–platelet aggregation
9-(Tetrahydro-2-furyl)adenine	100	90
9-Furfuryladenine	100	80
9-Cyclopentyladenine	100	35
9-Benzyladenine	100	30
PGE_1	0.5	10

A_2 (TXA_2) in inhibiting PGE_1-induced accumulation of cyclic AMP in platelets (Miller *et al.*, 1977). Londos and Wolff (1977) observed that adenosine and its analogues have opposing effects on adenylate cyclase activity of platelets. They found that adenosine derivatives having an intact ribose moiety were stimulatory, whereas those having only the purine moiety, like the adenine derivatives described here, are inhibitory. Others have reported that compounds of this type inhibit the cholera toxin-stimulated adenylate cyclase activity of intestinal epithelium (Zenser, 1976) and the PGE_1-, PGD_2- and adenosine-stimulated adenylate cyclase activity of human blood platelets (Haslam *et al.*, 1978; Salzman, 1978).

DISEASES ASSOCIATED WITH ABNORMAL ADENYLATE CYCLASE ACTIVITY. POTENTIAL OF INHIBITORS OF ADENYLATE CYCLASE AS THERAPEUTIC AGENTS

Some diseases are characterised by a hyperactivity of adenylate cyclase and thus by supranormal levels of cyclic AMP in the affected tissues. In table 8.7, discussed

Table 8.7 Diseases where inhibitors of adenylate cyclase may be potential therapeutic agents

Disease	Key reference
Highly probable	
1. Cholera and other enterotoxic diarrhoeal syndromes	Gill, 1977
2. Psychosis, schizophrenia	Kebabian, 1977
3. Duodenal ulcer	Jacobson and Thompson, 1976
4. Osteoporosis	Murad *et al.*, 1970
Probable	
1. Gastric ulcer	Bieck *et al.*, 1973
2. Hyperthyroidism (Graves disease)	Broadus, 1977
3. Osteogenesis imperfecta	Reite *et al.*, 1972
4. Muscular dystrophy	Gordon and Dowben, 1966
5. Zollinger–Ellison syndrome	Bieck *et al.*, 1973
6. Cushing's syndrome	Fleischer *et al.*, 1969
7. Anxiety	Margules, 1971; Tyrer and Lader, 1972

below, are listed some of the diseases where cyclic AMP concentrations are greater than normal which might possibly be treated with inhibitors of adenylate cyclase. As it was not possible to cover all the diseases falling into this category, the interested reader should consult the references given in table 8.7 for further information.

Cholera and other enterotoxic diarrhoeal syndromes

Cholera is a very serious problem in some developing parts of the world. Its symptoms are massive diarrhoea, with loss of water, bicarbonate, sodium and chloride, resulting in severe dehydration. Several laboratories have reported that bacterial enterotoxins cause an irreversible activation of intestinal adenylate cyclase in these lower bowel diseases (Adams, 1973; Sharp, 1973; Gill, 1977; Strewler and Orloff, 1977). The resulting elevation of cyclic AMP correlates in time and magnitude of effect with a hypersecretion of fluids and electrolytes into the intestinal lumen (Guerrant *et al.*, 1972). In addition, Gill (1977) reports that adenine, methylxanthine derivatives and papaverine at rather high concentrations (10^{-5}–10^{-3} M) are inhibitors of cholera toxin-activated intestinal adenylate cyclase, *in vitro*. We speculate that other inhibitors of bacterial enterotoxin-stimulated adenylate cyclase of intestinal origin would also be useful in reversing the biological effects of the enterotoxins.

Psychosis, schizophrenia

These disease conditions are characterised by bizarre behaviour, hallucinations and detachment from reality. Schizophrenia is thought by some to be a manifestation of hyperactive dopaminergic neuronal function in which the dopamine receptors become hypersensitive and the dopamine-sensitive adenylate cyclase overstimulated (Weiss and Greenberg, 1975; Kebabian *et al.*, 1972; Kebabian, 1977). The clinically effective antipsychotic (neuroleptic) drugs possess some actions which have been interpreted as central dopamine receptor antagonism and may conceivably exert their pharmacologic effect by inhibiting dopamine-sensitive cyclase activity in the brain. It has been shown (Clement-Cormier *et al.*, 1974, 1975) that the inhibitory potency of the phenothiazine antipsychotic agents as inhibitors of dopamine-sensitive adenylate cyclase correlates well with their clinical antipsychotic activity. Kebabian (1977) reported that CNS stimulants such as amphetamines and drugs which may have action on the dopamine receptor, such as morphine and heroin, can cause schizophrenia-like symptoms and are stimulators of the dopamine adenylate cyclase. Recently, Weiss and Levin (1978) proposed that antipsychotic drugs act by selectively inhibiting activation of the peak II phosphodiesterase isoenzyme, a form of the enzyme found largely in the brain. They suggested that the net effect of antipsychotic drugs may depend on the relative activity of dopamine-sensitive adenylate cyclase and peak II phosphodiesterase isoenzyme present in different parts of the brain. We suggest that new classes of central dopamine-sensitive adenylate cyclase inhibitors might provide a fresh approach to the treatment of psychosis.

Duodenal ulcer

This condition is characterised by hypersecretion of HCl from the stomach and is relieved by antacids and agents which block acid secretion. The role of cyclic AMP in gastric secretion has been controversial (Jacobson and Thompson, 1976) but there is substantial evidence that this compound mediates histamine-stimulated HCl secretion. Histamine increases the activity of adenylate cyclase in cell-free preparations of gastric tissues (Sung *et al.*, 1973), and increases cyclic AMP and HCl levels in gastric juice, an effect which is potentiated by an inhibitor of PDE (Bieck *et al.*, 1973). In addition, exogenous cyclic AMP and derivatives increase the gastric secretion of HCl (Jawaharlal and Berti, 1972). Experiments with the clinically-effective antiulcer agents, metiamide and burimamide (potent histamine (H_2) receptor antagonists), provide evidence that inhibitors of gastric adenylate cyclase could reduce the secretion of gastric acid. Burimamide and metiamide block gastrin- and histamine-stimulated secretion of HCl (Black *et al.*, 1972) and inhibit histamine-stimulated adenylate cyclase (Dousa and Code, 1974).

Osteoporosis

This condition is characterised by bone resorption and the loss of calcium from bone. The incorporation of calcium into bone is controlled by calcitonin, whereas its loss is controlled by parathyroid hormone (PTH). Normal bone structure maintenance therefore requires a balance between the activity of these two hormones. Both calcitonin and PTH raise cyclic AMP levels in bone, and these effects are additive (Murad *et al.*, 1970). Therefore agents which inhibit PTH-stimulated adenylate cyclase in bone, or can potentiate the effects of calcitonin, such as inhibitors of bone cyclic AMP PDE, might be possible therapeutic agents for osteoporosis.

Table 8.8 Diseases where inhibitors of cyclic AMP phosphodiesterase may be potential therapeutic agents

Disease	Key reference
Highly probable	
1. Asthma	Mathé *et al.*, 1977; Parker and Smith, 1973
2. Diabetes mellitus	Colwell *et al.*, 1976; Kuo *et al.*, 1975
3. Female fertility control	Singhal, 1973
4. Male infertility	Garbers *et al.*, 1978
5. Neoplasia and psoriasis	Strada and Pledger, 1975; Voorhees *et al.*, 1975
6. Thrombosis	Moncada and Vane, 1978
7. Anxiety	Beer *et al.*, 1972
8. Hypertension	Amer, 1975; Moncada and Vane, 1978
Probable	
1. Myocardial malfunction	Wollenberger *et al.*, 1973
2. Diabetes insipidus	Strewler and Orloff, 1977
3. Hypoparathyroidism	Broadus, 1977
4. Pseudohypoparathyroidism	Broadus, 1977
5. Parkinson's disease	Kebabian, 1977

DISEASES ASSOCIATED WITH ABNORMAL CYCLIC AMP PHOSPHODIESTERASE ACTIVITY. POTENTIAL OF INHIBITORS OF CYCLIC AMP PHOSPHODIESTERASE AS THERAPEUTIC AGENTS

Drugs that inhibit cyclic AMP PDE have been used as therapeutic agents for many years (Weinryb *et al.*, 1972; Amer and McKinney, 1973; Weiss and Hait, 1977). Although it is doubtful that all of these drugs work *in vivo* by inhibiting cyclic AMP PDE, at least some of their action could be due to blocking this vital enzyme. Table 8.8 lists some of the diseases that are associated with abnormal cyclic AMP metabolism and where inhibitors of cyclic AMP PDE might possibly ameliorate the condition. In the following paragraph, we discuss the use of inhibitors of cyclic AMP PDE in those diseases where it is very probable that there are abnormally high levels of cyclic AMP in the diseased tissue. For other diseases where the association is more doubtful, we leave it to the reader to consult the references cited in table 8.8.

Asthma

Asthmatic subjects display a diminished β-adrenergic responsiveness. There is a reduced responsiveness of urinary cyclic AMP in asthmatic children to epinephrine injections (Broadus, 1977). The disease may reflect a defect in β-adrenergic receptor–adenylate cyclase complex (Mathé *et al.*, 1977). Agents which elevate cyclic AMP, such as epinephrine, isoprenaline, theophylline and disodium cromoglycate, are also effective in treating asthma. In addition, the release of mediators of immediate hypersensitivity such as histamine, bradykinin and slow-reacting substance of anaphylaxis (SRS-A) from sensitised mast cells is inhibited by agents that raise cyclic AMP (Parker and Smith, 1973; Ignarro, 1975). Others have shown that agents which cause a rise in cyclic AMP levels inhibit antigen-induced release of mediators from human lung tissues (Kaliner and Austen, 1974). Taylor *et al.* (1974*a,b*) compared the antiallergic actions of disodium cromoglycate and other drugs known to inhibit cyclic AMP PDE. They found that these compounds inhibit passive cutaneous anaphylaxis and peritoneal mast cell degranulation reactions. These results, along with the demonstration (Roy and Warren, 1974) that disodium cromoglycate is an inhibitor of cyclic AMP PDE, led to the suggestion that its antianaphylactic action is the result of an inhibitory effect on PDE. Sullivan and Parker (1979), however, suggested in a recent report that steroids used in the treatment of asthma work indirectly via prostaglandins by stimulating adenylate cyclases in the target cells.

Diabetes mellitus

Brooker and Fichman (1971) reported that the oral antidiabetic drugs, chlorpropamide and tolbutamide, are inhibitors of cyclic AMP PDE of rat pancreas. They suggested that the hypoglycaemic action of the sulphonylurea compounds is caused by inhibition of pancreatic islet cell cyclic AMP PDE, leading to the release of insulin via elevation of islet cell cyclic AMP levels. Etazolate (SQ 20 009), a pyrazolopyridine inhibitor of cyclic AMP PDE, stimulated the release of insulin when it was perfused through the pancreas (Smith, 1974). Widstrom and Cerasi (1973) reported that the insulin release response to glucagon, an agent known to stimulate adenylate cyclase, is markedly potentiated by tolbutamide. Kuo *et al.*

(1975), however, found that all hormones and prostaglandins which bring about insulin secretion from beta cells also augment the synthesis of cyclic AMP by stimulating islet adenylate cyclase activity; tolbutamide potentiates adenylate cyclase activity and has no effect on cyclic AMP PDE. Thus it appears that agents which favour insulin release increase cyclic AMP levels and inhibitors of cyclic AMP PDE potentiate this action.

Female fertility control

There is substantial evidence favouring the involvement of cyclic AMP in modulating the action of steroid sex hormones, In a review article, Singhal (1973) concluded that cyclic AMP meets the criteria set forth by Sutherland to act as intracellular mediator of the action of both oestrogens and androgens. Exogenously administered cyclic AMP can mimic the anabolic effects of sex hormones in the respective target organ. In addition, inhibition of cyclic AMP PDE can potentiate the biological effect of submaximal doses of steroid sex hormones on their target tissues. Endogenous levels of cyclic AMP are depleted following oestrogenic and androgenic deprivation and are restored towards normal levels following administration of exogenous oestrogens and androgens. Finally, adenylate cyclase activity can be stimulated by sex hormones, *in vitro*, using isolated particulate preparations. Additional evidence to support the use of cyclic AMP in fertility control is provided by studies which show that cyclic AMP is effective as a contraceptive in female mice (Ryan and Coronel, 1969; Pearse and McClurg, 1971). New therapeutic contraceptives might well be numbered among the cyclic nucleotide analogues, cyclic AMP PDE inhibitors or adenylate cyclase stimulators.

Male infertility

One of the major causes of male infertility in animals and humans is lack of motility of otherwise viable sperm. There is considerable evidence and speculation that cyclic AMP mediates some of the actions of sperm cells; see for example Garbers *et al.* (1978). These investigators reported that spermatozoa contain some of the highest concentrations of cyclic AMP enzymes of any group of cells in the body. Exogenous cyclic AMP PDE inhibitors and cyclic AMP increase motility, respiration and metabolism of sperm cells (Garbers *et al.*, 1971; Li, 1972).

Neoplasia and psoriasis

These diseases are characterised by rapid, proliferative, uncontrolled cell growth (Strada and Pledger, 1975; Voorhees *et al.*, 1975). During the rapid cell growth phase, cyclic AMP levels are depressed. Exogenous cyclic AMP and agents which increase endogenous levels of cyclic AMP reduce the growth rate of malignant cells and cause a reversion of their biochemical and morphological characteristics toward those of normal cells. In a double blind study (Stawiski *et al.*, 1975), a cyclic AMP PDE inhibitor, papaverine, resulted in significant clinical benefit when topically applied to psoriatic lesions.

Arterial thrombosis

The role of blood platelets in haemostasis and thrombosis has been greatly clarified in recent years and has been the subject of several reviews (Marx, 1977; Moncada and Vane, 1978). Two major breakthroughs have helped in elucidating the role of prostaglandins in platelet function. First, there was the discovery that the major prostaglandin produced in platelets is the potent, unstable platelet aggregator, bronchoconstrictor and vasoconstrictor, thromboxane A_2 (TXA_2) (Hamberg and Samuelsson, 1973; Hamberg *et al.*, 1975; Nugteren and Hazelhof, 1973). Second, it was discovered that prostacyclin, a potent, unstable inhibitor of platelet aggregation, bronchodilator and vasodilator, is formed from arachidonic acid by endothelial cells of blood vessel walls (Gryglewski *et al.*, 1976; Johnson *et al.*, 1976; Moncada *et al.*, 1976*a,b*). Normally there is a homeostatic balance maintained between the production of thromboxane and prostacyclin. When there is damage to blood vessels, as in atherosclerosis, they become denuded of endothelium resulting in decreased prostacyclin synthesis. Thus the normal balance is upset, allowing platelet aggregation by the now predominant thromboxane. The end result is intravascular thrombosis. There is substantial evidence that inhibition of platelet aggregation is linked to an elevation of cyclic AMP levels (Haslam *et al.*, 1978; Salzman, 1977; Gorman *et al.*, 1977; Tateson *et al.*, 1977). Moreover, it was recently proposed that platelet aggregation is controlled by a balance between the cyclic AMP-lowering activity of TXA_2 (via inhibition of adenylate cyclase) and the cyclic AMP-elevating properties of prostacyclin (via stimulation of adenylate cyclase) (Gorman *et al.*, 1978). PGI_2 is the most potent stimulator of platelet adenylate cyclase thus far described, being 10 and 30 times more active than PGD_2 and PGE_1, respectively (Tateson *et al.*, 1977). Exogenous

Table 8.9 Diseases and conditions where inhibitors of platelet aggregation may be therapeutic agents

Disease or condition	Key references
1. Diabetes	Robertson and Chen, 1977; Peterson and Gormsen, 1978; Colwell *et al.*, 1976
2. Myocardial ischemia and infarction	Ogletree and Lefer, 1978; Lefer *et al.*, 1978
3. Atherosclerosis	Moncada and Vane, 1978; Vergroesen, 1977
4. Asthma	Mathé *et al.*, 1977
5. Stroke and transient ischemic attacks (TIA)	Rosenblum and El-Sabban, 1977
6. Use of prosthetic devices	Moncada and Vane, 1978
7. Depression	Abdulla and Hamadah, 1975
8. Hypertension	Moncada and Vane, 1978

PGI_2 and stable PGI_2 analogues or agents which potentiate the action of endogenous PGI_2 (inhibitors of cyclic AMP PDE) should act as antithrombotic agents. An example to prove this point is provided by the recent experiments of Moncada and Korbut (1978) who showed that dipyridamole and theophylline acted as antithrombotic agents by potentiating the disaggregation of platelets by circulating endogenous PGI_2. Other potential antithrombotic agents may be selective inhibitors of thromboxane synthetase and antagonists of the TXA_2 receptor. Some diseases and conditions where antiplatelet drugs might possibly find therapeutic utility are shown in table 8.9.

Anxiety
Antianxiety drugs such as diazepam and chlordiazepoxide are potent inhibitors of cyclic AMP PDE from rat brain (Beer *et al.*, 1972; Weinryb *et al.*, 1972). In addition Beer *et al.* (1972) reported that etazolate (SQ 20 009), a pyrazolopyridine derivative and a more potent inhibitor of cyclic AMP PDE, produced behavioural effects in experimental animals similar to those seen after the administration of antianxiety drugs. They also reported that dibutyryl cyclic AMP causes similar behavioural effects. Furthermore, etazolate, diazepam and chlordiazepoxide are more potent inhibitors of cyclic AMP PDE of rat brain than of enzymes of cat heart or human blood platelets (Hess *et al.*, 1975). Etazolate (Smith, 1974) and another pyrazolopyridine derivative, cartazolate (SQ 65 396) (Sakalis *et al.*, 1974) which is also a potent inhibitor of cyclic AMP PDE, have been evaluated in man and found to have anxiolytic activity without sedation. Recently, cartazolate was reported to enhance the binding of diazepam to its receptor (Beer *et al.*, 1978). It was suggested that this action of cartazolate may mediate its antianxiety effect by enhancing the binding of an endogenous modulator at the receptor.

Hypertension
In a series of studies using aorta and mesenteric artery of spontaneously stress induced-, DOCA- and neurogenic-hypertensive rats, hypertension was associated with an elevation of the ratio of cyclic GMP to cyclic AMP (Amer, 1973, 1975; Amer *et al.*, 1974; Namm and Leader, 1976). Based on these findings, Amer hypothesised that an aberration in the ratio plays a central role in the aetiology of hypertension. This aberration is due to an increase in cyclic AMP PDE activity and a decrease in activity of adenylate cyclase in the blood vessels of hypertensive animals. In three of the hypertensive models, there was a concurrent increase in activity of guanylate cyclase. Amer *et al.* (1974) also reported that antihypertensive drugs, such as guanethidine, diazoxide and minoxidil, lower the ratio of cyclic GMP to cyclic AMP of hypertensive animals towards that observed in normotensive animals. As supporting evidence of Amer's hypothesis, Rosman *et al.* (1976) reported that the hypertensive agent angiotensin II causes a twofold increase in plasma cyclic GMP level whereas cyclic AMP level rises only by 30 per cent. In a recent review, Diamond (1978), however, concluded that there is no correlation between increase in cyclic AMP level and smooth-muscle relaxation or increases in cyclic GMP levels and contraction. Prostacyclin is a very potent vasodilator produced by blood vessels whereas TXA_2 synthesised by platelets is a potent vasoconstrictor (Moncada and Vane, 1978). These unstable prostaglandin products may therefore play a role in controlling hypertension. Inhibitors of cyclic AMP PDE, stimulators of guanylate cyclase, cyclic AMP analogues, prostacyclin-like compounds, selective thromboxane synthetase inhibitors and TXA_2 receptor antagonists could all therefore be possible therapeutic agents for hypertension.

SUMMARY AND CONCLUSIONS

Some of the biochemical approaches we have used in our laboratory, utilising platelet aggregation to detect potential pharmacological and therapeutic agents, are described. A series of coumarin derivatives and etazolate inhibit platelet cyclic

AMP PDE and potentiate the effects of PGE_1 on platelet aggregation and platelet cyclic AMP concentrations. These compounds probably behave like dipyridamole and theophylline *in vivo* and act as antithrombotic agents by potentiating the inhibition of platelet aggregation via endogenous prostacyclin. A series of 9-substituted adenine derivatives inhibit PGE_1-stimulated adenylate cyclase activity in platelets. These compounds also reverse PGE_1-induced inhibition of platelet aggregation and the elevation of platelet cyclic AMP levels brought about by this prostaglandin. It is possible that compounds such as these will have therapeutic utility in such diverse diseases as cholera, psychosis and duodenal ulcers.

Cyclic AMP mediates a multitude of biological functions and abnormal concentrations of this cyclic nucleotide are associated with many diseases. In those we have discussed, the change derives from alterations of adenylate cyclase or cyclic AMP PDE. Thus, selective inhibition of these enzymes might help to ameliorate some of these diseases; indeed many current therapeutic agents appear to inhibit cyclic AMP PDE and adenylate cyclase (Weinryb *et al.*, 1972). An argument against such an approach to drug development is the difficulty in developing specific drugs, since cyclic AMP is involved in so many vital processes. Success in this respect, however, has been achieved at the Squibb Institute for Medical Research (Smith, 1974, 1978). Etazolate, a potent inhibitor of cyclic AMP PDE from many sources, potentiates the effects of agonists in many biological systems mediated by cyclic AMP (for example, fat cell lipolysis, adrenal cell steroidogenesis, inhibition of platelet aggregation). When etazolate was tested in animals, however, it was found only to affect the central nervous system. In clinical studies, etazolate failed to cause significant disturbance in a wide variety of biochemical parameters, but had anxiolytic activity without sedation. Although it seems likely that cyclic GMP, the cyclic nucleotide with opposing actions to cyclic AMP in many biological processes, is also involved in many diseases, this possibility has not been explored in the present work.

ACKNOWLEDGEMENT

The authors are indebted to Dr Nick S. Semenuk for bibliographic assistance and to Mrs Susan M. Smith for invaluable secretarial assistance. The data in figure 8.4 and in tables 8,4, 8.5 and 8.6 were used by permission of the authors, the *Journal of Cyclic Nucleotide Research* and Raven Press, New York, N.Y.

REFERENCES

Abdulla, Y. (1969). *J. Atheroscler. Res.*, **9**, 171
Abdulla, Y. and Hamadah, K. (1975). *Br. J. Psychiat.*, **127**, 591
Adams, M. (1973). *Science, N.Y.*, **179**, 552
Amer, M. (1973). *Science, N.Y.*, **179**, 807
Amer, M. (1975). *Life Sci.*, **17**, 1021
Amer, M. and Kreighbaum, W. (1975). *J. pharm. Sci.*, **64**, 1
Amer, M. and McKinney, G. (1973). *Life Sci.*, **13**, 753
Amer, M., Perhach, J., Jr, Gomoll, A. and McKinney, G. (1974). *Pharmacologist*, **16**, 288
Amer, M. Gomoll, A., Perhach, J., Jr, Ferguson, H. and McKinney, G. (1974). *Proc. natn. Acad. Sci. U.S.A.*, **71**, 4930
Beer, B., Klepner, C., Lippa, A. and Squires, R. (1978). *Pharmac. Biochem. Behav.*, **9**, 849
Beer, B., Chasin, M., Clody, D., Vogel, J. and Horovitz, Z. (1972). *Science, N.Y.*, **176**, 428

Bieck, P., Oates, J., Robison, G. and Adkins, R. (1973). *Am. J. Physiol.*, **224**, 158
Black, J., Duncan, W., Durant, C., Ganellin, G., and Parsons, E. (1972). *Nature, Lond.*, **236**, 385
Born, G. (1962). *Nature, Lond.*, **194**, 927
Broadus, A. (1977). *Adv. cyclic Nucleotide Res.*, **8**, 509
Brooker, G. and Fichman, M. (1971). *Biochem. Biophys. Res. Commun.*, **42**, 824
Brooker, G., Thomas, L. and Appleman, M. (1968). *Biochemistry*, **7**, 4177
Clement-Cormier, Y., Kebabian, J., Petzold, G. and Greengard, P. (1974). *Proc. natn. Acad. Sci. U.S.A.*, **71**, 1113
Clement-Cormier, Y., Parrish, R., Petzold, G., Kebabian, J. and Greengard, P. (1975). *J. Neurochem.*, **25**, 143
Colwell, J., Halushka, P., Sarji, K., Levine, J., Sagel, J. and Nair, R. (1976). *Diabetes*, **25**, 826
Diamond, J. (1978). *Adv. cyclic Nucleotide Res.*, **9**, 327
Dousa, T. and Code, C. (1974). *J. clin. Invest.*, **53**, 334
Fleischer, N., Donald, R. and Butcher, R. (1969). *Am. J. Physiol.*, **217**, 1287
Garbers, D., Lust, W., First, N. and Lardy, H. (1971). *Biochemistry*, **10**, 1825
Garbers, D., Watkins, H., Tubb, D. and Kopf, G. (1978). *Adv. cyclic Nucleotide Res.*, **9**, 583
Gill, D. (1977). *Adv. cyclic Nucleotide Res.*, **8**, 85
Gordon, P. and Dowben, R. (1966). *Am. J. Physiol.*, **210**, 728
Gorman, R., Bunting, S. and Miller, O. (1977). *Prostaglandins*, **13**, 377
Gorman, R., Fitzpatrick, F. and Miller, O. (1978). *Adv. cyclic Nucleotide Res.*, **9**, 597
Greengard, P. and Robison, G. (eds) (1975). *Adv. cyclic Nucleotide Res.*, **6**
Greengard, P., Robison, G. and Paoletti, R. (eds) (1972). *Adv. cyclic Nucleotide Res.*, **1**
Guerrant, R., Chen, L. and Sharp, G. (1972). *J. infect. Dis.*, **125**, 377
Gryglewski, R., Bunting, S., Moncada, S., Flower, R. and Vane, J. R. (1976). *Prostaglandins*, **12**, 685
Hamberg, M. and Samuelsson, B. (1973). *Proc. natn. Acad. Sci. U.S.A.*, **70**, 899
Hamberg, H., Svensson, J. and Samuelsson, B. (1975). *Proc. natn. Acad. Sci. U.S.A.*, **72**, 2994
Harris, D., Asaad, M., Phillips, M., Goldenberg, H. and Antonaccio, M. (1979). *J. cyclic Nucleotide Res.*, **5**, 125
Haslam, R. (1978). In *Platelet Function Testing* (ed. H. Day, H. Holmsen and M. Zucker), U.S. Department of Health, Education and Welfare, Publication No. (NIH) 78 1087, p. 487
Haslam, R. and Lynham, J. (1972). *Life Sci., Part II*, **11**, 1143
Haslam, R. J., Davidson, M., Davies, T., Lynham, J. and McClenaghan, M. (1978). *Adv. cyclic Nucleotide Res.*, **9**, 533
Hess, S. M., Chasin, M., Free, C. and Harris, D. (1975). In *Mechanism of Action of Benzodiazepines* (ed. E. Costa and P. Greengard), Raven Press, New York, p. 153
Ignarro, L. (1975). In *Cyclic Nucleotides in Disease* (ed. B. Weiss), University Park Press, Baltimore, p. 187
Jacobson, E. and Thompson, W. (1976). *Adv. cyclic Nucleotide Res.*, **7**, 199
Jawaharlal, K. and Berti, F. (1972). *Pharmac. Res. Commun.*, **4**, 143
Johnson, R., Morton, D., Kinner, J., Gorman, R., McGuire, J., Sun, F., Whittaker, N., Bunting, S., Salmon, J., Moncada, S. and Vane, J. (1976). *Prostaglandins*, **12**, 915
Kaliner, M. and Austen, K. (1974). *Biochem. Pharmac.*, **23**, 763
Kebabian, J. (1977). *Adv. cyclic Nucleotide Res.*, **8**, 421
Kebabian, J., Petzold, G. and Greengard, P. (1972). *Proc. natn. Acad. Sci. U.S.A.*, **69**, 2145
Kuo, J., Kuo, W. and Hodgins, D. (1975). In *Cyclic Nucleotides in Disease* (ed. B. Weiss), University Park Press, Baltimore, p. 211
Lefer, A., Ogletree, M., Smith, B., Silver, M., Nicolaou, K., Barnette, W. and Gasic, G. (1978). *Science, N.Y.*, **200**, 52
Li, T. (1972). *Life Sci., Part II*, **11**, 939
Londos, C. and Wolff, J. (1977). *Proc. natn. Acad. Sci. U.S.A.*, 74, 5482
McDonald, J. and Stuart, R. (1973). *J. Lab. clin. Med.*, **81**, 838
Margules, D. (1971). *Eur. J. Pharmac.*, **16**, 21
Marquis, N., Vigdahl, R. and Tavorima, P. (1969). *Biochem. biophys. Res. Commun.*, **36**, 965
Marx, J. (1977). *Science, N.Y.*, **196**, 1072
Mathé, A., Hedqvist, P., Strandberg, K. and Crystal, L. (1977). *New Engl. J. Med.*, **296**, 910
Miller, O., Johnson, R. and Gorman, R. (1977). *Prostaglandins*, **13**, 599

Mills, D. and Smith, J. (1971). *Biochem. J.*, **121**, 185
Moncada, S. and Korbut, R. (1978). *Lancet*, 1286
Moncada, S. and Vane, J. (1978). In *Platelets: A Multidisciplinary Approach* (ed. G. de Gaetano and S. Garattini), Raven Press, New York, p. 239
Moncada, S., Gryglewski, R., Bunting, S. and Vane, J. (1976*a*). *Nature, Lond.*, **263**, 663
Moncada, S., Gryglewski, R., Bunting, S. and Vane, J. (1976*b*). *Prostaglandins*, **12**, 715
Murad, F., Brewer, H. and Vaughn, M. (1970). *Proc. natn. Acad. Sci. U.S.A.*, **65**, 446
Namm, D. and Leader, J. (1976). *Blood Vessels*, **13**, 24
Nugteren, D. and Hazelhof, E. (1973). *Biochim. biophys. Acta*, **326**, 448
Ogletree, M. and Lefer, A. (1978). *Circulation Res.*, **42**, 218
Parker, C. and Smith, J. (1973). *J. clin. Invest.*, **52**, 48
Pearse, W. and McClurg, J. (1971). *Am. J. Obstet. Gynecol.*, **109**, 724
Peterson, H. and Gormsen, J. (1978). *Acta med. scand.*, **203**, 125
Reite, M., Davis, K., Solomons, C. and Ott, J. (1972). *Am. J. Psychiat.*, **128**, 1540
Robertson, P. and Chen, M. (1977). *J. clin. Invest.*, **60**, 747
Robison, G., Butcher, R. and Sutherland, E. (1971). In *Cyclic AMP*, Academic Press, New York, p. 36
Rosenblum, W. and El-Sabban, F. (1977). *Stroke*, **8**, 691
Rosman, P., Agrawal, R., Goodman, A. and Steiner, A. (1976). *J. clin. Endocrinol. Metab.*, **42**, 531
Roy, A. and Warren, B. (1974). *Biochem. Pharmac.*, **23**, 917
Ryan, W. and Coronel, D. (1969). *Am. J. Obstet. Gynecol.*, **105**, 121
Sakalis, G., Sathananthan, G., Collier, P. and Gerson, S. (1974). *Curr. Therap. Res.*, **16**, 861
Salzman, E. (1977). *Biochim. biophys. Acta*, **499**, 48
Salzman, E. (1978). In *Platelets: A Multidisciplinary Approach* (ed. G. de Gaettano and S. Garattini), Raven Press, New York, p. 227
Salzman, E. and Levine, L. (1971). *J. clin. Invest.*, **50**, 131
Schultz, J. and Gratzner, H. (eds) (1973). *The Role of Cyclic Nucleotides in Carcinogenesis*, Academic Press, New York
Sharp, G. (1973). *A. Rev. Med.*, **24**, 19
Singhal, R. (1973). *Adv. Pharmac. Chemother.*, **11**, 99
Smith, C. (1974). *Adv. Enzyme Regulation*, **12**, 187
Smith, C. (1978). In *Principles and Techniques of Human Research and Therapeutics*, Vol. 15, *Future Trends in Therapeutics* (ed. F. McMahon), Futura Publishing Company, Mount Kisco, N.Y., p. 55
Stawiski, M., Powell, J., Lang, P., Schork, M., Duell, E. and Voorhees, J. (1975). *J. invest. Dermatol.*, **64**, 124
Strada, S. and Pledger, W. (1975). In *Cyclic Nucleotides in Disease* (ed. B. Weiss), University Park Press, Baltimore, p. 3
Strewler, G. and Orloff, J. (1977). *Adv. cyclic Nucleotide Res.*, **8**, 311
Sullivan, T. and Parker, C. (1979). *J. Immunol.*, **122**, 431
Sung, C., Jenkins, B., Burns, L., Hackney, V., Spenney, J., Sachs, G. and Wiebelhaus, V. (1973). *Am. J. Physiol.*, **225**, 1359
Tateson, J., Moncada, S. and Vane, J. (1977). *Prostaglandins*, **13**, 389
Taylor, W., Francis, D., Sheldon, D. and Roitt, I. (1974*a*). *Int. Arch. Allergy appl. Immunol.*, **46**, 104
Taylor, W., Francis, D., Sheldon, D. and Roitt, I. (1974*b*). *Int. Arch. Allergy appl. Immunol.*, **47**, 175
Tyrer, P. and Lader, M. (1972). *Lancet*, **ii**, 542
Vergroesen, A. (1977). *Nutr. Rev.*, **35**, 1
Volicer, L. (ed.) (1977). *Clinical Aspects of Cyclic Nucleotides*, Spectrum Publications, New York
Voorhees, J., Duell, E., Stawiski, M., Creehan, P. and Harrell, E. (1975). In *Cyclic Nucleotides in Disease* (ed. B. Weiss), University Park Press, Baltimore, p. 79
Weinryb, I. (1972). In *Methods in Cyclic Nucleotide Research* (ed. M. Chasin), Marcel Dekker, New York, p. 39
Weinryb, I. and Michel, I. (1974). *Biochim. biophys. Acta*, **334**, 218
Weinryb, I., Chasin, M., Free, C., Harris, D., Goldenberg, H., Michel, I., Paik, V., Phillips, M., Samaniego, S. and Hess, S. (1972). *J. pharm. Sci.*, **61**, 1556

Weiss, H. (ed.) (1975). *Cyclic Nucleotides in Disease*, University Park Press, Baltimore
Weiss, B. and Greenberg, L. (1975). In *Cyclic Nucleotides in Disease* (ed. B. Weiss), University Park Press, Baltimore, p. 269
Weiss, B. and Hait, W. (1977). *Ann. Rev. Pharmac. Toxicol.*, **17**, 441
Weiss, B. and Levin, R. (1978). *Adv. cyclic Nucleotide Res.*, **9**, 285
Widstrom, A. and Cerasi, E. (1973). *Acta endocrinol.*, **72**, 532
Wollenberger, A., Babskii, E., Krouse, E., Genz, S., Blohm, D. and Bogdanova, E. (1973). *Biochem. Biophys. Res. Commun.*, **55**, 446
Zenser, T. (1976). *Proc. Soc. exp. Biol. Med.*, **152**, 126

9

Monoamine oxidase inhibitors: introduction

H. Blaschko (Department of Pharmacology, University of Oxford, South Parks Road, Oxford, OX1, UK)

Some time ago, when I was writing a review article on monoamine oxidase (Blaschko, 1952), I discovered that this enzyme had a very respectable pre-history in which many names can be found familiar to all pharmacologists. The story begins over a hundred years ago when Oswald Schmiedeberg (1877) described the metabolism of benzylamine and other monoamines, and a few years later his famous pupil, Otto Minkowski (1883), demonstrated the breakdown of benzylamine to benzoic acid in tissue preparations.

Sixty years after Schmiedeberg's first paper, we described the oxidative deamination of adrenaline by what is now known as monoamine oxidase (Blaschko *et al.*, 1937). Here for the first time an amine known to be present in animals was shown to be a substrate of the enzyme that we are interested in, in the following contributions. However, it still took a long time for the role of the oxidase in the biological inactivation of the catecholamines to be generally accepted. For instance, in a review article by Bacq (1949) entitled 'The metabolism of adrenaline', written twelve years after our first paper, we read that 'its deamination by amine oxidase is unlikely'.

There are good reasons why understanding of the biological role of monoamine oxidase progressed at such a slow pace. At the time Bacq wrote his review, the enzyme, catechol *O*-methyltransferase, had not been discovered. More importantly, the phenomenon of amine uptake was not appreciated. Students of amine metabolism are greatly indebted to Iversen (1967) for making us fully aware of the importance of uptake. Until then it remained a puzzle how an intracellular, mitochondrial enzyme could be responsible for the metabolism of substances like the catecholamines which have such evanescent pharmacological effects. People were conditioned to look for an analogy with the enzyme acetylcholinesterase, the action of which was much better understood: it was known to be located, at least partly, at the level of the cell membrane, where it probably destroyed its substrate in the extracellular compartment.

Monoamine oxidase received its name from Dr Albert Zeller, and it was he also who discovered the first inhibitor of the type that we are interested in today. I am afraid my own experimental acquaintance with these inhibitors is very limited. Over twenty years ago, when Dr Oleh Hornykiewicz came to Oxford I suggested to him that he might take an interest in dopamine. I had been thinking for some time that this amine might have some regulatory function of its own, and I was wondering if one could find some method of bioassay. At that time I knew of only one action of dopamine in which it differed from that of the other two naturally occurring catecholamines: Dr Peter Holtz and his colleagues (Holtz *et al.*, 1942) had found that dopamine lowered the arterial blood pressure of the guinea-pig whereas the other two amines had a pressor effect. Holtz thought that the depressor effect was caused not by dopamine itself but by an oxidation product formed under the action of monoamine oxidase.

Hornykiewicz (1958) confirmed Holtz's findings but showed that one had to look for a different interpretation. He found that iproniazid, the monoamine oxidase inhibitor studied by Zeller, caused a slight potentiation of the pressor response to noradrenaline. When this potentiation was fully established he tested the response to dopamine and found that under these conditions the depressor effect to dopamine was likewise enhanced. This made it unlikely that the depressor response was due to the presence of a compound produced through the action of monoamine oxidase.

I am afraid Dr Hornykiewicz did not continue with this problem when he returned to Vienna. By that time more convenient methods of dopamine assay had become available and you know that he put these to excellent use.

This is ancient history. The problems of the monoamine oxidase inhibitors keep us busy still. Four years ago I attended a Ciba Foundation Symposium on this subject (Wolstenholme and Knight, 1976). I shall be very interested to learn what new material has been accumulated in these last few years. In 1975 the importance of amine uptake was already fully realised and we also heard much about the various molecular forms of the enzyme. I feel sure these two aspects will again be discussed in the contributions which follow.

However, there are some new factors which have only been fully appreciated in the past few years. Most of these are connected with our knowledge of the receptors for catecholamines. This knowledge is important when we wish to consider the potential usefulness of monoamine oxidase inhibitors. These substances must obviously exert their action at the level of the synapse, in the area where the amines impinge on the receptors.

First of all, we now have to take into account the presence of receptors in two different locations: there are pre-synaptic and post-synaptic receptors. The pre-synaptic receptors, situated at the endings of the aminergic neurones, are important for us because their activation modifies the amounts of amine released. The post-synaptic receptors, of course, are important because their activation directly determines the response to release, its intensity and its duration. We have learnt to understand that the receptors are capable of lateral movement within the membrane. In addition, they may vary in number: the effector cell seems to respond to the amount of transmitter it is exposed to by modifying receptor synthesis. Such a phenomenon seems likely to be important in the context of the presentations in this volume, discussing situations when the concentration of free transmitter is modified by drug action over a prolonged period of time.

REFERENCES

Bacq, Z. M. (1949). *Pharmac. Rev.*, **1**, 1-26
Blaschko, H. (1952). *Pharmac. Rev.*, **4**, 415–58
Blaschko, H., Richter, D. and Schlossmann, H. (1937). *J. Physiol., Lond.*, **90**, 1–17
Holtz, P., Credner, K. and Koepp, W. (1942). *Naunyn-Schmiedeberg's Arch. exp. Path. Pharmak.*, **200**, 356–88
Hornykiewicz, O. (1958). *Br. J. Pharmac.*, **13**, 91–4
Iversen, L. L. (1967). *The Uptake and Storage of Noradrenaline in Sympathetic Nerves.* Cambridge University Press, Cambridge
Minkowski, O. (1883). *Naunyn-Schmiedeberg's Arch. exp. Path. Pharmak.*, **17**, 445–65
Schmiedeberg, O. (1877). *Naunyn-Schmiedeberg's Arch. exp. Path. Pharmak.*, **8**, 1–14
Wolstenholme, G. E. W. and Knight, J. (eds) (1976). *Monoamine Oxidase and its Inhibition*, Ciba Foundation Symposium 39 (new series), Elsevier, Amsterdam

10
Monoamine oxidase inhibitors: chemistry and pharmacology

J. Knoll (Semmelweis University of Medicine, Department of Pharmacology, 1445 Budapest, P.O.B. 370 Hungary)

INTRODUCTION

The fact that monoamines are deaminated in the organism has been known for a hundred years, dating back to a study by Schmiedeberg (1877) who found that benzylamine given orally to dogs is excreted as benzoylglycine (hippuric acid, urobenzoic acid). He also succeeded in demonstrating the presence of free benzoic acid in the urine (Schmiedeberg, 1881). These findings were corroborated by Minkowski (1883) who also proved that minced rabbit tissues catalyse the conversion of benzylamine to benzoic acid. Ewins and Laidlaw (1910) presented evidence that tyramine, an endogenous phenylalkylamine, and tryptamine, an endogenous indolealkylamine (Ewins and Laidlaw, 1913) are oxidised and deaminated in the liver, yielding corresponding acids. The same fact was demonstrated with regard to phenylethylamine by Guggenheim and Löffler (1916).

The specific enzyme for the catalysis of this type of conversion *in vivo* was discovered by Mary Hare, who identified 'tyramine oxidase' in 1928 as a new enzyme system in the liver for the oxidation of tyramine. The important study by Blaschko *et al.* (1937) on inactivation of adrenaline led to the concept of amine oxidase, an enzyme system which catalyses the oxidation of a large number of amines (cf. Blaschko, 1974). Zeller (1938) proposed the name 'monoamine oxidase' (MAO) for the class of enzymes responsible for the oxidative deamination of monoamines.

LOCALIZATION AND PHYSIOLOGICAL FUNCTIONS OF MAO

Mitochondrial MAO, a spherical protein with a density of 1.4 and a molecular weight of about 100 000, is widely distributed in different types of neurone, glia, liver, kidney, heart, intestine, lung, salivary gland, gonads and various kinds of smooth muscle. It is bound to the outer mitochondrial membrane (Schnaitman *et al.*, 1967) and plays an important role in the inactivation of both exogenous and endogenously-formed amines. The pressor amines of foodstuffs are inactivated by intestinal MAO, blood vessel MAO protects the organs from toxic effects of circulating amines and MAO in tissues helps to regulate the intracellular concentration of certain monoamines (e.g. phenylethylamine (PEA), phenylethanolamine, tyramine, noradrenaline, dopamine, octopamine, 5-hydroxytryptamine (5-HT), tryptamine, *N*-methylhistamine). The functional significance of non-mitochondrial amine oxidases, like spermine-oxidase and benzylamine-oxidase in the mammalian blood plasma is obscure, while connective tissue amine oxidases bring about cross-linking in elastic and collagen fibres (cf. Blaschko, 1972).

The concept that mitochondrial MAO plays a physiological role in the biological inactivation of transmitters like noradrenaline, dopamine and 5-HT is strongly supported by much experimental evidence. Substantial MAO activity is associated with synaptosomes (De Lores Arnaiz and De Robertis, 1962) and the activity of the enzyme is reduced after denervation of different organs (Snyder *et al.*, 1965; Klingman, 1966; Goridis and Neff, 1971). Inhibition of MAO leads to an increase in concentration of noradrenaline, dopamine and 5-HT in the brain to new steady-state levels (Brodie *et al.*, 1956; Spector *et al.*, 1957) and increases the stores of these amines in reserpinised animals. Important surveys regarding the localisation and physiological functions of MAO are to be found in the proceedings of two international meetings on the topic (Costa and Sandler, 1972; Wolstenholme and Knight, 1976).

MAO is a very stable enzyme. Even after death, changes are negligible for up to 120 h at 5 °C (MacKay *et al.*, 1978). Activity seems to depend on genotype. As platelet MAO activity appears to be a stable characteristic of the individual and activity in different parts of the human brain is highly intercorrelated in a particular subject, brain and platelet enzyme has been studied extensively under different psychiatric conditions (Murphy and Wyatt, 1972; Wyatt *et al.*, 1973; Friedmann *et al.*, 1974; Sandler *et al.*, 1974; Shaskan and Becker, 1975; Gottfries *et al.*, 1975; Buchsbaum *et al.*, 1976; Berger *et al.*, 1978). The view which has begun to emerge that low MAO activity in brain and platelets reflects a constitutional predisposition in the individual, making him vulnerable to suicidal be-

Table 10.1 The most important known physiological functions of mitochondrial MAO and the pharmacological consequences of enzyme inhibition

Physiological functions of MAO	Consequences of MAO inhibition
1. Intraneuronal inactivation of transmitter in noradrenergic, dopaminergic and serotonergic neurones	Transmitters increase to new steady-state levels in catecholaminergic and serotonergic neurones
2. Gastrointestinal inactivation of pressor amines present in foodstuffs	Facilitated access of pressor amines into circulation ('cheese effect')
3. Inactivation of a large variety of endogenously formed and exogenous amines in liver and blood vessels	Increased levels of endogenous 'trace amines' (e.g. phenylethylamine, octopamine, tryptamine, etc.). Endogenously formed pressor amines as well as those absorbed from the intestines (mainly tyramine) circulate and reach the organs without control

haviour, ethanol abuse or manic-depressive illness is a potentially important aspect of MAO research, which needs further experimental backing.

The discovery that iproniazid is an inhibitor of MAO activity (Zeller *et al.*, 1952) opened the way to the synthesis and pharmacological assessment of hundreds of irreversible inhibitors of the enzyme and fostered the development of present-day views on the physiological role of mitochondrial MAO. Table 10.1 shows what appear, on present knowledge, to be the most important physiological functions of mitochondrial MAO and the pharmacological consequences of irreversible inhibition of this enzyme.

BASIC STRUCTURE-ACTIVITY RELATIONSHIPS OF MAO INHIBITORS

Scrutiny of the potent irreversible inhibitors of MAO described in the literature during the past two decades shows that an aromatic ring, an amine group, a short carbon chain between them and a proper 'enzyme-killing' group are the essential structural requirements of an MAO inhibitor. One of the following three solutions to the problem is characteristically used for the formation of the 'killing group': (1) the amino group is changed to a hydrazide; (2) the short carbon chain between the aromatic ring and the nitrogen is changed to the cyclopropyl moiety; (3) a

Table 10.2 Prototypes of MAO inhibitors derived from phenylethylamine

	Structure	Name
Substrate	$C_6H_5-CH_2-CH_2-NH_2$	Phenylethylamine
Reversible inhibitor	$C_6H_5-CH_2-CH(CH_3)-NH_2$	Phenylisopropylamine (amphetamine)
Irreversible inhibitors	$C_6H_5-CH_2-CH_2-NH-NH_2$	Phenylethylhydrazine (phenelzine)
	$C_6H_5-CH_2-CH_2-NH-C_3H_5$ (cyclopropyl)	Phenylethylcyclopropylamine
	$C_6H_5-CH_2-CH_2-NH-CH_2-C{\equiv}CH$	Phenylethylpropargylamine
	$C_6H_5-CH(CH_2)CH-NH_2$ (cyclopropane ring)	Phenylcyclopropylamine (tranylcypromine)

propargyl or cyclopropyl group is attached to the nitrogen. Table 10.2 shows some prototypes of MAO inhibitors derived from PEA, one of the basic substrates of MAO. Table 10.3 exemplifies the importance of aromatic ring and unsubstituted two-carbon chain in binding inhibitor to enzyme.

TWO FUNCTIONALLY DIFFERENT FORMS OF MITOCHONDRIAL MAO

There is now general agreement that two main forms of mitochondrial MAO exist. The discovery of this situation and the development of our present knowledge concerning the dual nature of MAO is inseparable from the introduction of two

Table 10.3 The importance of the aromatic ring and the unsubstituted two-carbon chain in the binding of an MAO inhibitor to the enzyme

$R = -N(CH_3)-CH_2-C{\equiv}CH$	Code no.	ID_{50}† ($mg\ kg^{-1}$) i.v.	Relative potency
$C_6H_5-CH_2-CH_2-R$	TZ-650	0.1	1
$C_6H_5-C(CH_3)_2-CH_2-R$	TZ-996	>10	<0.01
$C_6H_5-CH_2-CH(CH_2-C_6H_5)-R$	J-504	>10	<0.01
$C_6H_{11}-CH_2-CH_2-R$	J-505	>10	<0.01

†*In vivo* inhibition of MAO in rat brain homogenate.
Substrate: [^{14}C]-PEA

substrate-selective highly potent irreversible inhibitors, deprenyl and clorgyline.

Deprenyl (phenylisopropylmethylpropargylamine) was developed by Knoll *et al.* in 1964 and noted to be a compound with a new spectrum of MAO inhibitory action (Knoll *et al.*, 1965; 1968; Magyar *et al.*, 1967). Detailed analysis of its pharmacological and biochemical effects (cf. Knoll, 1976*a,b*, 1978*a,b,c*, Knoll *et al.*, 1978) revealed that (–)-deprenyl is a selective inhibitor of a particular type of MAO which deaminates benzylamine and metaiodobenzylamine (Knoll and Magyar, 1972), as well as PEA (Yang and Neff, 1973).

Clorgyline (2,4-dichlorophenoxypropyl-*N*-methylpropargylamine), a compound similar in structure to deprenyl, was developed in 1968 by Johnston, who found it to be a selective inhibitor of that type of MAO which deaminates 5-HT. To distinguish the two forms of MAO, one highly sensitive to clorgyline and one relatively insensitive to it, he introduced the terms 'type A' and 'type B' MAO. This nomenclature has become widely accepted. MAO-A is selectively inhibited by clorgyline and MAO-B by (–)-deprenyl.

In 1968, another selective irreversible MAO-A inhibitor, *o*-chlorophenoxyethyl-*N*-cyclopropylamine (Lilly 51641) was described by Fuller. An oxygen bridge between aromatic ring and side chain and a chlorine atom in the *ortho* position are common to clorgyline and Lilly 51641. These features seem to be essential in determining selectivity of these inhibitors for MAO-A. An unsubstituted aromatic ring and a two-carbon chain between aromatic ring and nitrogen are common in the structure of potent and highly selective inhibitors of MAO-B (Knoll *et al.*, 1978). Pargyline, which is structurally similar to deprenyl but contains one carbon only between benzene ring and nitrogen, is a much less selective inhibitor of MAO-B than deprenyl.

Figures 10.1 and 10.2 illustrate schematically the working hypothesis that two functionally different forms of MAO exist which can be inhibited selectively: MAO-A is specialised for binding and metabolising the ethylamine side chain of a substrate if it is attached to a 5-hydroxyindole ring ('serotonin oxidase'), and

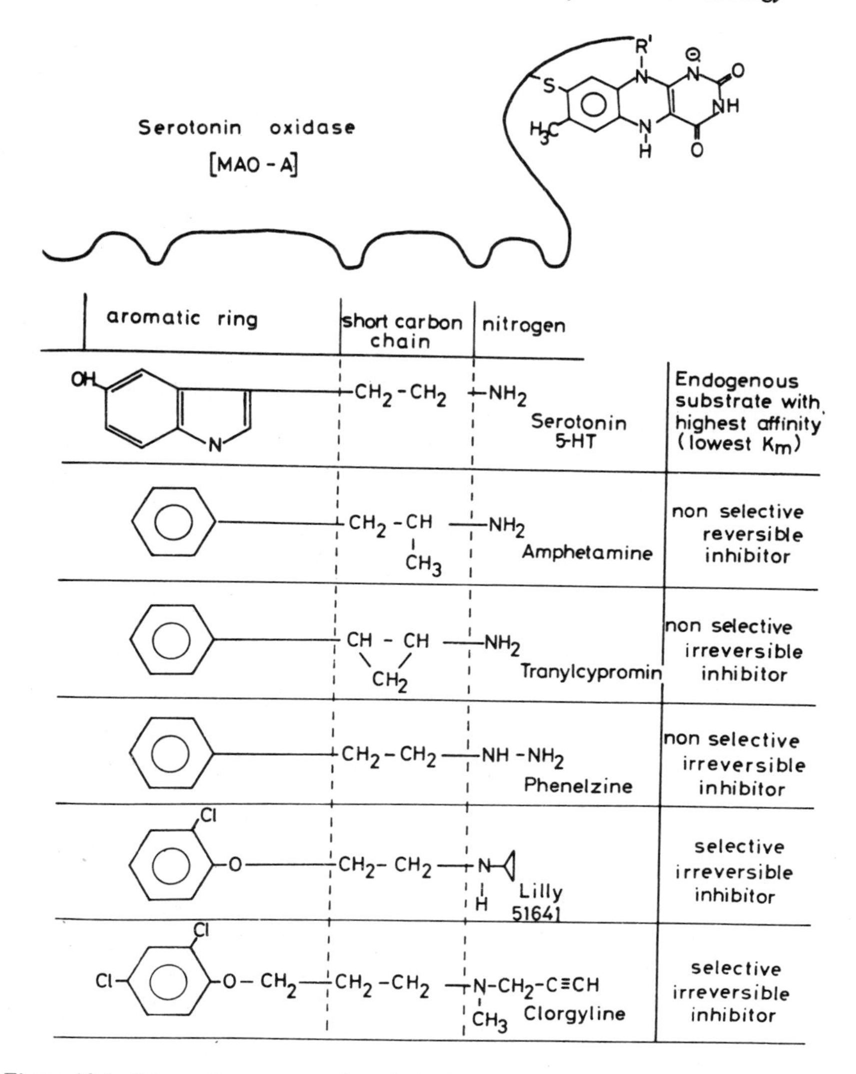

Figure 10.1 Schematic representation of the binding of serotonin (5-HT) and different inhibitors to 'serotonin oxidase' (MAO-A).

MAO-B is specialised for recognising and metabolising PEA ('phenylethylamine oxidase') (Knoll, 1976*a,b*, 1978*a,b,c*). There are many other amines which, because of appropriate structural similarities, are substrates of either MAO-A or MAO-B or are common substrates of both enzymes.

According to the models, the binding site for the nitrogen is in the vicinity of the covalently bound flavin group of the enzyme. This situation offers a good explanation for the destruction of MAO-A and B by their respective selective inhibitors, clorgyline and (–)-deprenyl (Knoll, 1978*a*). The proposed mechanism

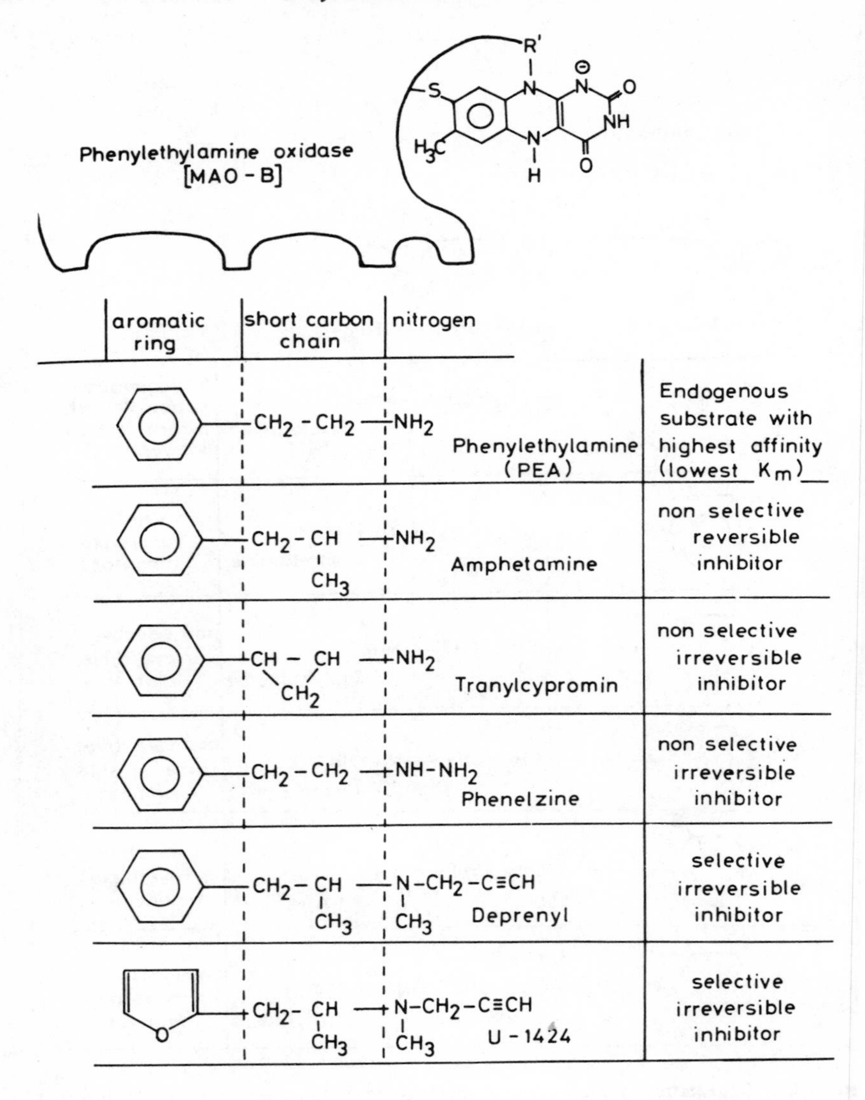

Figure 10.2 Schematic representation of the binding of phenylethylamine (PEA) and different inhibitors to 'phenylethylamine oxidase' (MAO-B).

of action of the 'enzyme-killing' group in clorgyline is visualised in figure 10.3. The propargyl group is known to be capable of reacting irreversibly with the covalently bound flavin (Maycock *et al.*, 1976) and it has been demonstrated that ^{14}C-deprenyl binds in stoichiometric fashion to the flavin active site of MAO in position 5 of riboflavin (Youdim, 1978).

Figure 10.4 demonstrates the selective inhibition of MAO-B *in vivo*. In this experiment, a method for continuous monitoring of MAO-B activity *in vivo* was

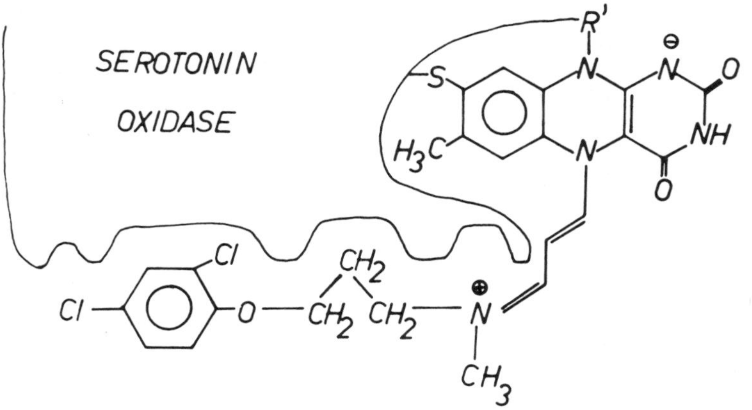

Figure 10.3 Schematic representation of the binding of clorgyline to MAO-A.

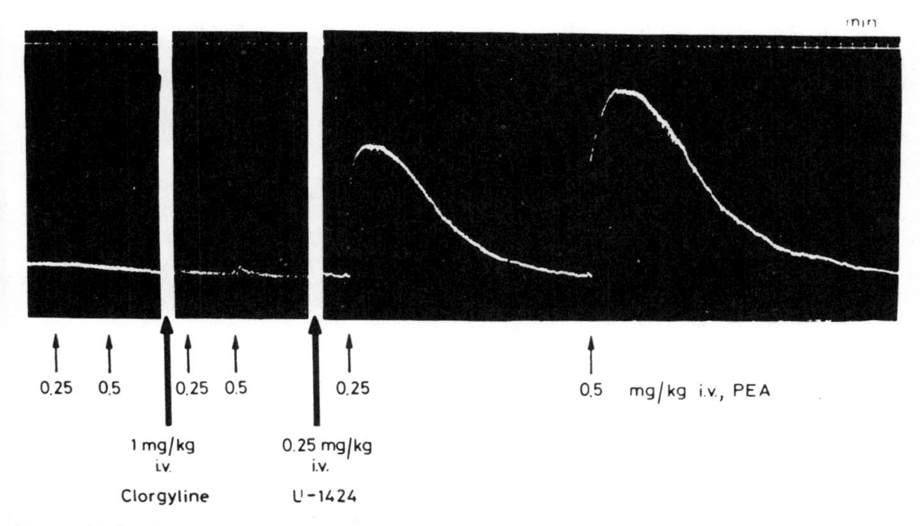

Figure 10.4 Phenylethylamine (PEA)-induced contractions of the nictitating membrane in a cat weighing 3.1 kg. Chloralose–urethane anaesthesia. Artificial respiration. Doses of PEA (0.25–0.5 mg kg^{-1}) insufficient to induce contractions of the nictitating membrane remained ineffective after MAO-A blockade by clorgyline (1 mg kg^{-1}, i.v.) but evoked powerful contractions after inhibition of MAO-B by U-1424 (0.25 mg kg^{-1}, i.v.).

employed (Knoll, 1976*a*). The nictitating membrane of the cat responds to intravenous injection of PEA, a specific substrate of MAO-B, with a dose-dependent contraction. The lowest dose of PEA which evoked a noticeable contraction of this organ is usually about 500 μg kg^{-1} and 2–3 mg kg^{-1} exert significant effects. Liver bypass experiments proved that the need for such high doses of PEA is a consequence of continuous degradation of this amine by MAO-B in the liver (Knoll, 1976*a*). Figure 10.4 shows that the effect of PEA, unchanged after

Table 10.4 The dose-dependency of the selectivity of U-1424 for MAO-B

Duration of treatment (days)		Rat brain MAO activity in per cent of control					
	Daily dose ($mg\ kg^{-1}$, s.c.):	0.05		0.25		1.0	
	Substrate:	[^{14}C]-PEA	[^{14}C]-5-HT	[^{14}C]-PEA	[^{14}C]-5-HT	[^{14}C]-PEA	[^{14}C]-5-HT
7		41.6	100.1	29.1	89.3	20.5	68.6
14		41.5	96.3	25.3	79.5	13.0	48.8
21		40.4	98.4	22.5	81.9	13.6	46.5

PEA = phenylethylamine; 5-HT = serotonin.
Radiometric assay: each value is the mean for 3 rats.

MAO-A blockade with a huge dose of clorgyline, is enormously potentiated after MAO-B inhibition with U-1424, a newly developed selective MAO-B inhibitor (Knoll *et al.*, 1978). Potent, selective MAO blockers of this type inhibit the enzyme with a remarkable safety margin. Less than 0.5 per cent of the LD_{50} should usually be sufficient to block MAO completely in the various organs of the body (Knoll, 1978*c*).

As MAO-A and MAO-B, which are otherwise similar (see figures 10.1 and 10.2), are thought to differ from each other in the parts complementary to the aromatic ring (Knoll, 1976*a*, 1978*a*), so 'selectivity' of an inhibitor to each enzyme variant is based on a relatively higher affinity of the compound to one form of the enzyme. However, none of the inhibitors blocks one type of MAO solely. In planning experiments with compounds possessing this property, or in making use of their selectivity in therapy, it must not be overlooked that they lose selectivity in higher doses.

It is of great practical importance that selective inhibitors of MAO-B do not lose selectivity even during long-term treatment if appropriate doses are selected (Ekstedt *et al.*, 1979). Table 10.4 is an example showing persistent selective inhibition of MAO-B in the brain of rats treated chronically with low doses of U-1424.

PHARMACOLOGICAL EFFECTS OF MAO INHIBITORS AND THERAPEUTIC IMPLICATIONS OF SELECTIVE INHIBITORS

A pharmacological effect is related to MAO inhibition if it can be produced by structurally different MAO inhibitors and has a time course paralleling MAO inhibition. As MAO inhibitors may have multiple effects on the action and economy of biogenic amines (Knoll and Magyar, 1972), it is sometimes not easy to prove the causal relationship between the observed pharmacological effect of an MAO inhibitor and its specific action on MAO.

An immense amount of data has accumulated from animal experiments showing that concentrations of catecholamines, 5-HT and other endogenous amines (e.g. tryptamine, PEA, octopamine, tyramine, phenylethanolamine, *N*-methylhistamine) are increased in the brain and in other tissues under the influence of MAO inhibitors. Central effects of MAO inhibitors are essentially related to an increase of these brain concentrations to new steady-state levels. There is, of course, a limiting value for each amine as its biosynthesis is regulated by negative feedback. It was known early that 5-HT always reaches the highest concentration in the brain after such treatment (Pletscher, 1957; Spector *et al.*, 1958; Brodie *et al.*, 1959). One of the consequences of such increased brain concentrations of 5-HT is a continuous, increased leakage of this transmitter from the neurones leading, presumably by stimulating presynaptic inhibitory autoreceptors, to inhibition of spontaneous firing of raphe neurones (Aghajanian *et al.*, 1970, 1972) and to suppression of REM sleep (Jouvet *et al.*, 1965).

As levels of 5-HT and catecholamines control seizure thresholds for electroshock and metrazol, depletors of these amines (reserpine, tetrabenazine) facilitate both types of seizure, as was first demonstrated with reserpine by Chen *et al.* (1954). MAO inhibitors antagonise these seizures in both naive and reserpinised rats (Prockop *et al.*, 1959). There are certain strains of mice which develop convulsions to acoustic stimuli: MAO inhibitors protect them from these audiogenic seizures (Plotnikoff *et al.*, 1963).

Non-selective MAO inhibitors prevent nidation and interrupt pregnancy (Poulson and Robson, 1963), block ovulation (Alleva *et al.*, 1966) and inhibit male copulatory behaviour (Tagliamonte *et al.*, 1971; Dewsbury *et al.*, 1972). The mechanism of action of these effects is poorly understood, so that a re-evaluation using selective MAO inhibitors seems timely. MAO inhibitors probably influence hormone production in the brain because of their complex effects on transmitter metabolism. It was found recently that (–)-deprenyl increases plasma prolactin concentration (Mendlewicz and Youdim, 1977). Precise analysis of the influence of selective inhibitors of MAO on brain hormone secretion is a promising and hitherto unexploited field of research.

MAO inhibitors potentiate the effect of many drugs which inhibit the uptake of monoamines. In humans, dangerous incompatibilities with imipramine (Luby and Domino, 1961; Brachfeld *et al.*, 1963) and pethidine (Mitchell, 1955; Shee, 1960; Goldberg, 1964) have been described. In the toxic interaction between pethidine and MAO inhibitors 5-HT, noradrenaline and PEA have all been invoked (Jounela, 1970; Jounela *et al.*, 1977). Non-selective and A-selective inhibitors of MAO strongly potentiate the effect of laevodopa on blood pressure. Clinical studies reveal that MAO inhibitors of this type are clearly contraindicated in patients taking laevodopa because of the risk of hypertensive crisis (Hunter *et al.*, 1970).

MAO inhibitors strongly potentiate the effect of tryptamine (e.g. Tedeschi *et al.*, 1959) and of the hallucinogenic indole-amines, e.g. dimethyltryptamine (Szara, 1956), diethyltryptamine (Szara, 1960), 5-methoxy-*N*-methyltryptamine (Taborsky and McIsaac, 1964), bufotenine (Gessner *et al.*, 1960), etc. Some PEA derivatives, identified in the urine of schizophrenics (3,4-dimethoxyphenylethylamine and *p*-methoxyphenylethylamine) are rapidly metabolised by MAO (Shulgin *et al.*, 1966) and their CNS effects can be demonstrated only after the administration of MAO inhibitors (Hallasmøeller *et al.*, 1973). In addition, the behaviour effects of PEA and phenylethanolamine are strongly potentiated by MAO inhibitors (Sabelli *et al.*, 1975).

It was observed early that MAO inhibition potentiates the behavioural effects of 5-hydroxytryptophan (Udenfriend *et al.*, 1957) and that L-tryptophan causes behavioural excitation in patients treated with MAO inhibitors (Sjoerdsma *et al.*, 1959). The 'tryptophan syndrome' was also observed in animals (Hess *et al.*, 1959) and simultaneous inhibition of both MAO-A and MAO-B was found to be prerequisite for full expression of the syndrome (Squires and Buus Lassen, 1975).

MAO inhibitors were found to exert a hypotensive effect in man but not in experimental animals (cf. Schoepke and Swett, 1967). Its mechanism remains obscure. Impaired release of noradrenaline because of accumulation of octopamine (a 'false transmitter') in noradrenergic neurones (Kopin *et al.*, 1965) and increased adrenergic activity in the n.tractus solitarii, one of the central inhibitory control mechanisms of blood pressure and heart rate (De Jong, 1974) are thought to be important components of this drug response. MAO inhibitors may provide symptomatic relief in about 50 per cent of patients suffering from angina pectoris (Cesarman, 1959). The mechanism of action of this effect too is unexplained.

The cardiovascular effects of the MAO inhibitors are now rarely employed therapeutically, although pargyline is still occasionally used as an antihypertensive agent.

Table 10.5 The low affinity of (–)-deprenyl, a selective inhibitor of MAO-B, for rat intestinal MAO *in vitro* and *in vivo*

Inhibitor	ID_{50} (M)†	Relative potency
(–)-Deprenyl	2×10^{-5}	1
Clorgyline	8×10^{-8}	250
	ID_{50} † (mg kg^{-1}, s.c.)	Relative potency
(–)-Deprenyl	15	1
Clorgyline	0.25	60

†Substrate: [^{14}C]-tyramine.

Potentiation of the pressor effect of tyramine is likely to be the main cause of the dangerous hypertensive reactions which supervene after the intake of certain food materials containing high amounts of the free amine (e.g. cheeses, yeast products, beans, Chianti wines, pickled herring, chicken liver, etc.) in patients treated with MAO inhibitors. This 'cheese reaction' seriously discredited the MAO inhibitors and restricted their therapeutic use, which needs careful medical control. The 'cheese effect', first described by Blackwell (1963), is thought to be primarily a consequence of inhibition of the intestinal enzyme (Blackwell *et al.*, 1967).

Magyar *et al.* (1973) found that (–)-deprenyl is a poor inhibitor of intestinal MAO in rat. Squires (1972) demonstrated that MAO-A is the predominant activity in the gut of many different species, providing a rational explanation for the weak inhibitory action of (–)-deprenyl and, conversely, why clorgyline is a highly effective inhibitor.

Table 10.5 illustrates that intestinal MAO belongs to the A type. Clorgyline was 250 times more potent than (–)-deprenyl in inhibiting mitochondrial MAO from rat intestine *in vitro* and 60 times more potent in blocking the intestinal enzyme *in vivo*. Thus the main barrier controlling the access of tyramine to the circulation by effecting its degradation is practically unaffected in (–)-deprenyl-treated animals but is blocked in clorgyline-treated ones. As in animals, MAO-A is also the main form of MAO in human intestine (Squires, 1972). This finding is in good agreement with the early observations of Varga and Tringer (1967) and Tringer *et al.* (1971); the appearance of a hypertensive reaction was never a problem during their clinical trials of racemic or (–)-deprenyl as an antidepressant. Neither Birkmayer, who administered (–)-deprenyl for years in combination with laevodopa (Birkmayer *et al.*, 1975, 1977; Birkmayer, 1978), nor other clinicians who worked with (–)-deprenyl (Lees *et al.*, 1977; Stern *et al.*, 1978; Rinne *et al.*, 1978; Csanda *et al.*, 1978; Mendlewicz and Youdim, 1978; Yahr, 1978) mentioned this type of side-effect.

In some volunteers treated with deprenyl, Varga tried to provoke the 'cheese reaction' by administering huge amounts of tyramine-rich cheeses, but no significant change in blood pressure could be detected (E. Varga, unpublished results). Elsworth *et al.* (1978) and Sandler *et al.* (1978) demonstrated recently that subjects were able to consume up to 200 mg of tyramine during (–)-deprenyl treatment before a rise in blood pressure and slowing of pulse occurred. This amount of tyramine is unlikely to be encountered during the course of a normal diet. All these experiments support the view (Knoll, 1976*a*) that the acute hypertensive reaction caused by MAO inhibitors is related to inhibition of MAO-A. Thus, selective MAO-B inhibitors appear to be safer than agents which block MAO-A. As the MAO inhibitors have a limited therapeutic role because of serious interactions with pressor amines, an effect related mainly to the inhibition of MAO-A, as described above, it seems justifiable to conclude that selective MAO-A inhibitors have no advantage over non-selective ones. Certainly, the 'cheese effect' in clorgyline-pretreated volunteers after tyramine challenge was as severe as that observed with non-selective inhibitors (Lader *et al.*, 1970).

The antidepressant effect of the MAO inhibitors has been thought by some to be mainly related to an increase in catecholamine levels. Noradrenaline is oxidised by MAO-A (cf. Tipton *et al.*, 1976). Dopamine was originally described as a substrate for both types of MAO (Neff and Yang, 1974), but more recent experiments support the view that it has much higher affinity for MAO-A in the rat brain (Maitre *et al.*, 1976; Sharman, 1976; Braestrup *et al.*, 1975). Braestrup *et al.* (1975) measured the three major catecholamine metabolites, homovanillic acid (HVA), dihydroxyphenylacetic acid (DOPAC) and 3-methoxy-4-hydroxyphenylglycol (MOPEG) in whole rat brain and compared clorgyline and (–)-deprenyl with regard to inhibitory effect on their formation. They found that clorgyline (ED_{50} = 0.2 mg kg^{-1}) was much more potent in inhibiting the metabolism of noradrenaline and dopamine than (–)-deprenyl (ED_{50} = 15 mg kg^{-1}), showing that these amines are mainly metabolised by MAO-A in whole brain. These figures are in good agreement with the conclusion that catecholamines are better substrates for MAO-A than MAO-B, but the result does not necessarily mean that rat brain MAO-A is the enzyme of the adrenergic neurones. Synaptosomal MAO is only a small (less than 5 per cent) portion of total brain activity. Changes in such a small proportion of the total are not detectable by standard methods of measuring MAO.

Racemic and (–)-deprenyl were found to be potent antidepressants in clinical trials (Varga and Tringer, 1967; Tringer *et al.*, 1971). In the light of accumulated data on the selectivity of (–)-deprenyl for MAO-B, which seems to be a highly specific 'phenylethylamine oxidase', a reevaluation and extension of clinical trials with (–)-deprenyl and an investigation of the combination of (–)-deprenyl and phenylalanine, a precursor of PEA, seems to be indicated in depressive illness: ample evidence now exists that PEA is an endogenous 'trace amine' within the brain (Fischer *et al.*, 1972; Saavedra, 1974; Willner *et al.*, 1974) and might play a role in affective behaviour (e.g. Sabelli and Mosnaim, 1974). Fischer *et al.* (1968) claimed that urinary excretion of free phenylethylamine is reduced in depressed patients, supporting the hypothesis that phenylethylamine deficit is one of the biochemical lesions in depression.

Other observers find good results with 5-hydroxytryptophan in depressed patients, supporting the hypothesis that 5-HT deficiency also plays a role in the genesis of depression (Van Praag, 1976). As (–)-deprenyl is the first MAO inhibitor without the cheese effect, it was safely combined with 5-hydroxytryptophan in patients with unipolar and bipolar depression and a good antidepressant response claimed (Mendlewicz and Youdim, 1978).

In contrast to the non-selective and A-selective MAO inhibitors which potentiate the effect of laevodopa, (–)-deprenyl is devoid of this effect and can be safely administered concurrently with laevodopa without special dietary care. In man, (–)-deprenyl provides good therapeutic benefit in Parkinson's disease. Birkmayer demonstrated that, when given together with L-dopa and a peripheral decarboxylase inhibitor, there were excellent results in patients on long-term L-dopa treatment who had developed the 'on–off' phenomenon (Birkmayer *et al.*, 1975, 1977; Birkmayer, 1978). In a two-year study on 233 patients, (–)-deprenyl produced a significant improvement in the therapeutic effect of L-dopa or L-dopa plus benserazide, and a smaller daily dose of L-dopa sufficed to enable reduction of functional disability to be achieved (Birkmayer *et al.*, 1977). Birkmayer and co-workers suggested that (–)-deprenyl may act not only as a selective MAO-B inhibitor but also as a dopamine-releasing agent in the striatum. This suggestion was based on a paper of Knoll *et al.* (1965) introducing racemic deprenyl (at that time under the code name E-250) and showing that the racemic form exerts an acute 'amphetamine-like' psychostimulant effect. Amphetamine is known to release noradrenaline and dopamine. In contrast to the (+) form however, (–)-deprenyl is devoid of acute psychostimulant effect (Magyar *et al.*, 1967) and neither the racemic form nor (–)-deprenyl was a releaser of noradrenaline (Knoll and Magyar, 1972) or dopamine (Knoll, 1978*c*). Although the possibility of a specific dopamine-releasing effect in human striatum cannot be excluded there is no precedent in animal experiments for such an action. Even taking into account some recent observations indicating that (–)-deprenyl is metabolised to methamphetamine and amphetamine in man (Reynolds *et al.*, 1978*a,b*) it seems unlikely that the amount of these metabolites generated from a daily dose of 10–15 mg (–)-deprenyl produce marked amphetamine-like effects. In any case the metabolites are likely to be (–) isomers, which are known to be less active psychostimulants than (+) isomers.

An important consideration in elucidating the nature of the therapeutic effect of (–)-deprenyl in Parkinson's disease is the predominant presence of MAO-B in the human striatum (Glover *et al.*, 1977). The fact that the dopaminergic nigrostriatal neurons contain MAO-A in the rat (Knoll, 1978*c*) and MAO-B in man (Glover *et al.*, 1977) shows once again the dangers in extrapolating from animal experiments to man.

The therapeutic benefit achieved with the drug may well stem from the fact that more than 80 per cent of MAO in human brain is type B (Youdim, 1976). Riederer *et al.* (1978) recently compared the sensitivity of human brain mitochondrial MAO to (–)-deprenyl and clorgyline in 15 brain areas *in vitro*. (–)-Deprenyl was many thousand times more powerful an inhibitor than clorgyline. For (–)-deprenyl the ID_{50} varied between 0.05 nM (hypothalamus) and 0.95 nM (caudate nucleus); values for clorgyline varied between 500 nM (pineal gland) and 3200 nM (raphe + reticular formation). They also demonstrated that deamination of dopamine was almost completely inhibited in the brains of seven patients with Parkinson's disease treated with 10 mg (–)-deprenyl daily, 6.0 ± 1.8 days before death, whilst 34 per cent MAO activity against 5-HT was still present.

As with all other known MAO inhibitors, (–)-deprenyl similarly exerts multiple effects on catecholaminergic and, in higher doses, on serotonergic transmission (Knoll and Magyar, 1972; Knoll, 1976*a*; Knoll, 1978*b,c*). Thus differences in the pharmacological effects of (–)-deprenyl and clorgyline depend not only on their selective inhibitory action but also on their ability to influence catecholaminergic and serotonergic transmission to a different degree. Three important differences between (–)-deprenyl and clorgyline have to be considered (cf. Knoll, 1976*a*) in explaining why the former is an MAO inhibitor without the cheese effect:

(1) (–)-Deprenyl leaves the gut wall MAO-A barrier intact, so that tyramine access to the circulation can be controlled; clorgyline inactivates the barrier (Magyar *et al.*, 1973; Squires, 1972; Knoll, 1976*a*, 1978*c*).

(2) (–)-Deprenyl reduces the pressor effect of tyramine by inhibiting its uptake (Knoll *et al.*, 1968; Knoll, 1976*a*, 1978*c*); clorgyline potentiates the pressor responses to tyramine (Knoll *et al.*, 1972; Knoll, 1976*a*).

(3) (–)-Deprenyl inhibits the release of noradrenaline from synaptosomes (Knoll and Magyar, 1972); clorgyline potentiates it (Knoll, 1976*a*).

A combination of these effects might well explain why prolonged treatment in parkinsonian volunteers with (–)-deprenyl still results in complete safety from the 'cheese effect' after challenge with increasing dosage of tyramine (Elsworth *et al.*, 1978; Sandler *et al.*, 1978).

When explaining the potentiation of the anti-akinetic effect of laevodopa treatment by (–)-deprenyl observed in parkinsonian patients, a number of effects other than MAO inhibition have to be considered. Laevodopa, the precursor of dopamine, penetrates easily into the brain and is converted in the striatum to dopamine, which stimulates postsynaptic dopaminergic receptors on the cholinergic caudate interneurons. As the essential lesion of Parkinson's disease is a loss of melanin-containing neurones in the pars compacta of the substantia nigra leading to a deficiency of striatal dopamine (Ehringer and Hornykiewicz, 1960), laevodopa administration is a kind of substitution therapy (Cotzias *et al.*, 1969; Yahr *et al.*, 1969; Birkmayer, 1969). Exogenous dopamine, however, also stimulates presynaptic 'autoreceptors' (cf. Carlsson, 1975). This means that it suppresses the activity of those nigrostriatal dopaminergic neurones which still remained intact, an unwanted side-effect.

(–)-Deprenyl acts primarily on intact nigrostriatal neurones, and in an opposite manner to that of a laevodopa. As these neurones probably contain mainly type B MAO (cf. Glover *et al.*, 1977; Riederer *et al.*, 1978) (–)-deprenyl presumably increases the dopamine content of the nerve terminals. As a potent inhibitor of reuptake, it intensifies the activity of the nigrostriatal dopaminergic neurones (Knoll, 1978*c*). As (–)-deprenyl, in high concentration, blocks postsynaptic dopamine receptors (Knoll, 1978*c*) and as presynaptic receptors tend to be much more sensitive to both agonists and antagonist than postsynaptic, it is possible that the low doses of (–)-deprenyl employed counteract the stimulatory effect of laevodopa on the presynaptic 'autoreceptors'. It is even possible that the (–)-deprenyl metabolites (Reynolds *et al.*, 1978*a,b*), (–)-amphetamine and (–)-methamphetamine, exert some slight dopamine-releasing action on the nigrostriatal terminal. Thus, all the effects of (–)-deprenyl on the action and economy of the endogenous dopamine in the striatum envisaged here work towards a primary effect on the remaining intact nigrostriatal dopaminergic neurones, whereas laevodopa acts primarily on the postsynaptic cholinergic caudate interneurone.

If this working hypothesis be accepted, it may be of great importance to start administering the drug in the earliest stages of the disease, in order to preserve, as far as possible, the physiological control of acetylcholine release in the striatum, thereby reducing the amount of laevodopa required to the lowest possible level. Even the known slow degradation of nigrostriatal dopaminergic neurones, a frequent concomitant of ageing, might conceivably be counteracted by starting long-term treatment with low doses of (–)-deprenyl at a suitable time. This possibility might be worth investigation in the future.

It is always difficult to decide what proportion of the total pharmacological effect to assign to selective MAO inhibition in evaluating the experimental or clinical actions of (–)-deprenyl or clorgyline. Detailed studies with (–)-deprenyl have revealed that whilst it acted as selective inhibitor of MAO-B in a dose of about 0.25 per cent of the LD_{50} in mice, dogs and cats, it only inhibited amine reuptake in much higher concentration (Knoll, 1978*c*).

PEA, like tyramine (cf. Knoll, 1978*b,c*), is an indirectly-acting sympathomimetic agent on the cat nictitating membrane. (–)-Deprenyl inhibits the uptake of both amines but in much higher doses than are needed to block MAO-B activity selectively. Figure 10.5 shows the effect of this inhibitor on the nictitating membrane of the cat *in vivo*, administered in a selective hepatic MAO-B blocking dose. Blocking PEA metabolism allows the accumulation of sufficient amine circulation to release noradrenaline from sympathetic nerve terminals in the detector organ.

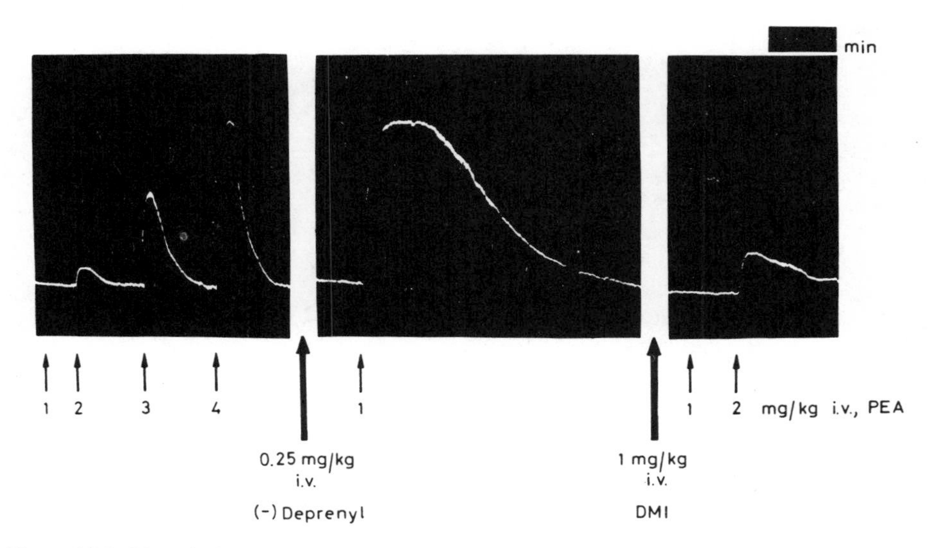

Figure 10.5 Phenylethylamine (PEA)-induced contractions of the nictitating membrane in a cat weighing 3.3 kg. The potentiation of the PEA effect by (–)-deprenyl (0.25 mg kg^{-1}, i.v.) and complete inhibition of PEA effects by desmethylimipramine (DMI, 1 mg kg^{-1}, i.v.). Chloralose–urethane anaesthesia. Artificial respiration.

Table 10.6 Selective inhibitors of MAO-B with different actions on the reuptake and release of amines

Structure		Relative potency in inhibiting brain MAO-B (–)-deprenyl = 1	Capability of	
			inhibiting tyramine uptake	releasing noradrenaline
$C_6H_5-CH_2-CH_2-NH_2$	Phenylethylamine	Substrate	–	+
$C_6H_5-CH_2-CH_2-N(CH_3)-CH_2-C{\equiv}CH$	TZ-650	1	–	+
$C_6H_5-CH_2-CH(CH_3)-N(CH_3)-CH_2-C{\equiv}CH$	(–)-Deprenyl	1	+	–
$C_4H_3O-CH_2-CH_2-N(CH_3)-CH_2-C{\equiv}CH$	TZ-945	0.6	–	–
$C_{10}H_{11}-N(H_3C)-CH_2-C{\equiv}CH$ (tetrahydronaphthyl)	J–512	0.6	–	–

Thus, as the method is based on PEA-induced release of noradrenaline, it is evident that 0.25 mg (–)-deprenyl kg^{-1} does not interfere with the PEA uptake, which can easily be blocked by the selective noradrenaline reuptake inhibitor, desmethylimipramine.

Structure–activity relationship studies (Knoll *et al.*, 1978) have revealed the prerequisites for any deprenyl-derived new selective MAO-B inhibitors. (–)-Deprenyl and several such new derivatives affected adrenergic transmission differently. Table 10.6 demonstrates that of four selective inhibitors of MAO-B, one (TZ-650) is a releaser of noradrenaline, one ((–)-deprenyl) is an inhibitor of tyramine uptake, and two (TZ-945 and J-512) are devoid of these side-effects. The usefulness of such compounds as pharmacological research tools seems obvious. Clinical studies will then be needed to determine the therapeutic potential of selective MAO-B inhibitors with different pharmacological profiles.

SUMMARY

Irreversible inhibitors of mitochondrial monoamine oxidase (MAO) are potentially important drugs because the enzyme plays a crucial role in intraneuronal inactivation of transmitter in catecholaminergic and serotoninergic neurones. However, as intestinal MAO inactivates the pressor amines of foodstuffs, hepatic enzyme controls pressor amine blood levels and blood vessel enzyme protects organs from the toxic effects of circulating monoamines, inhibitor-treated patients have always been open to the risk of hypertensive reactions ('cheese effect') which has limited the therapeutic usefulness of this group of drugs. The discovery of two forms of MAO and the development of (–)-deprenyl, the first highly potent irreversible MAO inhibitor without the 'cheese effect', may change this situation.

There are two main forms of mitochondrial MAO, type A and B. Inhibitors selective for each are known and are appropriate tools for their analysis, clorgyline preferentially inhibiting the A form and (–)-deprenyl the B.

Structure–activity relationship studies reveal that the binding of the aromatic ring and the CH_2-CH_2 chain between the aromatic ring and nitrogen moiety and the attraction of the nitrogen localised in the vicinity of the covalently-bound flavin group of MAO determine the affinity of substrates for the enzyme and their ability to be metabolised on its surface. MAO-A and MAO-B are thought to differ from each other in complementary parts of the aromatic ring. In MAO-A this part of the enzyme is adapted for the attachment of the 5-hydroxyindole ring of serotonin ('serotonin oxidase'), whereas in MAO-B it is a simpler, flat surface for the binding of the benzene ring of phenylethylamine ('phenylethylamine oxidase'). The other sites of the two enzymes for interacting with the CH_2-CH_2 chain and nitrogen are the same.

An aromatic ring, an amino group, a short carbon chain between them and the introduction of a proper 'enzyme-killing' group in such a structure are essential requirements of an MAO inhibitor. Both (–)-deprenyl and clorgyline contain a propargyl group attached to the nitrogen which reacts irreversibly with the covalently bound flavin in MAO. (–)-Deprenyl, a selective inhibitor of MAO-B, is closely related to phenylethylamine, the main endogenous substrate for type B MAO. The essential difference between (–)-deprenyl and clorgyline is an oxygen bridge between the aromatic ring and the side chain and a chlorine atom in the *ortho* position. These differences seem to determine the selectivity of clorgyline towards MAO-A.

As MAO-A and MAO-B are unevenly distributed in the organs of the body there are striking differences in pharmacological activity between their selective inhibitors. Because MAO inhibitors have multiple effects in the action and economy of biogenic amines, differences between them influence the pharmacological spectrum of each type of selective inhibitor.

In contrast to clorgyline, (–)-deprenyl is an MAO inhibitor without the 'cheese effect' because:

(1) intestinal mucosa, the first protective barrier, which by degrading tyramine in foodstuffs prevents an excess of this pressor amine gaining access to the circulation, predominantly contains MAO-A in man;

(2) MAO-A inhibitors potentiate the effect of tyramine, whereas (–)-deprenyl inhibits it,

(3) MAO-A inhibitors increase the outflow of noradrenaline from synaptosomes, whereas (–)-deprenyl inhibits it.

Because of the safety of (–)-deprenyl and (–)-deprenyl-derived selective MAO-B inhibitors, the antidepressant effect of these compounds is open to careful clinical scrutiny. MAO-B is the predominant form of the enzyme in human brain; phenylethylamine, its specific substrate, is an endogenous stimulatory amine, a deficit of which might be one of the biochemical lesions in certain types of depression.

MAO inhibitors are normally contraindicated in parkinsonian patients taking laevodopa, because hypertensive crises will result. (–)-Deprenyl is an exception. It can safely be combined with laevodopa, and dietary care need not be exercised.

The selective MAO-A inhibitory agents have no advantage over previously used non-selective inhibitors, but selective MAO-B inhibitors because of the absence of 'cheese effect' show greater promise.

REFERENCES

Aghajanian, G. K., Graham, A. W. and Sheard, M. H. (1970). *Science, N.Y.*, **169**, 1100
Aghajanian, G. K., Haigler, H. J. and Bloom, F. E. (1972). *Life Sci.*, **11**, 615
Alleva, J. J., Overpeck, J. G. and Umberg, E. J. (1966). *Life Sci.*, **5**, 1557
Berger, P. A., Ginsburg, R. A., Barchas, J. D., Murphy, D. L. and Wyatt, R. J. (1978). *Am. J. Psychiat.*, **135**, 95
Birkmayer, W. (1969). *Wien. klin. Wschr.*, **81**, 677
Birkmayer, W., Riederer, P., Youdim, M. B. H. and Linauer, W. (1975). *J. neural Transmiss.*, **36**, 303
Birkmayer, W., Riederer, P., Ambrosi, L. and Youdim, M. B. H. (1977). *Lancet*, **i**, 439
Birkmayer, W. (1978). *J. neural. Transmiss.*, **43**, 239
Blackwell, B. (1963). *Lancet*, **ii**, 849
Blackwell, B., Marley, E., Price, J. and Taylor, D. (1967). *Br. J. Psychiat.*, **113**, 349
Blaschko, H., Richter, D. and Schlossmann, H. (1937). *J. Physiol. Lond.*, **90**, 1
Blaschko, H. (1972). *Adv. Biochem. Psychopharmac.*, **5**, 1
Blaschko, H. (1974). *Rev. physiol. biochem. Pharmac.*, **70**, 83
Brachfeld, J., Wirtschafter, A. and Wolfe, S. (1963). *J. Am. med. Assoc.*, **186**, 1172
Braestrup, C., Andersen, H. and Randrup, A. (1975). *Eur. J. Pharmac.*, **34**, 181
Brodie, B. B., Pletscher, A. and Shore, P. A. (1956). *J. Pharmac. exp. Ther.*, **116**, 9
Brodie, B. B., Spector, S. and Shore, A. (1959). *Ann. N.Y. Acad. Sci.*, **80**, 609
Buchsbaum, M. S., Coursey, R. D. and Murphy, D. L. (1976). *Science, N.Y.*, **194**, 339

Carlsson, A. (1975). In *Pre- and Postsynaptic Receptors* (ed. E. Usdin and W. E. Bunney, Jr), Marcel Dekker, New York, pp. 49–66

Cesarman, T. (1959). *Ann. N.Y. Acad. Sci.*, **80**, 988

Chen, G., Ensor, C. R. and Bohner, B. (1954). *Proc. Soc. exp. Biol. Med.*, **86**, 507

Costa, E. and Sandler, M. (eds) (1972). *Adv. biochem. Psychopharmac.*, **5**

Cotzias, G. C., Papavasiliou, P. S. and Gellene, R. (1969). *New Engl. J. Med.*, **280**, 337

Csanda, E., Antal, J., Antony, M. and Csanaky, A. (1978). *J. neural Transmiss.*, **43**, 263

De Jong, W. (1974). *Eur. J. Pharmac.*, **29**, 179

De Lores Arnaiz, G. R. and De Robertis, E. D. P. (1962). *J. Neurochem.*, **9**, 503

Dewsbury, D. A., Davis, H. N. and Jansen, P. E. (1972). *Psychopharmacology*, **24**, 209

Ehringer, H. and Hornykiewicz, O. (1960). *Klin. Wschr.*, **38**, 1236

Ekstedt, B., Magyar, K. and Knoll, J. (1979). *Biochem. Pharmac.*, 28, 919

Elsworth, J. D., Glover, V., Reynolds, G. P., Sandler, M., Lees, A. J., Phuapradit, P., Shaw, K. M., Stern, G. M. and Kumar, P. (1978). *Psychopharmacology*, **57**, 33

Ewins, A. J., and Laidlaw, P. P. (1910). *J. Physiol. Lond.*, **41**, 78

Ewins, A. J. and Laidlaw, P. P. (1913). *Biochem. J.*, **7**, 18

Fischer, E., Heller, B. and Miro, A. H. (1968). *Arzneimittel-Forsch.*, **18**, 1486

Fischer, E., Spatz, H., Heller, B. and Reggiani, H. (1972). *Experientia*, **15**, 307

Friedmann, E., Shopsin, B., Sathananthan, G. and Gershon, S. (1974). *Am. J. Psychiat.*, **131**, 1392

Fuller, R. W. (1968). *Biochem. Pharmac.*, **17**, 2097

Gessner, P. K., Khairalla, P. A., McIsaac, W. M. and Page, I. H. (1960). *J. Pharmac. exp. Ther.*, **130**, 126

Glover, V., Sandler, M., Owen, F. and Riley, G. (1977). *Nature, Lond.*, **265**, 80

Goldberg, L. I. (1964). *J. Am. med. Assoc.*, **190**, 456

Goridis, C. and Neff, N. H. (1971). *Neuropharmacology*, **10**, 557

Gottfries, C. G., Oreland, L., Wiberg, A. and Winblad, B. (1975). *J. Neurochem.*, **25**, 667

Guggenheim, M. and Löffler, W. (1916). *Biochem. Z.*, **72**, 325

Hallasmøeller, T., Vizi, E. S. and Knoll, J. (1973). In *Symposium on Pharmacological Agents and Biogenic Amines in the Central Nervous System* (ed. J. Knoll and E. S. Vizi), Akadémiai Kiadó, Budapest, pp. 81–6

Hare, M. L. C. (1928). *Biochem. J.*, **22**, 968

Hess, S. M., Redfield, B. G. and Udenfriend, S. (1959). *J. Pharmac. exp. Ther.*, **127**, 178

Hunter, K. R., Boakes, A. J., Laurence, D. R. and Stern, G. M. (1970). *Br. med. J.*, **iii**, 388

Johnston, J. P. (1968). *Biochem. Pharmac.*, **17**, 1285

Jounela, A. J. (1970). *Ann. med. exp. fenniae*, **48**, 261

Jounela, A. J., Mattila, M. J. and Knoll, J. (1977). *Biochem. Pharmac.*, **26**, 806

Jouvet, M., Vimont, P. and Delmore, F. (1965). *C. r. Séanc. Soc. Biol.*, **159**, 1595

Klingman, G. I. (1966). *Biochem. Pharmac.*, **15**, 1729

Knoll, J. (1976*a*). In *Monoamine Oxidase and Its Inhibition* (eds G. E. W. Wolstenholme and J. Knight), Ciba Foundation Symposium 39 (new series). Elsevier, Amsterdam, pp. 135–61

Knoll, J. (1976*b*). In *Neuron Concept Today* (ed. J. Szentágothai, J. Hámori and E. S. Vizi), Akadémiai Kiadó, Budapest, pp. 109–17

Knoll, J. (1978*a*). *Horizons Biochem. Biophys.*, **5**, 37

Knoll, J. (1978*b*). *Enzyme Activated Irreversible Inhibitors* (ed. N. Seiler, M. J. Jung and J. Koch-Weser), Elsevier, Amsterdam, pp. 253–69

Knoll, J. (1978*c*). *J. neural Transmiss.*, **43**, 177

Knoll, J. and Magyar, K. (1972). *Adv. biochem. Psychopharmac.*, **5**, 393

Knoll, J., Ecsery, Z., Nievel, J. G. and Knoll, B. (1964). *MTA V. Oszt. Közl.*, **15**, 231

Knoll, J., Ecsery, Z., Kelemen, K., Nievel, J. G. and Knoll, B. (1965). *Arch. int. Pharmacodyn. Ther.*, **155**, 154

Knoll, J., Vizi, E. S. and Somogyi, G. (1968). *Arzneimittel-Forsch.*, **18**, 109

Knoll, J., Vizi, E. S. and Magyar, K. (1972). In *Recent Developments of Neurobiology in Hungary*, Vol. III, *Results in Neuroanatomy, Neurophysiology, Neuropathophysiology and Neuropharmacology* (ed. K. Lissák), Publishing House of the Hungarian Academy of Sciences, Budapest, pp. 167–217

Knoll, J., Ecsery, Z., Magyar, K. and Sátory, E. (1978). *Biochem. Pharmac.*, **27**, 1739

Kopin, I. J., Fischer, J. E., Musacchio, J. M., Horst, W. D. and Weise, V. K. (1965). *J. Pharmac. exp. Ther.*, **147**, 186

Lader, M. H., Sakalis, G. and Tansella, M. (1970). *Psychopharmacologia*, **18**, 118

Lees, A. J., Shaw, K. M., Kohout, L. J., Stern, G. M., Elsworth, J. D., Sandler, M. and Youdim, M. B. H. (1977). *Lancet*, **ii**, 791

Luby, E. D. and Domino, E. F. (1961). *J. Am. med. Assoc.*, **177**, 68

MacKay, A. V. P., Davies, P., Dewar, A. J. and Yates, C. M. (1978). *J. Neurochem.*, **30**, 827

Maitre, I., Delini-Stula, A. and Waldmeier, P. C. (1976). In *Monoamine Oxidase and its Inhibition* (eds G. E. W. Wolstenholme and J. Knight) Ciba Foundation Symposium 39 (new series), Elsevier, Amsterdam, pp. 247–70

Magyar, K., Vizi, E. S., Ecsery, Z. and Knoll, J. (1967). *Acta physiol. hung.* **32**, 377

Magyar, K., Sátory, E. and Knoll, J. (1973). In *Symposium on Pharmacological Agents and Biogenic Amines in the Nervous System* (general ed. J. Knoll; ed. K. Magyar), Akadémiai Kiadó, Budapest, pp. 107–22

Maycock, A. L., Abeles, R. H., Salach, J. I. and Singer, T. O. (1976). In *Monoamine Oxidase and Its Inhibition* (eds G. E. W. Wolstenholme and J. Knight) Ciba Foundation Symposium 39 (new series). Elsevier, Amsterdam, pp. 33–47

Mendlewicz, J. and Youdim, M. B. H. (1977). *Lancet*, **i**, 507

Mendlewicz, J. and Youdim, M. B. H. (1978). *J. neural Transmiss.*, **43**, 279

Minkowski, O. (1883). *Naunyn-Schmiedeberg's Arch. exp. Path. Pharmak.*, **17**, 445

Mitchell, R. S. (1955). *Ann. intern. Med.*, **42**, 417

Murphy, D. L. and Wyatt, R. J. (1972). *Nature, Lond.*, **238**, 225

Neff, N. H. and Yang, H. Y. T. (1974). *Life Sci.*, **14**, 2061

Pletscher, A. (1957). *Schweiz. med. Wschr.*, **87**, 1532

Plotnikoff, N., Huang, J. and Havens, P. (1963). *J. pharm. Sci.*, **52**, 172

Poulson, E. and Robson, J. M. (1963). *J. Endocrinol.*, **30**, 205

Prockop, D. J., Shore, P. A. and Brodie, B. B. (1959). *Experientia*, **15**, 145

Reynolds, G. P., Riederer, P., Sandler, M., Jellinger, K. and Scheemann, D. (1978*a*). *J. neural Transmiss.*, **43**, 271

Reynolds, G. P., Elsworth, J. D., Blau, K., Sandler, M., Lees, A. J. and Stern, G. M. (1978*b*). *Br. J. clin. Pharmac.*, **6**, 542

Riederer, P., Youdim, M. B. H., Rausch, W. D., Birkmayer, W., Jellinger, K. and Seemann, D. (1978). *J. neural Transmiss.*, **43**, 217

Rinne, U. K., Siirtola, T. and Sonninen, V. (1978). *J. neural Transmiss.*, **43**, 253

Saavedra, J. M. (1974). *J. Neurochem.*, **22**, 211

Sabelli, H. C. and Mosnaim, A. D. (1974). *Am. J. Psychiat.*, **131**, 695

Sabelli, H. C., Vazquez, A. J. and Flavin, D. (1975). *Psychopharmacology*, **42**, 117

Sandler, M., Youdim, M. B. H. and Hanington, E. (1974). *Nature, Lond.*, **250**, 335

Sandler, M., Glover, W., Ashford, A. and Stern, G. M. (1978). *J. neural Transmiss.*, **43**, 209

Schmiedeberg, O. (1877). *Naunyn-Schmiedeberg's Arch. exp. Path. Pharmak.*, **8**, 1

Schmiedeberg, O. (1881). *Naunyn-Schmiedeberg's Arch. exp. Path. Pharmak.*, **14**, 379

Schnaitman, C., Erwin, V. G. and Greenwalt, J. W. (1967). *J. Cell Biol.*, **34**, 719

Schoepke, H. G. and Swett, L. R. (1967). In *Antihypertensive Agents* (ed. E. Schlitter), Academic Press, New York and London, pp. 393–428

Sharman, D. F. (1976). In *Monoamine Oxidase and Its Inhibition* (ed. G. E. W. Wolstenholme and J. Knight), Ciba Foundation Symposium 39 (new series). Elsevier, Amsterdam, pp. 203–29

Shaskan, E. G. and Becker, R. E. (1975). *Nature, Lond.*, **253**, 659

Shee, J. C. (1960). *Br. med. J.* **ii**, 507

Shulgin, A. T., Sargent, T. and Naranjo, C. (1966). *Nature, Lond.*, **212**, 1606

Sjoerdsma, A., Lovenberg, W., Oates, J. A., Crout, J. R. and Udenfriend, S. (1959). *Science, N.Y.*, **130**, 225

Snyder, S. H., Fischer, J. and Axelrod, J. (1965). *Biochem. Pharmac.*, **14**, 363

Spector, S., Prockop, D. and Shore, P. A. (1957). *Science, N.Y.*, **127**, 704

Spector, S., Prockop, D., Shore, P. A. and Brodie, B. B. (1958). *Science, N.Y.*, **128**

Squires, R. F. (1972). *Adv. biochem. Psychopharmac.*, **5**, 355

Squires, R. F. and Buus Lassen, J. (1975). *Psychopharmacology*, **41**, 145

Stern, G. M., Lees, A. J. and Sandler, M. (1978). *J. neural Transmiss.*, **43**, 245

Szara, S. (1956). *Experientia*, **12**, 441

Szara, S. (1960). *Arch. gen. Psychiat.*, **15**, 320

Taborsky, R. G. and McIsaac, W. M. (1964). *Biochem. Pharmacol.*, **13**, 531
Tagliamonte, A., Tagliamonte, P. and Gessa, G. L. (1971). *Nature, Lond.*, **230**, 244
Tedeschi, D. H., Tedeschi, R. E. and Fellows, E. J. (1959). *J. Pharmac. exp. Ther.*, **126**, 223
Tipton, K. F., Houslay, M. D. and Mantle, T. J. (1976). In *Monoamine Oxidase and its Inhibition* (ed. G. E. W. Wolstenholme and J. Knight), Ciba Foundation Symposium 39 (new series). Elsevier, Amsterdam, pp. 5–31
Tringer, L., Haits, G. and Varga, E. (1971). In *Societies Pharmacologica Hungarica. Vth Conferentia Hungarica pro Therapis et Investigations in Pharmacologia* (ed. G. P. Leszkovszky), Akadémiai Kiadó, Budapest, pp. 111–4
Udenfriend, S., Weissbach, H. and Bogdanski, D. F. (1957). *J. biol. Chem.*, **224**, 803
Van Praag, H. M. (1976). In *Neuroregulators and Psychiatric Disorders* (ed. E. Usdin, D. A. Hamburg and J. D. Barchas), Oxford University Press, New York, pp. 163–76
Varga, E. and Tringer, L. (1967). *Acta med. acad. sci. hung.*, **23**, 289
Willner, J., LeFevre, H. P. and Costa, E. (1974). *J. Neurochem.*, **23**, 857
Wolstenholme, G. E. W. and Knight, J. (eds) (1976). *Monoamine Oxidase and its Inhibition*, Ciba Foundation Symposium 39 (new series), Elsevier, Amsterdam
Wyatt, R. J., Murphy, D. L., Belmaker, R., Cohen, S., Donnelly, C. H. and Pollin. W. (1973). *Science, N.Y.*, **179**, 916
Yahr, M. D., Duvoisin, R. C. and Sheard, M. J. (1969). *Arch. Neurol., Chicago*, **21**, 343
Yahr, M. D. (1978). *J. neural Transmiss.*, **43**, 227
Yang, H. Y. and Neff, N. N. (1973). *J. Pharmac. exp. Ther.*, **187**, 365
Youdim, M. B. H. (1976). In *Neuroregulators and Psychiatric Disorders* (ed. E. Usdin, D. A. Hamburg and S. D. Barchas) Oxford University Press, New York, pp. 57–67
Youdim, M. B. H. (1978). *J. neural Transmiss.*, **43**, 199
Zeller, E. A. (1938). *Helv. chim. Acta*, **21**, 881
Zeller, E. A., Barsky, J., Fouts, J. R., Kirchheimer, W. F. and Van Orden, L. S. (1952). *Experientia*, **8**, 349

11
Monoamine oxidase inhibition: some clinical dimensions

M. Sandler, Vivette Glover, J. D. Elsworth, Rachel Lewinsohn and M. A. Reveley
(Bernhard Baron Memorial Research Laboratories and Institute of Obstetrics and Gynaecology, Queen Charlotte's Maternity Hospital, Goldhawk Road, London W6 0XG, UK)

INTRODUCTION

Monoamine oxidase (MAO) is an insoluble enzyme located on the outer membrane of the mitochondrion (Schnaitman and Greenawalt, 1968) and probably forms an intrinsic part of the structure of this membrane (Sawyer and Greenawalt, 1979). Several converging lines of evidence have led to the inescapable conclusion that MAO exists in more than one form. Indeed, in the late 1960s, considerable effort was expended in attempting physically to separate and characterise these multiple forms (see Sandler and Youdim, 1972). However, the rigorous solubilisation procedures employed, demanding the use of sonication or detergent before apparent physical separation could be achieved, probably also led to conformational changes in enzyme structure. Thus, even though multiplicity was clearly demonstrated during *in vitro* testing and these effects were highly reproducible, doubt still exists as to whether the phenomena observed are truly representative of the *in vivo* situation: if the extent to which these *in vitro* findings reflect *in vivo* action be considered as criteria then the question of whether the phenomena reviewed by Sandler and Youdim (1972) are sophisticated artefacts cannot be ignored.

The extensive literature on MAO is riddled with inconsistencies. However, the working hypothesis of Johnston (1968), even while generating a new crop of problems (e.g. Fowler *et al.*, 1978; Owen *et al.*, 1979), has provided a framework on which to hang further experiments and has brought a semblance of order to an otherwise chaotic situation. Johnston's classification is based on the selective inhibitory action of the drug, clorgyline: by definition, that proportion of enzyme activity sensitive to relatively small concentrations of this compound is termed MAO A whilst the fraction resistant to it is given the name MAO B. The claim by Williams *et al.* (1975) that a further form exists, MAO C, localised to the circumventricular region of the rat brain, has recently been shown to be unfounded (Kim *et al.*, 1979; Toyoshima *et al.*, 1979; Böhm *et al.*, 1979*a*). MAO A and MAO B may well owe their distinguishing physicochemical properties to the intimate proximity of different types and proportions of phospholipid (Houslay and Tipton, 1973; Dennick and Mayer, 1977) although there is no general agreement on this point. Broadly, MAO A selectively oxidises the so-called neurotransmitter monoamines, 5-hydroxytryptamine, noradrenaline and adrenaline, although in some organs of the pig, at least, 5-hydroxytryptamine is also vigorously metabolised by MAO B (Ekstedt and Oreland, 1976). There exists a 'central bloc' of substrates which are generally thought of as showing little preference for either form–tyramine, octopamine and, according to some, dopamine (Yang and Neff, 1974)–although dopamine must be considered very much a special case: its properties as a preferred substrate vary considerably from species to species and organ to organ. Thus, in rat brain it is preferentially metabolised by MAO A (Waldmeier *et al.*, 1976) but in human brain and platelet, by MAO B (Glover *et al.*, 1977). In human gastro-intestinal mucosa (Squires, 1972; Elsworth *et al.*, 1978) and endometrium (R. Mazumder, V. Glover and M. Sandler, in preparation), however, dopamine is selectively oxidised by MAO A. Even tyramine is predominantly oxidised by MAO B in the pig (Ekstedt and Oreland, 1976). Certain trace amines, e.g. phenylethylamine (Yang and Neff, 1973) and phenylethanolamine are preferred substrates for MAO B. However, the selectivity observed can depend on the substrate concentration present. At high substrate concentrations, phenylethylamine is an active substrate for MAO A (R. Lewinsohn, V. Glover and M. Sandler, in preparation) as shown by its active metabolism by the solely MAO A-containing organ, the human placenta (Egashira, 1976). The best substrate of all for MAO B and the best discriminator in practice is benzylamine (R. Lewinsohn, V. Glover and M. Sandler, in preparation) although the physiological significance of this finding is obscure. It should be noted in passing that benzylamine is also vigorously oxidized by another enzyme, benzylamine oxidase (Bergeret *et al.*, 1957; Lewinsohn *et al.*, 1978), which tends to have a somewhat different distribution to that of MAO B, being particularly active in the walls of blood vessels (R. Lewinsohn, V. Glover and M. Sandler, in preparation). Anatomically, the two different forms of MAO and benzylamine oxidase may conveniently be distinguished by a recently developed peroxidatic histochemical approach (Ryder *et al.*, 1979; Williams *et al.*, 1979).

An important substrate of MAO B both in animals (Waldmeier *et al.*, 1977; Hough and Domino, 1979) and man (J. D. Elsworth, V. Glover and M. Sandler, in preparation), is *tele-N*-methylhistamine. The human brain possess relatively high, discontinuously distributed concentrations of histamine but no diamine oxidase (Kapeller-Adler, 1970). Thus, in this site, histamine is only metabolised by *tele-N*-methylation with subsequent oxidation by MAO B which, in this way, assumes a key role in its further metabolism. Whether the clinical lightening of affect achieved by MAO inhibition stems in any way from a build up of the *N*-methylated metabolite of histamine is yet another matter requiring further exploration.

MAO inhibitors have been known to be useful antidepressant agents since the early 1950s when the irreversible 'suicide' inhibitor, iproniazid, appeared on the scene (Zeller *et al.*, 1955). Although the therapeutic benefit achieved with their aid has not always been obvious (Medical Research Council, 1965), the consensus of opinion now holds that they may be useful in certain kinds of depressive illness (e.g. Pare and Sandler, 1959; Robinson *et al.*, 1973; Tyrer *et al.*, 1973). It has always been assumed that the efficacy of this treatment depends on actual inhibition of MAO, with consequent accumulation of one or more monoamines (rather than by some secondary action of the drug on some other system) but it has not been possible so far to pinpoint the particular amine responsible. Whether the regulation of MAO activity by unknown endogenous mechanisms plays any role in determining the amount of amine which is to be available at particular sites is another unsolved problem although evidence has recently started to emerge for the first time which bears on this possibility, e.g. the selective potentiation of endometrial MAO A by progesterone (R. Mazumder, V. Glover and M. Sandler, in preparation), of platelet MAO B by adrenaline injection (Gentil *et al.*, 1975) or exercise (Gawel *et al.*, 1977) or the existence of endogenous inhibition (see below).

ENDOGENOUS MAO INHIBITORS?

In 1962, Phillips and his colleagues demonstrated a significant increase of MAO activity in the liver of germ-free chicks (Phillips *et al.*, 1962). We have recently confirmed and extended this observation (Böhm *et al.*, 1979*b*): despite a doubling in activity of the hepatic enzyme, in which both MAO A and B were involved proportionately, there was no change in activity in the brain of these animals. It thus seemed feasible that the establishment of gut flora in the conventional controls brought in its train the elaboration and release of an MAO inhibitory substance into the portal system and hence to the liver.

Whether such an inhibitory substance is generated similarly from human gut flora or, indeed, from some other site in the body, is quite unknown. However, during some recent experiments on platelet activity in patients under treatment with tricyclic antidepressant drugs, data have come to light compatible with an interpretation in terms of the presence of a circulating endogenous inhibitor (Reveley *et al.*, 1979).

Tricyclic antidepressants are known to be competitive inhibitors of MAO B, a property which has been well demonstrated on the human platelet enzyme (Edwards and Burns, 1975) which only contains activity of this type (Donnelly and Murphy, 1977). One would thus expect that platelets from individuals under treatment with a drug belonging to this group would manifest some degree of *in vivo* inhibition but that the process of blood collection and platelet harvesting would result in a diluting out of the competitive inhibitor so that no inhibition would be detectable on *in vitro* assay. Nevertheless, Sullivan *et al.* (1977) claim to have identified an inhibitory effect *in vitro* in platelets from patients so treated although the mechanism of its origin was not immediately obvious. We attempted to replicate this work but, surprisingly, observed the opposite effect (Reveley *et al.*, 1979). Platelet MAO activity in a washed preparation correlated directly and significantly with the concentration of amitriptyline in the blood sample from which the platelet preparation had been obtained. The most parsimonious explanation for this finding, it seemed to us, is that some endogenously-produced inhibitor, perhaps deriving from gut flora, is constantly coming into contact with platelet MAO but that the presence of the competitive inhibitor, amitriptyline, has a masking effect, protecting the enzyme from the endogenously-produced inhibitor. These data are preliminary and further observations are obviously needed to clarify the point. It is of interest, however, that Berrettini and Vogel (1978) have recently claimed to find a small molecule inhibitor, not yet further characterised, in the plasma of schizophrenic subjects.

If a small molecule inhibitor of the type we envisage really exists, the best place to seek for its presence might be in urine. Using a crude rat liver homogenate and, later, a human platelet preparation as enzyme source, we were immediately able to demonstrate a substantial inhibitory effect (V. Glover, M. Reveley and M. Sandler, in preparation). Although individual urinary constituents, e.g. urea (Giordano *et al.*, 1962) are known to manifest some degree of competitive inhibitory ability when tested in high concentration, experimental reasons soon emerged to indicate that we were seeking for something more specific. A 'mock urine' sample, made up of a number of those urinary solutes normally present in highest concentration, gave rise to a moderate degree of inhibition in its undiluted state: however, when diluted two- or fourfold, its inhibitory ability became minimal and an eightfold dilution resulted in its complete disappearance. By contrast, inhibitory activity was still demonstrable in normal urine samples even after 32-fold dilution. When urine samples were incubated with a sulphatase-glucuronidase mixture, there was some increase in inhibitory activity. Accordingly, a wide variety of endogenously-occurring substrates and products of MAO, which are often excreted as sulphate conjugates, were screened but, of those tested, none had sufficient inhibitory potential to account for the observed findings. We have not yet tested certain carbolines (tryptolines) which have recently been claimed to be formed endogenously (Elliott and Holman, 1977) and have a structure sufficiently close to a known MAO inhibitor of plant origin, harmaline (Udenfriend *et al.*, 1958) to make them of interest in this regard. Certainly, their molecular weight is of the right order of magnitude although harmaline itself, unlike the present unknown agent, is a specific inhibitor of MAO A (Nelson *et al.*, 1979*a, b*). When urine samples were fractionated on Sephadex, maximum inhibitory ability was identified in the fraction eluting at about the same time as dextrose which possesses a molecular weight of 180.

The effect of the urinary inhibitor was partly reversed by washing the rat liver enzyme preparation. Competitive inhibition was evident when 5-hydroxytryptamine was employed as substrate but a mixed type of inhibition was observed with phenylethylamine. Inhibitor was present in rat urine also but despite our earlier experiments with germ-free chicks, described above, urine samples from germ-free rats contained as much inhibitor as samples from conventional controls.

As we still have to identify this unknown inhibitory compound, we are not yet in a position to pronounce on any role it may play in the physiological regulation of MAO activity. This possibility is obviously a tempting one and we hope to investigate it in the future. It is of particular interest, in this connection, that urine samples from subjects with severe depression contain significantly less inhibitor than control samples (V. Glover, M. Reveley and M. Sandler, in preparation).

MAO INHIBITORS AS DRUGS

The realisation that multiple forms of MAO exist opened up the possibility of tailoring inhibitors able selectively to prevent the breakdown of a particular monoamine whilst allowing the metabolism of others to proceed at a normal rate (Youdim *et al.*, 1971). Thus if, for example, a noradrenaline deficit turned out to be the fundamental lesion in depressive illness, as mooted by some (e.g. Schildkraut, 1965), it would obviously be advantageous to prevent its further breakdown and promote a build-up within the synaptic cleft whilst allowing the degradation of tyramine to proceed. Dietary tyramine has largely been responsible for the 'cheese reaction' (Marley and Blackwell, 1970) which has brought MAO inhibitor therapy into such disrepute since the middle 1960s and curtailed its subsequent application. Because of these prevailing attitudes, the selective inhibitors of MAO A and B came on the scene rather late for exhaustive clinical trial even though they have roused considerable academic interest. It should be pointed out, however, that early clinical experience with the first of the selective inhibitors, clorgyline, provided little encouragement. Although the drug gives rise to significant lightening of affect (Heard, 1969; Wheatley, 1970; Lipper *et al.*, 1979), it is just as prone to provoke the 'cheese reaction' as any of the more standard MAO inhibiting drugs (Lader *et al.*, 1970). However, interest has shifted more recently to a selective inhibitor of MAO B, deprenyl, which Knoll discusses in depth in the preceding paper (p. 151). Professor Knoll very kindly arranged for us to obtain access to generous supplies of the drug from the Hungarian drug company, Chinoin.

(—)-Deprenyl is a particularly powerful 'suicide' MAO inhibitor and a dose of 10 mg, the standard clinical dose, results in complete inhibition of platelet enzyme within a matter of hours (Elsworth *et al.*, 1978). It is of interest that as soon as something of the order of 80–90 per cent whole body MAO B inhibition has been achieved, there is a massive increase of urinary phenylethylamine output (Elsworth *et al.*, 1978) and, indeed, the gas chromatographic measurement of this substance (Blau *et al.*, 1979) can be a useful clinical guide to the degree of MAO B inhibition. When deprenyl administration ceases, urinary phenylethylamine excretion drops to normal within a few days implying that a relatively small degree of enzyme regeneration is adequate to cope with overspill and restrict its urinary output. Platelet enzyme gradually rises to pretreatment levels over a period of approximately three weeks (Elsworth *et al.*, 1978).

Although the drug is powerful, its action is to some extent non-specific and by no means restricted to the target enzyme defining the therapeutic group into which it falls, MAO. Even apart from other pharmacological effects of the drug itself, however, there is also the question of pharmacologically-active metabolites to be reckoned with: we have recently been able to show that deprenyl is metabolized almost quantitatively to a mixture of methamphetamine and amphetamine (Reynolds *et al.*, 1978). We have not yet monitored the optical rotation of these metabolites but it seems not unreasonable to suppose that, deriving as they do from (—)-deprenyl, they will turn out to possess the (—) configuration themselves. The (—)-amphetamines are not as potent pharmacologically as the (+) isomers (Innes and Nickerson, 1977) but the fact that they may play a role during the adjunct treatment of parkinsonism with this drug, for example, has still to be ruled out. It must be remembered that a combination of (+)-amphetamine, at least, and L-dopa may in certain circumstances, be more effective in the treatment of this disease than L-dopa alone (Parkes *et al.*, 1975).

Perhaps the most important attribute of deprenyl is its freedom from the 'cheese effect'. This property was first identified by Knoll and his colleagues (see Knoll and Magyar, 1972; Knoll, 1976) extrapolating from pharmacological data and subsequently confirmed in man by our group. When normal subjects are challenged with doubling-up doses of tyramine approximately every half hour, their blood pressure begins to rise and bradycardia supervenes when the dose rises to 400 mg (Elsworth *et al.*, 1978). After pretreatment with an 'orthodox' MAO inhibitor, this 'cheese effect' is observed following a dose of tyramine as low as 25 mg (Lader *et al.*, 1970). A series of volunteers on full deprenyl dosage were able to tolerate a dose of at least 200 mg and some the full 400 mg. A number of these volunteers had been treated with deprenyl for a period of up to 2 months so that even long-term administration rendered the subject no more likely to suffer an exaggerated pressor response after tyramine dosage than acute treatment.

Knoll's original explanation for this freedom from the 'cheese effect' was in terms of an intact gut MAO A barrier being preserved (Knoll and Magyar, 1972). Indeed, high MAO A activity is present in both animal and human small intestinal mucosa (Squires, 1972; Elsworth *et al.*, 1978). Such a claim, however, needs more careful examination. If a standard dose of 10 mg of deprenyl be dissolved in the body water of an average-size man, a drug concentration of approximately 10^{-6} M will be achieved. As will be seen from figure 11.1, which demonstrates the selective inhibitory effect of deprenyl on MAO in a brain homogenate using 5-hydroxytryptamine and phenylethylamine as substrates, 10^{-6} M is optimal for selectivity and achieves something of the order of 95 per cent inhibition of phenylethylamine oxidation with only about 15-20 per cent of 5-hydroxytryptamine oxidation inhibited. Knoll and his group (p. 151) have shown that selective inhibition can be maintained in rats on low doses of (−)-deprenyl. However, at higher doses, the selectivity may be abolished (Waldmeier and Felner,

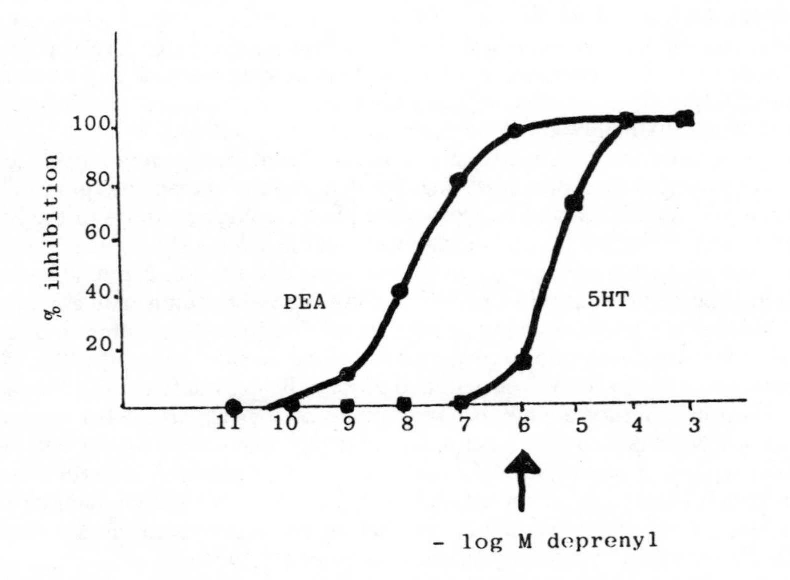

Figure 11.1 Inhibitory effect of (−)-deprenyl on phenylethylamine (PEA) compared with 5-hydroxytryptamine (5-HT) oxidation; showing that a 10^{-6} M concentration is optimal for selectivity.

1978). Whether 10 mg^{-1} d in man maintains selectivity is still an open question although the cumulative inhibition of 5-hydroxytryptamine oxidation of about 15–20 per cent per day is probably more rapid than the MAO regenerative process. Some exiguous and rather inconclusive evidence from post-mortem human brains suggests that while MAO A can be as much as 65 per cent inhibited functional selectivity is maintained, as dopamine concentrations are increased above controls although those of 5-hydroxytryptamine are not (Riederer *et al.*, 1978). However, the question must remain open, at present.

For a more likely explanation, one must turn to other properties of the drug and even the finding to be reported below can perhaps be extrapolated from an early pharmacological observation of Knoll and his group (Knoll and Magyar, 1972) who showed that deprenyl, at least in high concentration (although the doses employed in man are, by comparison, minuscule) prevents liberation of noradrenaline from its synaptic storage vesicles in response to tyramine challenge. In many tissues of the pig, including liver and brain, tyramine oxidation is effected predominantly by MAO B and this reaction is inhibited by deprenyl (Ekstedt and Oreland, 1976). The pig gut is an exception, for there, tyramine is predominantly oxidized by MAO A. It therefore seemed of interest to pretreat a group of pigs with a selective inhibitory dose of deprenyl and challenge them with tyramine *intravenously* in order to bypass the gut (Sandler *et al.*, 1978). It was predicted that this group should then undergo an exaggerated pressor response. Another group was treated with selective inhibitory doses of clorgyline, whilst a third was employed as control. All animals were killed after pressor tests and the predicted degree of MAO inhibition detected in every case. As anticipated, a relatively low inhibitory effect was observed after clorgyline but after deprenyl, inhibition was substantial.

The blood pressure response to intravenous tyramine challenge was quite paradoxical. In a manner opposite to prediction, the deprenyl group did not differ from control animals. The clorgyline group, on the other hand, despite minimal inhibition of tyramine-oxidizing ability, manifested with a very marked pressor reaction.

In order to interpret these responses, it is necessary to postulate the existence of two types of effect deriving from an irreversible MAO-inhibiting drug, MAO inhibition itself and, perhaps, facilitation of noradrenaline release from its storage vesicles. Alternatively, deprenyl might prevent the egress of noradrenaline out of its storage vesicle. If this interpretation be accepted, the 'cheese effect' is more a consequence of the second type of response than of MAO inhibition itself. It follows that this freedom from the 'cheese reaction' is not necessarily a concomittant of selectivity. Indeed, new compounds might, with confidence, be sought which possess such immunity from a tyramine pressor response and are yet able to inhibit MAO A. What little evidence exists suggests that MAO A inhibitors are potentially better antidepressants than those of MAO B (Lipper *et al.*, 1979). We have recently carried out a double-blind clinical trial of deprenyl in depressive illness (N. Mendis, C. M. B. Pare, M. Sandler and V. Glover, in preparation) and although some therapeutic benefit was identified, it was relatively small compared with the effect achieved with more orthodox MAO inhibitors.

If the 'cheese reaction' stems from an effect qualitatively different from MAO inhibition, it follows that such a property might be identified in certain compounds which are not known to be MAO inhibitors. And indeed, the antidepressant drug, isoniazid, which has no MAO inhibitory action, has been noted to be responsible on occasion for the 'cheese effect' by at least two groups of workers (Robinson *et al.*, 1968; Smith and Durack, 1978).

Although clinical research into the MAO inhibitors has been slow to develop because of the unfortunate occurrence of the 'cheese reaction', the experiments cited above might be considered sufficiently heartening to warrant a fresh start (e.g. Fuller, 1978).

REFERENCES

Bergeret, B., Blaschko, H. and Hawes, R. (1957). *Nature, Lond.*, **180**, 1127

Berrettini, W. H. and Vogel, W. H. (1978). *Am. J. Psychiat.*, **135**, 605

Blau, K., Claxton, I. M., Ismahan, G. and Sandler, M. (1979). *J. Chromatogr.*, **163**, 135

Böhm, K.-H., Glover, V., Sandler, M., Petty, M. and Reid, J. L. (1979*a*). *J. Neurochem.*, **33**, 607

Böhm, K.-H., Glover, V., Sandler, M. and Coates, M. E. (1979*b*). *Biochem. Pharmac.*, in press

Dennick, R. G. and Mayer, R. J. (1977). *Biochem. J.*, **161**, 161

Donnelly, C. H. and Murphy, D. L. (1977). *Biochem. Pharmac.*, **26**, 853

Edwards, D. J. and Burns, M. O. (1975). *Life Sci.*, **15**, 2045

Egashira, T. (1976). *Jap. J. Pharmac.*, **26**, 493

Ekstedt, B. and Oreland, L. (1976). *Arch. int. Pharmacodyn.*, **222**, 157

Elliott, G. R. and Holman, R. B. (1977). In *Neuroregulators and Psychiatric Disorders* (ed. E. Usdin, D. A. Hamburg and J. D. Barchas), Oxford University Press, New York, p. 220

Elsworth, J. D., Glover, V., Reynolds, G. P., Sandler, M., Lees, A. J., Phuapradit, P., Shaw, K. M., Stern, G. M. and Kumar, P. (1978). *Psychopharmacology*, **57**, 33

Fowler, C. J., Callingham, B. A., Mantle, T. J. and Tipton, K. F. (1978). *Biochem. Pharmac.*, **27**, 97

Fuller, R. W. (1978). *Progr. Neuro-Psychopharmac.*, **2**, 303

Gawel, M., Glover, V., Park, D., Rose, F. C. and Sandler, M. (1977). *Clin. Sci. molec. Med.*, **52**, 32P

Gentil, V., Greenwood, M. H. and Lader, M. H. (1975). *Psychopharmacologia*, **44**, 187

Giordano, C., Bloom, J. and Merrill, J. P. (1962). *J. Lab. clin. Med.*, **59**, 396

Glover, V., Sandler, M., Owen, F. and Riley, G. J. (1977). *Nature, Lond.*, **265**, 80

Heard, J. A. (1969). *Clin. Trials J.*, **6**, 219

Hough, L. B. and Domino, E. F. (1979). *J. Pharmac. exp. Ther.*, **208**, 422

Houslay, M. D. and Tipton, K. F. (1973). *Biochem. J.*, **135**, 173

Innes, I. R. and Nickerson, M. (1975). In *The Pharmacological Basis of Therapeutics*, 5th ed. (ed. L. S. Goodman and A. Gilman), Macmillan, New York, p. 477

Johnston, J. P. (1968). *Biochem. Pharmac.*, **17**, 1285

Kapeller-Adler, R. (1970). *Amine Oxidases and Methods for their Study*, Wiley-Interscience, New York

Kim, J. S., Hirano, M., Uchimura, H., Saito, M., Matsumoto, T. and Nakahara, T. (1979). *J. Neurochem.*, **32**, 253

Knoll, J. (1976). In *Monoamine Oxidase and its Inhibition* (ed. G. E. W. Wolstenholme and J. Knight), Elsevier, Amsterdam, p. 135

Knoll, J. and Magyar, K. (1972). *Adv. Biochem. Psychopharmac.*, **5**, 393

Lader, M. H., Sakalis, G. and Tansella, M. (1970). *Psychopharmacologia*, **18**, 118

Lewinsohn, R., Böhm, K.-H., Glover, V. and Sandler, M. (1978). *Biochem. Pharmac.*, **27**, 1857

Lipper, S., Murphy, D. L., Slater, S. and Buchsbaum, M. S. (1979). *Psychopharmacology*, **62**, 123

Marley, E. and Blackwell, B. (1970). *Adv. Pharmac. Chemother.*, **8**, 186

Medical Research Council (1965). *Br. med. J.*, **i**, 881

Nelson, D. L., Herbet, A., Glowinski, J. and Hamon, M. (1979*a*). *J. Neurochem.*, **32**, 1829

Nelson, D. L., Herbet, A., Pétillot, Y., Pichat, L., Glowinski, J. and Hamon, M. (1979*b*). *J. Neurochem.*, **32**, 1817
Owen, F., Cross, A. J., Lofthouse, R. and Glover, V. (1979). *Biochem. Pharmac.*, **28**, 1077
Pare, C. M. B. and Sandler, M. (1959). *J. Neurol. Neurosurg. Psychiat.*, **22**, 247
Parkes, J. D., Tarsy, D., Marsden, C. D., Bovill, K. T., Phipps, J. A., Rose, P. and Asselman, P. (1975). *J. Neurol. Neurosurg. Psychiat.*, **38**, 232
Phillips, A. W., Newcomb, H. R., Rupp, F. A. and Lachapelle, R. (1962). *J. Nutr.*, **76**, 119
Reveley, M., Glover, V., Sandler, M. and Coppen, A. (1979). *Br. J. clin. Pharmac.*, **8**, 375
Reynolds, G. P., Elsworth, J. D., Blau, K., Sandler, M., Lees, A. J. and Stern, G. M. (1978). *Br. J. clin. Pharmac.*, **6**, 542
Riederer, P., Youdim, M. B. H., Rausch, W. D., Birkmayer, W., Jellinger, K. and Seemann, D. (1978). *J. neural Transmiss.*, **43**, 217
Robinson, D. S., Lovenberg, W., Keiser, H. and Sjoerdsma, A. (1968). *Biochem. Pharmac.*, **17**, 109
Robinson, D. S., Nies, A., Ravaris, C. L. and Lamborn, K. R. (1973). *Arch. gen. Psychiat.*, **29**, 407
Ryder, T. A., MacKenzie, M. L., Pryse-Davies, J., Glover, V., Lewinsohn, R. and Sandler, M. (1979). *Histochemistry*, **62**, 93
Sandler, M. and Youdim, M. B. H. (1972). *Pharmac. Rev.*, **24**, 331
Sandler, M., Glover, V., Ashford, A. and Stern, G. M. (1978). *J. neural Transmiss.*, **43**, 209
Sawyer, S. T. and Greenawalt, J. W. (1979). *Biochem. Pharmac.*, **28**, 1735
Schildkraut, J. (1965). *Am. J. Psychiat.*, **122**, 509
Schnaitman, C. A. and Greenawalt, J. W. (1968). *J. Cell Biol.*, **38**, 158
Smith, D. K. and Durack, D. T. (1978). *Ann. intern. Med.*, **88**, 520
Squires, R. F. (1972). *Adv. biochem. Psychopharmac.*, **5**, 355
Sullivan, J. L., Dackis, C. and Stanfield, C. (1977). *Am. J. Psychiat.*, **134**, 188
Toyoshima, Y., Kinemuchi, H. and Kamijo, K. (1979). *J. Neurochem.*, **32**, 1183
Tyrer, P., Candy, J. and Kelly, D. (1973). *Psychol. Med.*, **3**, 120
Udenfriend, S., Witkop, B., Redfield, B. G. and Weissbach, H. (1958). *Biochem. Pharmac.*, **1**, 160
Waldmeier, P. C. and Felner, A. E. (1978). *Biochem. Pharmac.*, **27**, 801
Waldmeier, P. C., Delini-Stula, A. and Maitre, L. (1976). *Naunyn-Schmiedeberg's Arch. Pharmac.*, **292**, 9
Waldmeier, P. C., Feldtrauer, J. J. and Maitre, L. (1977). *J. Neurochem.*, **29**, 785
Wheatley, D. (1970). *Br. J. Psychiat.*, **117**, 573
Williams, D., Gascoigne, J. E. and Williams, E. D. (1975). *Histochem. J.*, **7**, 585
Williams, D., Gascoigne, J. E. and Williams, E. D. (1979). *Histochem. J.*, **11**, 83
Yang, H.-Y. T. and Neff, N. H. (1973). *J. Pharmac. exp. Ther.*, **187**, 365
Yang, H.-Y. T. and Neff, N. H. (1974). *J. Pharmac. exp. Ther.*, **189**, 733
Youdim, M. B. H., Collins, G. G. S. and Sandler, M. (1971). *Biochem. J.*, **121**, 34P
Zeller, E. A., Barsky, J. and Berman, E. R. (1955). *J. biol. Chem.*, **214**, 267

12

Mode of action of β-lactam antibiotics at the molecular level

J. M. Frère,† C. Duez, J. Dusart, J. Coyette, M. Leyh-Bouille and J. M. Ghuysen
(Service de Microbiologie, Faculté de Médecine, Institut de Botanique,
Université de Liège, Sart Tilman, B-4000 Liège, Belgium),
O. Dideberg (Laboratoire de Cristallographie approfondie,
Institut de Physique, Université de Liège, Sart Tilman, B-4000 Liège, Belgium)
and J. Knox (Biological Sciences Group and Institute of Materials Science,
University of Connecticut, Storrs, Connecticut 06268, USA)

INTRODUCTION

In recent years, two main approaches have been utilised for unravelling the mode of action of β-lactam antibiotics. One consists of treating whole bacteria or isolated cytoplasmic membranes with ^{14}C-labelled penicillin G and separating the so-called penicillin-binding proteins (PBPs) by polyacrylamide gel electrophoresis. In some mutants, the absence of one of the PBPs can be correlated with impaired function. Moreover, some β-lactams exhibit a strong preference for one of the PBPs and induce morphological changes similar to those observed with mutants lacking this same PBP (Spratt, 1978). This is mainly a physiologist's approach and, not surprisingly, it has yielded little information about the interaction between the antibiotics and their target(s) at the molecular level. This type of information has been obtained by an alternative approach, which might be thought of as a chemist's approach: it involves the isolation of penicillin-sensitive enzymes (D,D-carboxypeptidases and transpeptidases) (PSEs) and study of the interaction between purified enzymes and antibiotics as bimolecular reactions. Sometimes, for good measure, substrate is also thrown in: this makes the system a little more complicated but still manageable.

†To whom correspondence should be addressed.

In this paper, we will restrict our discussion to results obtained by this latter approach. The long-term goal of these studies is establishing correlations between structure and function so that, ideally, new molecules can be designed for inhibiting a given enzyme. We are still very far away from performing such a *tour de force*, but interesting features are beginning to emerge.

Transpeptidases and D,D-carboxypeptidases are the only bacterial enzymes which seem to be specifically inhibited by β-lactam antibiotics. At high concentrations, penicillins are known to acylate many proteins but, although this phenomenon might be of practical importance (such as in determining the allergenic effect of penicillins), it is not involved in the antibacterial effect itself. Many bacterial D,D-carboxypeptidases and transpeptidases have been described and some purified. All of them are sensitive to penicillins, with the notable exception of the D,D-carboxypeptidase from *Streptomyces albus* G (albus G. enzyme) which, as will be discussed below, displays a very low sensitivity. Are there other enzymes which specifically recognise the β-lactam ring? Only a limited number of enzyme classes catalyse reactions involving β-lactam antibiotics. The amide bond which connects the side-chain to the 6-aminopenicillanic acid (6-APA) or 7-aminocephalosporanic acid (7-ACA) nuclei can be hydrolysed by an amidase (Savidge and Cole, 1975; Vanderhaeghe, 1975) and the group on C_3 of some cephalosporins modified by an esterase (Abraham and Fawcett, 1975; Abbott and Fukuda, 1975) (figure 12.1). These enzymes usually decrease the biological

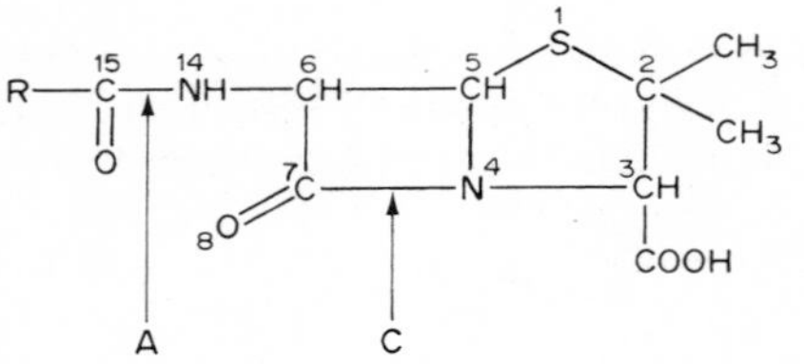

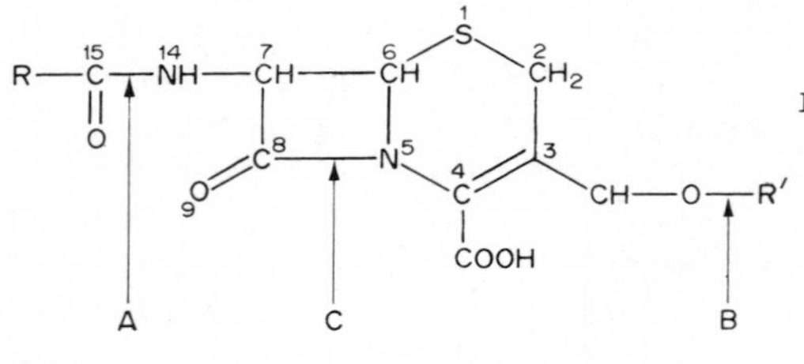

Figure 12.1 General structures of penicillins (I) and cephalosporins (II) showing the site of action of amidases (A), esterases (B) and β-lactamases (C). The group on C_3 of II does not necessarily bear an ester function; compounds such as cephalexin are thus not susceptible to esterases.

activity of the antibiotic (see below) but they do not alter its characteristic bicyclic structure. In contrast, however, penicillin-sensitive enzymes (PSEs) appear specifically to break open the four-membered ring. They share this property with the β-lactamases, a class of penicillin-destroying enzymes which, for therapeutic reasons, has received a lot of attention. Understandably, β-lactamase-producing bacteria exhibit an increased resistance to β-lactam antibiotics. Is there a relationship between the sensitivity of penicillins and cephalosporins to β-lactamase and their ability to inhibit PSEs? At first sight, the answer might be yes. Very generally, β-lactams with high biological activity are substrates for β-lactamases. An evolutionary relationship between β-lactamases and PSEs is thus a distinct possibility.

The existence of enzymes the only role of which would be to destroy penicillins has seemed puzzling to some authors (Saz, 1970; Ozer and Saz, 1970). However, some β-lactamases are among the enzymes exhibiting the highest turnover numbers and it seems unlikely that this high β-lactamase activity would only be a 'secondary activity'. Moreover, it was recently demonstrated that many PSEs (if not all of them) are endowed with a low, but significant (clearly a 'secondary activity') penicillin-destroying activity implying in all cases the hydrolysis of the β-lactam amide bond (Frère *et al.*, 1975*b*; Hammarström and Strominger, 1975; and references on tables 12.2 and 12.3). That penicillinases would be derived from PSEs (by gene duplication, for instance) would thus appear to be a reasonable hypothesis.

CHARACTERISTIC STRUCTURAL FEATURES OF β-LACTAM ANTIBIOTICS

This hypothesis suggests that the 'activity' of penicillins and cephalosporins as substrates of β-lactamases or inhibitors of PSEs could have a common structural basis. Unfortunately, only substances active as antibacterial agents have been tested as substrates for β-lactamases and, more rarely, as inhibitors for D,D-carboxypeptidases and transpeptidases. Several well documented reviews have been published, exploring the relationship between structure and antibacterial activity (Nayler, 1971; Sweet, 1972; Gorman and Ryan, 1972; Jászberényi and Gunda, 1975; Gunda and Jászberényi, 1977). These data are difficult to use in an analysis of the relationship between structure and activity as inhibitor of the target enzyme(s) since the measured value (minimum inhibitory concentration) depends upon at least three phenomena: penetration of the antibiotic into the cell, its sensitivity to the β-lactamase(s) which the cell might produce and, finally, its interaction with the target enzyme(s). However, from the data accumulated so far by X-ray crystallography and infrared and nuclear magnetic resonance spectroscopy, the following features appear to be of major importance:

(1) In active compounds, the β-lactam amide is more unstable than in normal amides. The main reason for this instability is not the strain inherent in a four-membered ring but the fact that the presence of a second, fused ring renders the β-lactam nitrogen pyramidal and not planar as in a normal amide or a peptide bond. The non-planarity of the β-lactam amide bond decreases the amide resonance, which makes the $\mathrm{N{-}C({=}O)}$ bond longer and more unstable than in normal amides (Sweet, 1972; Boyd, 1973). Compounds with a monocyclic β-lactam and many other bicyclic systems have a planar nitrogen and are not active. Striking examples of the last group of compounds are the Δ^2-cephems (Figure 12.2). These substances are inactive although, geometrically, they are more closely

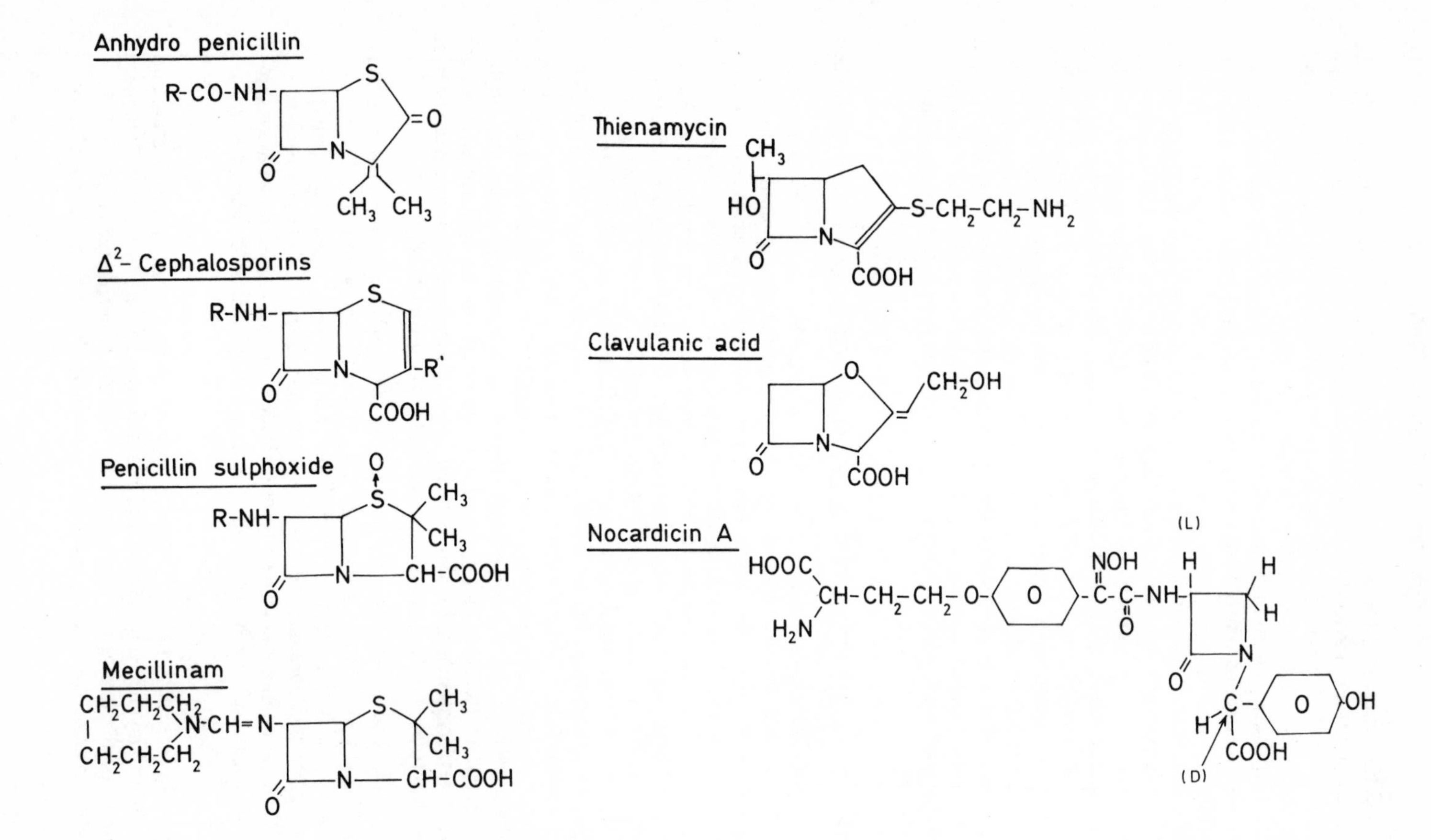

Figure 12.2 Structures of some 'non-classical' β-lactams

related to penicillins than are active cephalosporins. In contrast, the non-planarity of the β-lactam nitrogen does not appear to be a sufficient condition for activity: anhydropenicillin and penicillin sulphoxide have a pyramidal nitrogen atom but no antibacterial activity.

(2) The four atoms of the β-lactam ring do not lie exactly in the same plane: C_6 (C_7 in cephalosporins) is displaced and the lactam O_8 (O_9) is bent towards the S_1. The same bending is observed in anhydropenicillin and Δ^2-cephems. Again, the non-coplanarity of the four atoms of the β-lactam ring is not sufficient to determine antibiotic activity (Sweet, 1972).

(3) The R–NH– group on C_6 of penicillins (C_7 of cephalosporins) must be on the β-face of the β-lactam ring. The β-face is the upper face when the β-lactam amide is viewed in a direction parallel to the O_8, C_7, N_4 plane, with the O on the front left, the C on the back left and the N on the right. Inversion of the substituents on C_6 (6-epipenicillins) yields inactive compounds. R must be an acyl group.

(4) In general, replacement of the hydrogen atom on C_6 of penicillins or C_7 of cephalosporins by a larger group strongly decreases antibacterial activity. Only if the substituent is a methoxy group might activity be retained or even increased (Ho *et al.*, 1973).

(5) In cephalosporins, the presence of a good leaving group on the substituent of C_3 further decreases the stability of the β-lactam amide (Boyd *et al.*, 1975). Antibacterial activity is generally also increased by the presence of a better leaving group (Gorman and Ryan, 1972).

(6) Although esterification of the carboxyl on C_3 of penicillins strongly decreases the activity, lactonisation of the carboxyl on C_4 with an alcohol function on the substituent on C_3 of cephalosporins yields active compounds (Jászberényi and Gunda, 1975).

(7) No active compound has yet been found in the crystals of which the β-lactam C=O was engaged in a hydrogen bond (Sweet, 1972). The meaning of this observation is still obscure.

(8) Recently, various β-lactam antibacterial agents have been isolated or synthesised which do not present the structure of classical penicillins and cephalosporins: mecillinam, an amidino penicillin, still presents the usual fused-ring system of penicillins but nocardicin (Aoki *et al.*, 1976), clavulanic acid (Brown *et al.*, 1976), thienamycin and other derivatives of olivanic acid (Corbeth *et al.*, 1977) are quite different (figure 12.2). The mode of action of these substances is under current study in various laboratories. Thienamycin and clavulanic acid compete with penicillin G for the PBPs of *E. coli* (Spratt *et al.*, 1977) and mecillinam, nocardicin and clavulanic acid compete for the penicillin binding site of the D,D-carboxypeptidase-transpeptidase from *Actinomadura* R39 (J. Kelly, M. Noël, and J. M. Frère, unpublished results).

CHEMICAL AND β-LACTAMASE-CATALYSED OPENING OF THE β-LACTAM RING

As stated above, only a very limited number of substances not active as antibacterial agents have been tested as substrates for β-lactamases. From the scant information available, the following points appear to be of importance in relation to the preceding paragraphs.

(1) Δ^2-Cephalosporins are not substrates of the β-lactamases (O'Callaghan *et al.*, 1968). In contrast, penicillin G sulphoxide is a good substrate for β-lactamase I from *Bacillus cereus* (Davies *et al.*, 1974) and penicillin G sulphone, a good substrate for the exocellular β-lactamase from *Staphylococcus aureus* (Cartwright and Coulson, 1979). The sulphone of penicillanic acid inhibits various β-lactamases (English *et al.*, 1978) while that of 6-α-chloropenicillanic acid is a good inhibitor of the *S. aureus* β-lactamase (Cartwright and Coulson, 1979), but the real inactivator could be the β-chloro isomer, as in the case of the bromopenicillanic acids (Knott-Hunziker *et al.*, 1979*a, b*).

(2) 6-Epipenicillin G is not a substrate for the β-lactamase of *Pseudomonas aeruginosa* (Furth, 1975).

(3) 6-α-Methylpenicillin G is not a substrate for β-lactamases I and II from *B. Cereus*; 7-phenoxyacetamido-7-α-methyldesacetoxycephalosporanic acid and 7-methoxycephalothin are not substrates for β-lactamase II from *B. Cereus* (Davies *et al.*, 1974), an enzyme which is usually a good cephalosporinase.

(4) A cephalosporin with a good leaving group on the substituent on C_3 is generally a better substrate than the corresponding substance with a methyl group.

(5) The lactone from desacetoxycephalosporin C is a good substrate for β-lactamase I but not a substrate for β-lactamase II from *B. cereus* (Kuwabara and Abraham, 1969).

(6) The new, non-classical β-lactams do not appear to be substrates for the β-lactamases. However, some of them behave as rather specific inhibitors of these enzymes (Brown *et al.*, 1976; Maeda *et al.*, 1977).

(7) If one ignores the sulphoxide and the sulphone, the four-membered β-lactam ring must be condensed with a thiazoline or a dihydrothiazine ring (Richmond and Sykes, 1973).

Discounting two important exceptions (the sulphoxide–sulphone group and 7-methoxycephalothin), a relationship between antimicrobial activity and sensitivity to β-lactamases might exist with a common set of requirements responsible for the 'activity' of a β-lactam as antibacterial agent and as substrate (or inhibitor) of β-lactamases. However, as will be seen below, the diversity of β-lactamases is so great that it is extremely imprudent to draw conclusions from the sparse experimental data presented above.

In the microbial world, β-lactamases are very common enzymes. In Gram-negative bacteria, enzymes with so many different activity profiles have been described that Richmond and Sykes (1973) identified five classes and 15 subclasses. Class I enzymes are mainly cephalosporinases. They hydrolyse some cephalosporins up to 80 times faster than penicillin G. Some are active against ampicillin, others are not. Class II enzymes are predominantly active against penicillins. They have little or no activity against cephalosporins. Enzymes from classes III, IV and V hydrolyse both penicillins and cephalosporins but can be distinguished on the basis of their sensitivity to *p*-chloromercuribenzoate (PCMB) and interaction with penicillins, such as cloxacillin, which present a bulky, sterically-hindered side-chain directly attached to the exocyclic amide C atom (C_{15} in figure 12.1). Class IV enzymes are sensitive to PCMB and resistant to cloxacillin–some hydrolyse it, others simply do not recognise it. Classes III and V enzymes are resistant to PCMB. Cloxacillin is an inhibitor for class III enzymes, but a substrate for class V. In Gram-positive strains, the situation is also very

Table 12.1 Rate of hydrolysis of β-lactam antibiotics by OH^- ions and β-lactamases from various Gram-positive strains†

Hydrolysing agent		β-Lactamases									
	OH^- ions	*Bacillus cereus* 569/H		*Bacillus licheniformis*		*Staphylococcus aureus*			Actinomycetes		
Antibiotic		I	II	749/C	6346/C	A+B	C	E_3	*Actinomadura* R39	*Streptomyces albus* G	*Streptomyces* K11
Penicillin G	100	100	100	100	100	100	100	100	100	100	100
Penicillin V	170	153				110	100	110	220	80	116
Carbenicillin	100	22							33	10	27
Ampicillin	115	194	47	68	120	185		198	520	210	82
Methicillin	90	4	120	0.5	1.0	1.5	0.6	1.0	28	1.5	0
Oxacillin	100	5	89			4.5	1		320	19	0
Cloxacillin	100	0.5	89	0		2	1		30	2	0
6-APA		12	10	5	13	10		1	76	75	111
Cephalosporin C	47	0.1	14	1.1	1.0	1.1	1.2		6	6	5
Cephalothin	150	< 0.1	89						55	2.4	116
Cephaloglycine	180	< 0.1	41	1	42				215	7.5	88
Cephalexin	20		1.3						17	1.2	0
Nitrocefin	2000			20					230	23	
References‡		a, b	a, b	c	c	d	e		f	f	f

‡In each case, maximum velocities were calculated with penicillin G = 100. The rate of hydrolysis by OH^- ions was measured by incubating the substance at 37°C (at a 1 mM concentration) in 1 M potassium phosphate buffer, pH 12.0 and estimating the product by the iodine method. The concentration of nitrocefin was different (10^{-4} M), and the product was estimated by direct spectrophotometric reading at 482 nm. The value of the first-order rate constant for the hydrolysis of penicillin G by OH^- ions was 0.5×10^{-3} s^{-1}. For the cephalosporins, similar values were also obtained at lower concentrations (10^{-4} M) by following the variation of absorbance at 260 nm (Waley, 1974; Fuad *et al.*, 1976; J. M. Frère, unpublished).

‡References: a, Thatcher, 1975*a*; b, Kubarawa and Abraham, 1967; c, Thatcher, 1975*b*; d, Richmond, 1975; e, Citri, 1971; f, Johnson *et al.*, 1973.

complicated. Table 12.1 summarises data obtained with the most widely studied of these Gram-positive β-lactamases. The antibiotics have been divided into seven groups, according to their chemical structure: penicillins V and G have uncharged, sterically unhindered side-chains; carbenicillin and ampicillin have charged side-chains while methicillin, oxacillin and cloxacillin present a sterically hindered side-chain; 6-APA, with no side-chain at all, but an uncharged amino group at neutral pH, is in a group by itself. Cephalosporins are distinguished by the nature of the substituent on C_3: cephalosporin C, cephalothin and cephaloglycine have a $-CH_2-O-CO-CH_3$ group; cephalexin, a methyl group, and nitrocefin, a chromogenic cephalosporin, a 2, 4-dinitrostyryl group. The inhibitory effect of cloxacillin and of related compounds on some of these enzymes has received a lot of attention. In the presence of an antibiotic from this group (type A substrates), β-lactamase I from *B. cereus* undergoes a slow transconformation into a less active configuration (Citri, 1973). Addition of a saturating amount of penicillin G (or another type S substrate) restores the more active conformer of the enzyme. This mechanism thus appears to follow Frieden's hysteresis model (Frieden, 1970) and the two forms clearly differ by several of their physical or chemical properties (Csanyi *et al.*, 1970; Samuni and Citri, 1975; Kiener and Waley, 1977; Citri *et al.*, 1976; Strom *et al.*, 1976). A similar behaviour has been described for the β-lactamase from *S. aureus* (Sagai and Sato, 1973). An attractive hypothesis can be built on the basis of these data: β-lactamases use the side-chain of penicillins and cephalosporins as a handle further to destabilise the β-lactam amide. This deformation of the antibiotic is only possible if the side chain has a certain degree of rotational freedom. With sterically hindered side-chains, the enzyme itself must 'adapt' to the substrate and change into a less active conformation (P. Blanpain, G. Laurent, B. Nagy and F. Durant, in preparation). The hypothesis can easily be modified to account for enzymes which do not 'recognise' oxacillin (the side-chain just does not fit into the enzyme site) or for enzymes which hydrolyse it (they are somehow capable of overcoming the steric hindrance). More surprising is the fact that some enzymes hydrolyse 6-APA rather well although they may have no 'handle' to use. After examination of numerous specificity profiles, one might wonder whether it is reasonable to search for a unified hypothesis of β-lactamases activity. It is somewhat disappointing to realise that, although a large amount of information is available about the chemistry of these enzymes (four sequences are known (Meadway, 1969; Thatcher, 1975*c*; Ambler and Scott, 1978; Ambler, 1975) and X-ray crystallographic data have been obtained on three of them (Aschaffenburg *et al.*, 1978; Knox *et al.*, 1976)), very little is known about their respective active sites. The involvement of a tyrosine and a histidine residue has been suggested (Scott, 1973) but little direct experimental evidence has been accumulated in favour of this hypothesis. Recently, a serine residue of β-lactamase I from *B. cereus* has been shown to react with β-bromopenicillanic acid. This Ser residue is conserved in the amino acid sequences of all known β-lactamases. The same reagent effectively inactivates the β-lactamases from *S. aureus, E. coli* W3310, *B. licheniformis* and *Pseudomonas aeruginosa* but not β-lactamase II from *B. cereus*. The fact that treatment with dilute alkali results in the release of the label suggests the possibility of an ester or ether linkage (Knott-Hunziker *et al.*, 1979*a, b*). Recent experiments (J. F. Fisher and J. R. Knowles, chapter 13 in this volume) indicate the possibility of formation of a transient acyl-enzyme, involving a serine residue, in the catalytic pathway of β-lactamases.

The reactivity of the β-lactam amide bond in a given compound determines its properties as an acylating agent. Approximate relative values of the acylating power of various antibiotics were obtained by measuring their susceptibility to hydroxyl ions (see table 12.1). In addition, cefoxitin, a cephalosporin with a methoxy group on C_7, was found to be twice as stable as benzylpenicillin. It is not a substrate for various β-lactamases (from *E. coli, S. aureus, Klebsiella, Pseudomonas*, . . .)

Examination of table 12.1 shows that, with the exception of the particularly unstable nitrocefin, the range of 'chemical' stability of the β-lactam antibiotics barely spans one order of magnitude. Interestingly the most stable (cephalexin) and the most unstable (cephaloglycine) differ only by the nature of the substituent on C_3: cephalexin has a methyl group which contains no leaving group, while cephaloglycine has a $-CH_2-O-CO-CH_3$ group, containing the acetoxy moiety, a good leaving group. In contrast, sensitivity to β-lactamases sometimes extends over more than three orders of magnitude, and in all cases, over two orders. In consequence, there does not appear to be any clear indication of a correlation between sensitivity to OH^- ions and to β-lactamases although, in the particular case of the pair cephaloglycine–cephalexin, the chemically less stable compound (cephaloglycine) consistently behaves as a better substrate (with the exception of Gram-negative class I β-lactamases which hydrolyse cephalexin faster than cephaloglycine!).

In the absence of steric hindrance, the attack of the β-lactam ring by OH^- ions can occur on the β- or the α-face (Boyd *et al.*, 1975). The fact that epipenicillins and β-lactams with hindered α-faces are poor substrates for β-lactamases indicates that enzymatic hydrolysis is probably limited to an attack on the α-face.

Inhibition of the β-lactamases from *E. coli* (RTem factor) and *S. aureus* by clavulanate involves the formation of a 1:1 complex (Fisher *et al.*, 1978; Cartwright and Coulson, 1979). Spectroscopic properties of the complex are in agreement with an intermediate of structure

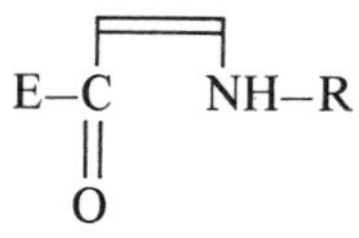

E–C   NH–R
   ||
   O

where E is the enzyme. In further steps, reactivation of the enzyme can occur, but the interaction appears to be rather complicated and, for the *E. coli* enzyme, a branched pathway has been proposed (Charnas *et al.*, 1978).

PENICILLIN-SENSITIVE ENZYMES: D,D-CARBOXYPEPTIDASES AND TRANSPEPTIDASES

As stated above, penicillin-sensitive enzymes exhibit D,D-carboxypeptidase activity, transpeptidase activity, or both (figure 12.3). Some transpeptidases never use water as acceptor, some D,D-carboxypeptidases only utilise water. Many enzymes can utilise both water and an aminated molecule as acceptors but, in most cases, this latter substance must be a simple amino acid, such as glycine. The D,D-carboxypeptidases–transpeptidases excreted by *Streptomyces* R61 and *Actinomadura* R39, on the contrary, catalyse reactions which really mimic the

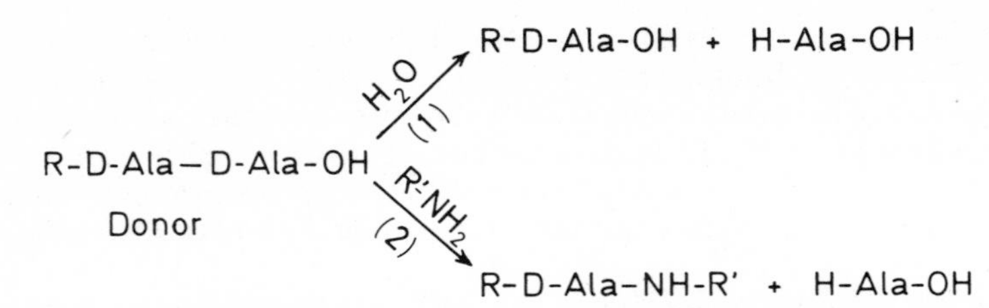

Figure 12.3 D,D-Carboxypeptidase (1) and transpeptidase (2) activities depend upon the nucleophilic acceptor on to which the R-Ala moiety of the donor is transferred: water in the first case, an amino group in the second

cross-linking of the peptide subunits of the nascent peptidoglycan (Ghuysen *et al.*, 1974; Zeiger *et al.*, 1975). In this article, we will use the term transpeptidase only for two classes of enzymes: those which do not use water as acceptor and those which can catalyse the dimerisation of donor–acceptor substrates.

The three-step mechanism which has been proposed for the interaction between the exocellular D,D-carboxypeptidase-transpeptidase from *Streptomyces* R61 and β-lactam antibiotics (Frère *et al.*, 1975*a*) appears to be valid for most penicillin-sensitive enzymes. This mechanism involves the reversible formation of a first stoichiometric complex, EI, which irreversibly transforms into a second, rather stable complex, EI*. This second complex irreversibly decays into a fully reactivated enzyme and a degraded, biologically inert antibiotic:

$$E + I \underset{}{\overset{K}{\rightleftharpoons}} EI \xrightarrow{k_3} EI^* \xrightarrow[\text{Nucleophilic agent}]{k_4} E + P(s)$$

The constants K, k_3 and k_4, respectively, represent the dissociation constant of EI, and the first-order rate constants for the two irreversible steps. The structure of the degradation product(s), P(s), is a function of three variables: the enzyme E, the antibiotic I and the nucleophilic agent involved in the degradation of EI*. For instance, with the exocellular D,D-carboxypeptidase-transpeptidase from *Streptomyces* R61, benzylpenicillin and water, the antibiotic is fragmented into phenylacetylglycine and an unidentified intermediate which spontaneously decays into *N*-formyl-D-penicillamine. In the presence of oxygen, this latter product can in turn give rise to the corresponding disulphide (Frère *et al.*, 1975*b*, 1976*a*). Hammarström and Strominger (1976) reported a similar mechanism for the membrane-bound D, D-carboxypeptidase from *Bacillus stearothermophilus* and claimed they had identified D-5,5-dimethyl-Δ^2-thiazoline-4-carboxylate as the possible intermediate, but it was later shown that the compound had been mischaracterised (Adriaens *et al.*, 1978). With the R61 enzyme (i.e. exocellular D,D-carboxypeptidase-transpeptidase from *Streptomyces* R61), a suitable amine ($R'–NH_2$), if present, can also accept the phenylacetylglycyl moiety (Figure 12.4). However, in the presence of various alcohols, the corresponding penicilloyl esters are also formed but penicilloic acid has never been detected (Marquet *et al.*, 1979). This enzyme similarly fragments penicillin V but only hydrolyses the β-lactam

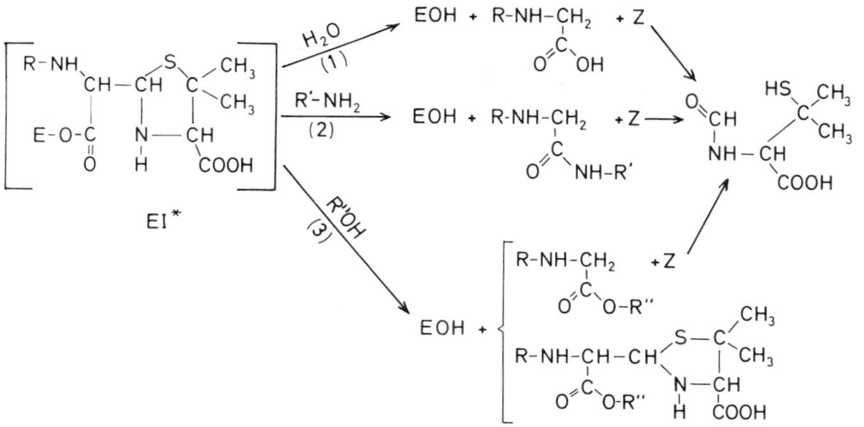

Figure 12.4 Degradation of benzylpenicillin (R = $C_6H_5-CH_2-CO-$) by the exocellular D,D-carboxypeptidase-transpeptidase from *Streptomyces* R61. Complex EI* is probably a penicilloyl-enzyme. Pathways 2 and 3, occurring in the presence of water are always in competition with pathway 1; Z is the unidentified intermediate which spontaneously decays into *N*-formyl-D-penicillamine. The structure of the complex is discussed below.

amide bond of nitrocefin (Frère *et al.*, 1978*a*; J. M. Frère, unpublished data). With other cephalosporins, the situation is still obscure for two reasons: the efficiency of the enzyme as a cephalosporinase is exceedingly low (half-lives of complex EI* amount to several days (Frère *et al.*, 1975*a*)) and the relative difficulty in obtaining radioactively labelled cephalosporins. The problem of the fate of cephalosporins is currently under study in our laboratory. Other enzymes, such as the exocellular D, D-carboxypeptidase from *S. albus* G (J. M. Frère, C. Duez and M. Noel, unpublished data), or the membrane-bound D,D-carboxypeptidase from *Proteus mirabilis* (Schilf *et al.*, 1978), transform benzylpenicillin into benzylpenicilloic acid (tables 12.2 and 12.3).

The fact that the three-step model appears to be valid for many enzymes does not, however, completely eliminate the possibility of a simple, reversible interaction between a β-lactam antibiotic and a PSE. For instance, no stable complex between benzylpenicillin and D,D-carboxypeptidase IB could be isolated from *E. coli* (Tamura *et al.*, 1976). In addition, the values of the constants for the interaction between the D,D-carboxypeptidase from *S. albus* G and cephalosporin C are such (Frère *et al.*, 1978*b*) that, for a short contact time (1 h or less), more enzyme will be immobilised in complex EI than in complex EI*. It should be noted here that enzymes exhibiting a similar behaviour pattern might easily escape detection by the PBP method.

Mainly for technical reasons, the individual constants k_3 and K have only been obtained for a rather limited number of enzyme–antibiotic interactions (table 12.2). However, values of the ratio k_3/K are available for some other enzymes (table 12.4). This ratio represents the efficiency of the bimolecular reaction $E + I \rightarrow EI^*$ when $[I] \ll K$ so that the amount of complex EI remains negligible. The most striking feature which appears upon examination of these tables is the

Table 12.2 Constants and products for the interaction between PSEs and β-lactams

	K (mM)	k_3 (s^{-1})	k_3/K (M^{-1} s^{-1})	k_4 (s^{-1})	Product(s)	References
D,D-Carboxypeptidase-transpeptidase *Streptomyces* R61 (exocellular)						
Benzylpenicillin	13 (25 °C)	180 (25 °C)	13 000 (25 °C)	1.4×10^{-4}	Fragments	
Penicillin V	> 1	> 1	1500	2.8×10^{-4}	Fragments	
Carbenicillin	0.11	0.09	800	1.4×10^{-4}	?	
Ampicillin	7.2	0.8	100	1.4×10^{-4}	?	Frère *et al.*, 1975*a*, *b*
Cephalosporin C	> 1	> 1	1100	1×10^{-6}	?	
Cephaloglycine	0.4	0.009	21	3×10^{-6}	?	
Nitrocefin	0.2 (10 °C)	0.1 (10 °C)	460 (10 °C)	3×10^{-4}	Hydrolysed β-lactam amide	
D,D-Carboxypeptidase *Streptomyces albus* G (exocellular)						
Penicillin V	150	8×10^{-4}	5×10^{-3}	9×10^{-5}	Penicilloic acid	
Cephalosporin C	1.5	1×10^{-4}	6×10^{-2}	8×10^{-5}	?	Frère *et al.*, 1978*b*
Cephalothin	9	5×10^{-4}	6×10^{-2}	3×10^{-5}	?	
D,D-carboxypeptidase-transpeptidase *Actinomadura* R39 (exocellular)						
Cephalosporin C	0.2 (20 °C)	10 (20 °C)	50 000 (20 °C)	0.3×10^{-6}	?	Fuad *et al.*, 1976
D,D-Carboxypeptidase *Streptococcus faecalis* ATCC 9790 after detergent-solubilisation of membrane-bound enzyme						
Benzylpenicillin	0.024	0.025	1050	4.4×10^{-5}	Fragments	Coyette *et al.*, 1978

Values were measured at 37 °C unless otherwise indicated (in parentheses).

Table 12.3 Products of the interaction between PSEs and benzylpenicillin

Enzymic activity	Strain	Physical state	k_4 (s^{-1})†	Products	References
D,D-Carboxypeptidase-transpeptidase	*Actinomadura* R39	Exocellular	2.8×10^{-6}	Fragments	Fuad *et al.*, 1976
Transpeptidase	*Streptomyces* R61	Membrane-bound	1.1×10^{-4}	Penicilloic acid‡	Marquet *et al.*, 1974
Transpeptidase	*Streptomyces* R61	Detergent-solubilised from membranes	0.1×10^{-4}	Fragments	Dusart *et al.*, 1977
Transpeptidase	*S. rimosus*	Detergent-solubilised from membranes	0.6×10^{-4}	Fragments	Dusart *et al.*, 1977
D,D-Carboxypeptidase	*Proteus mirabilis*	Detergent-solubilised from membranes	3.3×10^{-3}	Penicilloic acid	Schilf *et al.*, 1978
D,D-Carboxypeptidase	*Bacillus stearothermophilus*	Membrane-bound	1×10^{-3} (55 °C)	Fragments	Hammarström and Strominger, 1975
D,D-Carboxypeptidase	*B. stearothermophilus*	Detergent-solubilised from membranes	1×10^{-3} (55 °C)	Fragments	Blumberg *et al.*, 1974
D,D-Carboxypeptidase	*B. subtilis*	Detergent-solubilised from membranes	0.6×10^{-4}	Fragments	Blumberg *et al.*, 1974
D,D-Carboxypeptidase IA	*E. coli* H 2143	Detergent-solubilised from membranes	2×10^{-3}	Penicilloic acid§	Tamura *et al.*, 1976
D,D-Carboxypeptidase	*S. aureus*	Detergent-solubilised from membranes	0.3	Penicilloic acid§	Kozarich and Strominger, 1978

†Values of k_4 were obtained at 37 °C unless otherwise stated.
‡Product released from membranes containing several PBPs.
§No precaution was taken to exclude the possibility of contamination by a β-lactamase.

Table 12.4 Values of k_3/K for the interaction between some D,D-carboxypeptidases (D,D-Cbases) or transpeptidases (Tases) and various β-lactams

Origin of enzyme:	Exocellular *Actinomadura* R39†	Exocellular *Streptomyces albus* G	Exocellular *Streptomyces* R61†	Membrane enzymes, *Streptomyces* R61	Membrane enzymes, *Streptomyces* Detergent-solubilised R61	K15	*rimosus*	Membrane-bound *Streptococcus faecalis*	Membrane-bound *Streptococcus faecalis* Solub.‡	Membrane-bound *Bacillus subtilis* (solub.)‡§
Type of activity:	D,D-Cbase-Tase	D,D-Cbase	D,D-Cbase-Tase	Tase	Tase	Tase	Tase	D,D-Cbase	D,D-Cbase	D,D-Cbase
Penicillin G	300 000	$< 2 \times 10^{-3}$	12 000 (25 °C)	53	10	30	340	445	1050	1200 (4 °C)
Penicillin V	[>70 000]	5×10^{-3}	1500	140	15	30	400	560	1220	600 (4 °C)
Carbenicillin	6000	$< 2 \times 10^{-3}$	800	15	8	5	140	19	27	
Ampicillin	70 000 (20 °C)	$< 10^{-3}$	100	400	15	15	380	230	600	100
Methicillin	1000 (20 °C)	$< 2 \times 10^{-3}$	[15]					1.9	3.6	7
Oxacillin	[40 000]		[130]			0.6		4.5	4.6	5
Cloxacillin	[15 000]	$< 2 \times 10^{-3}$	[30]	13				0.8	0.8	8
6-APA	[900]		0.2							
Cephalosporin C	50 000 (20 °C)	6×10^{-2}	1100	5						200
Cephalothin	[>70 000]	6×10^{-2}	[3000]							4
Cephaloglycine	70 000 (20 °C)	*c.* 0.01	21							
Cephalexin	3000 (20 °C)	$< 10^{-3}$	[4]							
Nitrocefin 87/312	2 600 000 (10 °C)		460 (10 °C)							
References	Fuad *et al.*, 1976	Frère *et al.*, 1978*b*	Frère *et al.*, 1975*a*	Dusart *et al.*, 1977				Coyette *et al.*, 1978		Umbreit and Strominger, 1973

Values are given in $M^{-1}\ s^{-1}$ at 37 °C unless otherwise stated in parentheses.
†Values given in square brackets were calculated from the ID_{50} values given by Dusart *et al.* (1973).
‡Solub = after solubilisation with a detergent;
§Similar values were obtained with intact membranes except for cephalosporin C for which the membrane value was 10 $M^{-1}\ s^{-1}$;
After solubilization with a detergent, the less sensitive D,D-carboxypeptidase from *Proteus mirabilis* reacts with penicillin G with a k_3/K of 30 000–80 000 $M^{-1}\ s^{-1}$ (Schilf *et al.*, 1978).

major importance of k_3 in determining the inhibitory power of an antibiotic. With the possible exception of the D,D-carboxypeptidase from *Streptococcus faecalis*, antibiotics with a good inhibitory power have high k_3 values; an unfavourable, high K value is never, *per se*, an obstacle to good activity. For instance, benzylpenicillin is a very good inhibitor of the R61 enzyme, although the K value is very high. Conversely, the resistance of the *albus* G enzyme to cephalosporins is mainly due to very low k_3 values, while some K values are no worse than in the cases of good interactions. The value of K accounts for a certain degree of 'fine tuning' of the activity, but it generally remains secondary. The k_4 values appear to have a rather limited importance: increasing k_4 will increase the level of enzyme still active after the system has reached the steady state. With values of k_4 below 1×10^{-4} s^{-1}, the third step will increase the steady-state level of free enzyme (E) to a substantial proportion of total enzyme only if [I] is sufficiently low so that

$$k_a = \frac{k_3\,[\mathrm{I}]}{K} \approx k_4$$

Under these conditions, the time required to reach the steady state will be so long (several hours) that the antibiotic will be inefficient anyway. With values above 1×10^{-4} s^{-1}, the third step will be more important but, if one excludes the 0.3 s^{-1} value obtained for the D,D-carboxypeptidase from *Staphylococcus aureus*, the values of k_4 do not exceed the threshold value of 1×10^{-4} s^{-1} by a factor of more than 30, which is small compared with variations of k_3 by four or six orders of magnitude. Although the high k_4 value certainly confers a much higher resistance to the D,D-carboxypeptidase from *Staphylococcus aureus*, the important factor governing the interaction with good inhibitors will often be the k_3/K ratio which determines the rate of immobilisation of the enzyme in the EI* complex.

The structure of this complex has not been directly established but a large amount of evidence has been accumulated, indicating the penicilloylation of the hydroxyl group of a serine residue:

(1) As shown by deuterium fixation during degradation of the R61 enzyme-penicillin V complex, the C_5-C_6 bond of penicillins is intact in the EI* complex (Frère *et al.*, 1978*a*).

(2) Binding to the R61 and R39† enzymes and hydrolysis by a β-lactamase alter the ultraviolet spectra of cephalosporin C, cephaloglycine and cephalexin in a similar manner (Fuad *et al.*, 1976).

(3) The visible spectra of nitrocefin after binding to the R39 enzyme or hydrolysis by a β-lactamase are identical; binding to the R61 enzyme yields a somewhat different spectrum, but the product released upon reactivation of the enzyme is indistinguishable from β-lactamase-opened nitrocefin (Frère *et al.*, 1974*a, b*).

(4) The nuclear magnetic resonance spectrum of the denatured benzylpenicillin-R61 enzyme complex presents a proton resonance characteristic of α-methylpenicilloate (Degelaen *et al.*, 1979).

†R39 enzyme is the exocellular D,D-carboxypeptidase-transpeptidase from *Actinomadura* R39.

(5) After denaturation and digestion of a R61 enzyme-[^{14}C]-benzylpenicillin complex by pronase, the radioactive label is found on a tripeptide; the sequence of the tripeptide is H-Val-Gly-Ser-OH and the penicilloyl moiety can only be attached to the hydroxyl group of the Ser residue (Frère *et al.*, 1976*b*). The label is spontaneously but slowly released from the peptide as penicilloic acid (C. Duez, unpublished results). A serine residue is also involved in the binding of penicillin to the R39 enzyme (C. Duez, unpublished results) and the membrane-bound D,D-carboxypeptidase from *B. subtilis* (Georgopapadakou *et al.*, 1977).

Although a migration of the penicillin molecule during the denaturation process still remains a remote possibility (a comparable migration was observed by Virden *et al.* (1978) with quinacillin and the β-lactamase from *S. aureus*), it seems reasonable to assume that the structure of the EI* complex is as shown in figure 12.4.

It should also be noted that, with the D,D-carboxypeptidase from *S. albus* G, the interaction is so poor that a non-specific acylation similar to that which occurs with serum proteins remains quite possible.

In consequence, since penicillins seem to inactivate sensitive enzymes by acylating the hydroxyl group of a serine residue, one might be tempted to try to correlate the acylating power (as determined above) with the k_3 value for one given enzyme. It is evident that no such correlation exists. If it were true, the k_3 values for the antibiotics listed in table 12.1 should not, with a given enzyme, differ by a factor larger than 10: the ratio of the k_3 values for the reactions of the R61 enzyme with benzylpenicillin and cephaloglycine is 2×10^4! In the penicillin group, the only structure on which the enzyme might have a specific action is the acyl side-chain. Apparently, this side-chain is used as a handle by the enzyme to destabilise the β-lactam amide bond and increase its acylating properties. The interaction between the side-chain and the enzyme site must be extremely specific, since large variations of k_3 are observed for relatively small structural modifications.

Another possible explanation of these observations is that the nature of the side-chain influences the exact localisation of the C=O of the β-lactam in the complex, thus determining the efficiency of the acylation reaction for purely steric reasons. Both hypotheses could also partially contribute in establishing the value of k_3.

COMPARISON BETWEEN β-LACTAMASES AND PENICILLIN-SENSITIVE ENZYMES

Tables 12.1–12.5 supply elements for this comparison; assuming a close relationship between the two types of enzymes, one can formulate two alternative hypotheses:

(1) β-Lactamases have somehow replaced the serine hydroxyl group by a water molecule or a hydroxyl ion. Relative hydrolysis velocities or relative physiological efficiencies (V_M/K_m) for β-lactamases should be compared respectively to the k_3 or k_3/K values for the PSEs.

(2) β-Lactamases have developed a very efficient third step (high k_4) in the interaction pathway. Relative hydrolysis velocities for β-lactamases should be compared to the k_4 values for PSEs.

Table 12.5 Relative values for constants involved in the interaction of β-lactams with PSEs and β-lactamases

	Actinomadura R39				*Streptomyces* K11–R61			*Streptomyces albus* G			
	β-Lactamase		D,D-Cbase-Tase		β-Lactamase K11	D,D-Cbase-Tase (R61)		β-Lactamase		D,D-Cbase	
	V_M	Physiological efficiency	k_3/K	k_4	Relative velocity	k_3/K	k_4	V_M	Physiological efficiency	k_3/K	k_4
Penicillin G	100	100	100	100	100	100	100	100	100	$<$ 40	–
Penicillin V	220	52	$\geqslant$ 25	–	116	6	200	80	95	100	100
Carbenicillin	33	11	2	200	27	0.4	100	10	7	$<$ 40	–
Ampicillin	520	59	90	150	82	3.0	100	210	170	$<$ 20	–
Methicillin	28	6	1.3	950	0	0.1	–	1.5	1.4	$<$ 40	–
Oxacillin	320	40	13	–	0	0.5	–	19	41	–	–
Cloxacillin	30	5	5	–	0	0.2	–	2	5	$<$ 40	–
6-Aminopenicillanic acid	80	70	0.3	–	111	$<$ 0.01	–	75	93	–	–
Cephalosporin C	6	0.8	67	10	5	5	0.7	6	1.3	1200	90
Cephalothin	55	30	$\geqslant$ 25	–	116	13	–	2.4	1.4	1200	36
Cephaloglycine	215	4	90	29	88	0.1	2.1	7.5	1.4	*c.* 200 $<$	–
Cephalexin	17	1.3	4	85	0	0.02	–	1.2	0.2	$<$ 20	–
Nitrocefin	230	–	6000	53	–	16	220	23	–	–	–

The values for penicillin G were arbitrarily set to 100 in all cases except for the D,D-carboxypeptidase from *Streptomyces albus* G for which penicillin V was chosen as a reference. Inclusion of this latter enzyme in the table is somewhat arbitrary since it is not, strictly speaking, a PSE. Since *Streptomyces* R61 excretes only very minute amounts of β-lactamase, values of relative velocities of hydrolysis (with $[S] = 2 \times 10^{-4}$ M) are given for the β-lactamase of a closely related strain, *Streptomyces* K11. It should be noted that the sensitivities of the D,D-carboxypeptidase-transpeptidase from this latter strain to penicillins closely parallels those of the D,D-carboxypeptidase-transpeptidase from *Streptomyces* R61 (Dusart *et al.*, 1973). The values for the PSEs were computed from those of table 12.4 and, when k_3/K values were determined at a lower temperature, they were normalised to 37 °C assuming a Q_{10} of 2 (which is close to the average Q_{10} value measured for the interaction between the exocellular D,D-carboxypeptidase-transpeptidase from *Streptomyces* R39 and carbenicillin, nitrocefin and penicillin G).

Particularly interesting are the data presented in table 12.5, since they compare results obtained for PSEs and β-lactamases from the same or closely related strains of bacteria. For instance, if one ignores 6-aminopenicillanic acid, cephalosporin C and cephaloglycine, the correlation between physiological efficiencies of β-lactamases and k_3/K values for PSEs is rather good for enzymes of *Actinomadura* strain R39. In a very general way, methicillin, cloxacillin and oxacillin are relatively poor substrates for β-lactamases and poor inhibitors of PSEs. But it is also clear that there is no close relationship between kinetic parameters of the two types of enzymes. Alternatively, the second hypothesis would be in good agreement with the observed covalent binding of penicillins to a serine residue of some PSEs and some β-lactamases; but no correlation appears between k_4 values for PSEs and the activity of the various β-lactamases. This is hardly surprising for PSEs which fragment penicillin, since values of k_4 do not appear to be determined by release of the acylglycyl moiety but by breaking of the C_5-C_6 bond. This latter observation makes a strict comparison very hazardous for strains R39 and R61-K11. Furthermore, peptide substrates for the exocellular D,D-carboxypeptidase-transpeptidases from *Actinomadura* R39 and *S. albus* G were assayed as substrates or inhibitors of the corresponding β-lactamases; no interaction was observed (Johnson *et al.*, 1973). Nonetheless, the β-lactamases from these two strains are under current study in our laboratory, although difficult problems are being encountered in purifying sufficient quantities of these latter enzymes: 500-litre cultures usually yield 2 or 3 mg of enzyme purified to protein homogeneity. So far, no structural relationship has been observed between the β-lactamase and the D,D-carboxypeptidase from the same strain (C. Duez and J. M. Frère, unpublished results). Recently, Ohya *et al.* (1979) obtained antisera against β-lactamases from *Proteus morganii* and *P. rettgeri*. These antisera had no effect on the binding of penicillin by the PBPs of the same strains or of *P. mirabilis*. In this latter strain and in *P. vulgaris*, PBPs 4, 5 and 6 are known to be D,D-carboxypeptidases (W. Schilf and H. Martin, personal communication; A. Rousset, personal communication). Since the PBPs patterns in these four *Proteus* species are very similar, it is likely that no immunological relationship exists here between the D,D-carboxypeptidases and the β-lactamases.

STRUCTURAL ANALOGY OR NO STRUCTURAL ANALOGY?

The original hypothesis of Tipper and Strominger (1965) implied a structural analogy between the β-lactam antibiotic and the D-alanyl-D-alanine C-terminal of donor substrates of D,D-carboxypeptidases and transpeptidases. Formation of an acyl-enzyme was proposed which, in the case of the substrate, would in turn react with the acceptor. However, acylation by β-lactams was assumed to be irreversible, an hypothesis which is not completely true as we have seen above. Virudachalam and Rao (1977) have calculated that the conformations of penicillins and cephalosporins fall within the range of allowed conformations for X-D-alanyl-D-alanine peptide, *if one first assumes the nitrogen of the peptide to be pyramidal as in penicillins or cephalosporins*. A second assumption was that 'the activity of the β-lactam antibiotics was not particularly dependent on the nature of the substituent beyond the carbonyl group of the amino acyl side-chain'. The analysis given above clearly disproves this latter assumption. The former assumption can be accepted by assuming that β-lactam antibiotics are analogues of the transition state reached by the peptide in the active site of the enzyme (Lee, 1971). In fact, Boyd *et al.* (1975) have calculated that the substrate–enzyme interaction must supply 8.5 or 3 kcal to strain the peptide bond into a conformation similar to that of the β-lactam amide of penicillins and cephalosporins, respectively. Boyd (1977) has also computed that the tetrahedral adducts formed during nucleophilic attack by a hydroxyl ion on the carbonyl carbon of a R-Gly-Gly peptide are very similar to the antibiotic structure, if one assumes the attack to take place on the α-face of the peptide. There are two objections to this attractive hypothesis:

(1) Transition-state analogues are known to exhibit a particularly high affinity towards their target enzymes: the K_is of these inhibitors are lower than the K_is of true analogues of the substrate in the ground state (Jencks, 1969). As was seen above, this is obviously not the case with β-lactam antibiotics.

(2) The β-lactam antibiotic, upon binding to the enzyme, must itself be activated into a transition state to become a powerful acylating agent. This transition state would be the one which might be common to antibiotic and substrate, but this would request the enzyme to supply more free energy to strain the amide bond further.

Thus the hypothesis must be further adapted as follows: in reaching the enzyme–substrate transition state, the enzyme also undergoes a conformation change. Recognition of the antibiotic is not excellent because, although it presents a structure close to the correct one, the enzyme does not; the conformation of the enzyme then switches to the 'transition-state' structure but, in doing so, must further strain the antibiotic molecule because the analogy remains imperfect. The two phenomena could be represented as follows, assuming that $[I] \simeq [S^*]$:

$$E + S \rightleftharpoons ES \rightleftharpoons E^*S^* \rightarrow E + P$$

$$E + I \underset{}{\overset{K_1}{\rightleftharpoons}} EI \underset{}{\overset{K_2}{\rightleftharpoons}} E^*I^* \xrightarrow{k'} (E\text{–}I)^*$$

This latter scheme would be in good agreement with Rando's (1975) proposal that β-lactam antibiotics act as 'suicide' substrates; after being activated by the enzyme, these compounds rapidly react with some essential group. The data which have been discussed above appear to support this hypothesis.

It is also fundamental to decide whether substrate and inhibitor bind to the same enzyme site. The catalytic pathway of transpeptidases and D,D-carboxypeptidases could point to the acylation of a serine residue accompanied by the release of the C-terminal D-Ala, followed by the transfer of the R-D-Ala moiety to the nucleophilic acceptor (water or $R'NH_2$, figure 12.2). With penicillins, the thiazolidine moiety of the molecule would occupy the acceptor site and seriously impair the second part of the process.

Kinetic data have not provided convincing evidence. Competitive kinetics were obtained in several cases (Izaki and Strominger, 1968; Yocum *et al.*, 1974; Umbreit and Strominger, 1973; Leyh-Bouille *et al.*, 1971, 1972) but it was later demonstrated that the results had not been obtained under steady-state conditions and that the transient formation of a ternary enzyme–inhibitor–substrate complex could not be excluded (Frère *et al.*, 1975*c*). In contrast, with the D,D-carboxypeptidase from *Proteus mirabilis* L-form, reaction velocities could be measured after the steady state had been established in the interaction with the antibiotic (Schilf *et al.*, 1978). The results were indicative of non-competitive behaviour.

When the degradation of enzyme–penicillin complexes was studied, it appeared that the rate-limiting step of the phenomenon preceded the release of the thiazolidine moiety of the antibiotic (Frère *et al.*, 1978*a*); the acylglycyl-enzyme was then rapidly attacked by nucleophilic agents (water or other acceptors). The following experimental results suggested that the acceptor site was really occupied by the thiazolidine ring:

(1) The intact penicilloyl moiety was never transferred to compounds behaving as good acceptors in the transpeptidation reaction (Marquet *et al.*, 1979).
(2) After the thiazolidine part of the inhibitor had been released, the acylglycyl moiety could be transferred to an aminated acceptor (Marquet *et al.*, 1979).

However, the specificity profile for nucleophilic acceptors in this transfer did not closely parallel the profile for the same compounds in the transpeptidation reactions. Moreover, a good acceptor, D-Ala, had no effect on the formation of EI* complexes (Fuad *et al.*, 1976), and steady state studies of the concomitant transpeptidation and hydrolysis reactions indicated that the transpeptidation pathway followed an ordered mechanism in which an enzyme–acceptor complex was formed first (Frère *et al.*, 1973).

With some D,D-carboxypeptidases, the acyl-D-Ala moiety of the donor can be transferred to some simple nucleophiles, mainly hydroxylamine and glycine. The results obtained upon studying the partitioning of the donor substrate between water and the other nucleophile indicated that, in these cases, the donor was first bound to the enzyme (and not, as stated by the authors (Nishino *et al.*, 1977), that acylation of the enzyme was taking place). Complexes were also obtained between several D,D-carboxypeptidases and peptide or depsipeptide substrates which were ^{14}C-labelled in the R-D-Ala moiety. The fact that the radioactivity remained linked to the protein after denaturation was, in this case, a clear indication of the formation of a covalent bond (Rasmussen and Strominger, 1978; Kozarich and Strominger, 1978). One should note, however, that these enzymes were rather inefficient D,D-carboxypeptidases (specific activity lower than 1 IU mg^{-1} of protein).

Similar experiments were performed (B. Joris, personal communication) with the D,D-carboxypeptidases-transpeptidases from *Streptomyces* R61 and *Actinomadura* R39, which behave as true and efficient transpeptidases in the presence of acceptors the structures of which closely resemble those of the natural cell-wall peptides. Only a very low labelling of the proteins was recorded which in the case of the R61 enzyme completely disappeared in the presence of glycylglycine, a good acceptor in the transpeptidation reaction. Moreover, a very low labelling was also obtained when the enzymes were replaced by bovine serum albumin at a similar concentration.

To reconcile these conflicting data, one can assume that 'efficient' transpeptidases and 'inefficient' carboxypeptidases do not use the same reaction pathway. In the first case, the R-D-Ala moiety of the donor would be transferred directly to the amino group of the acceptor previously bound to its own site; in the second case, in the absence of an acceptor site, the substrate would acylate the enzyme, the C-terminal D-alanine would be released and only then would a nucleophile be able to enter the catalytic site to attack the acyl-enzyme. This, however, does not answer the basic question about the relationship between the penicillin and the substrate binding sites. Only X-ray crystallography is finally capable of solving this problem. Crystals suitable for structure determination have now been obtained for two enzymes: the penicillin-sensitive D,D-carboxypeptidase-transpeptidase from *Streptomyces* R61 (Knox *et al.*, 1979) and the penicillin-resistant D,D-carboxypeptidase from *Streptomyces albus* G (Dideberg *et al.*, 1979) (Figure 12.5).

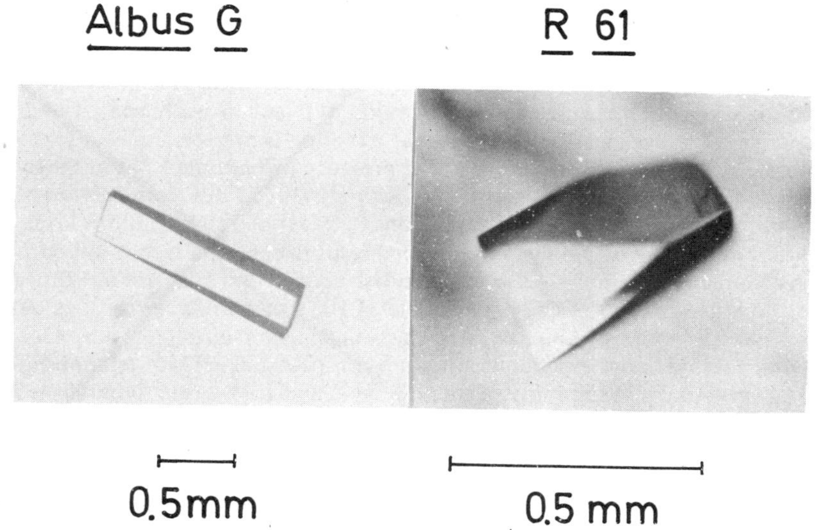

Figure 12.5 Crystals of the D,D-carboxypeptidase of *Streptomyces albus* G and of the D,D-carboxypeptidase-transpeptidase of *Streptomyces* R61.

The properties of this latter enzyme are rather puzzling: it binds and hydrolyses peptide substrates quite well but does not recognise most penicillins or cephalosporins; binding does occur in very few cases, but the enzyme then seems unable to activate the β-lactam. In these cases, moreover, the reversible binding is clearly non-competitive with the binding of the substrate. This behaviour thus appears to be in conflict with the hypothesis discussed above (Leyh-Bouille *et al.*, 1970; Frère *et al.*, 1978*b*). Recently, the presence of one tightly bound Zn^{2+} ion per molecule of enzyme has been discovered (O. Dideberg; unpublished results). It is thus possible that the *Streptomyces albus* G D,D-carboxypeptidase might be a Zn^{2+} enzyme, utilising a catalytic pathway very different from that of the R61 or R39 D,D-carboxypeptidase-transpeptidases.

The theoretical studies which are presently available suffer greatly from the simplifying assumptions which have been made: the side-chain R (figure 12.1) of penicillins was replaced by a methyl group (Virudachalam and Rao, 1977), the D-Ala-D-Ala-OH C-terminus of the peptide substrate was replaced by -Gly-Gly-OH

(Boyd, 1977) and the nature of the residue preceding the penultimate D-Ala was never considered. However, it has been clearly demonstrated that the structure of this residue is of vital importance in determining whether a peptide could be a substrate for the D,D-carboxypeptidases and transpeptidases (Ghuysen *et al.*, 1972). For instance, the peptide acetyl-D-Ala-D-Ala-OH is neither a good substrate nor a good inhibitor of various enzymes (Nieto *et al.*, 1973); similarly, the replacement of only one of the D-Ala residues in the standard substrate Ac_2-L-Lys-D-Ala-D-Ala by a glycine residue drastically reduces the efficiency of the *Streptomyces* and *Actinomadura* exocellular enzymes. (For complete reviews of the substrate specificity problem, see Ghuysen *et al.*, 1972 and Ghuysen *et al.*, 1979.)

CONCLUSIONS

Fourteen years ago, the hypothesis of Tipper and Strominger (1965) seemingly solved the problem of the mechanism of action of β-lactam antibiotics. The discussion presented above clearly shows that the situation is not so simple. A further degree of complexity is introduced by the presence of multiple PBPs in the cytoplasmic membrane of bacteria. Among these, only three PBPs from *Salmonella typhimurium* (Shepherd *et al.*, 1977) and one from various strains of *Streptomyces* (J. Dusart and M. Leyh-Bouille, unpublished results) exhibit a true transpeptidase activity. Some exhibit a D,D-carboxypeptidase activity but they are not thought to be the killing targets. The enzymic role of the other PBPs remains, as yet, undetermined. Moreover, the new, non-classical β-lactam antibiotics have recently shattered accepted concepts about structure–antibacterial activity relationships in the β-lactam family. Preliminary results (M. Noël and J. M. Frère, unpublished results) indicate these substances to be poor inhibitors of the exocellular D,D-carboxypeptidase-transpeptidases from *Streptomyces* R61 and *Actinomadura* R 39. Interestingly, one of the branches of the interaction between the *E. coli* and *S. aureus* β-lactamases and clavulanic acid could be very similar to the three-step model described above for transpeptidases and D,D-carboxypeptidases.

Many previous discussions have tried to establish correlations between structure and activity of β-lactam antibiotics. A factor which has certainly been neglected so far is the crucial importance of the target enzyme itself. This point can be stressed by comparing the values of the parameters k_3/K and k_4 for the interaction of one enzyme with several antibiotics: there is certainly no correlation between the two values. If one changes enzymes, prediction of even the relative values of these parameters clearly appears impossible. Considering the values of k_4 which have so far been measured, one can only hope that pathogenic bacteria will not develop transpeptidases with higher k_4 values!

A crude parallelism can be established between the activity of β-lactam antibiotics as substrates of β-lactamases and inhibitors of PSEs. It seems reasonable to assume that, in both cases, the nucleophilic attack on the β-lactam amide bond happens from the α-face of the ring. In consequence, bulky substituents on this α-face appear to decrease both inhibitory activity and susceptibility to β-lactamases. However, one should not be too optimistic. For instance, many β-lactamases appear to make a very clear distinction between the penam and the 3-cephem nuclei. Most PSEs do not. There is only one good example of a clear correlation between chemical stability, susceptibility to β-lactamase and inhibitory power against PSEs: cephalexin is chemically more stable, more resistant to most β-lactamases and less active against PSEs than cephaloglycine, differences which can probably be attributed to the absence of a leaving group on the substituent on C_3 of cephalexin. From a purely therapeutic point of view, one can however only be happy that such a good correlation remains an exception.

Note added in proof: While this article was in press, Yocum, Waxman, Rasmussen and Strominger (*Proc. natn. Acad. Sci. U.S.A.*, **76**, 2730–4, 1979) obtained results indicating that substrate and penicillin could bind to the same serine residue of the D,D-carboxypeptidases from *B. stearothermophilus* and *B. subtilis*. Moreover, these authors have found significant sequence homology around the active site serine of these enzymes and that of the β-lactamases of known sequence. These results shed new light on some of the problems left unsolved in the present paper. However, many important questions still remain unanswered.

REFERENCES

Abbott, B. J. and Fukuda, D. S. (1975). *Meth. Enzymol.*, **43**, 731–4

Abraham, E. P. and Fawcett, P. (1975). *Meth. Enzymol.*, **43**, 728–31

Adriaens, P., Meesschaert, B., Frère, J. M., Vanderhaeghe, H., Degelaen, J., Ghuysen, J. M. and Eyssen, H. (1978). *J. biol. Chem.*, **253**, 3660–5

Ambler, R. P. (1975). *Biochem. J.*, **151**, 197–218

Ambler, R. P. and Scott, G. K. (1978). *Proc. natn. Acad. Sci. U.S.A.*, **75**, 3732–6

Aoki, H., Sakai, H., Kohsaka, M., Konomi, T., Hosada, J., Kubachi, Y. and Iguchi, E. (1976). *J. Antibiot.*, **29**, 492–500

Aschaffenburg, R., Phillips, D. C., Sutton, B., Baldwyn, G., Kiener, P. A. and Waley, S. G. (1978). *J. molec. Biol.*, **120**, 447–9

Blumberg, P. M., Yocum, R. R., Willoughby, E. and Strominger, J. L. (1974). *J. biol. Chem.*, **249**, 6828–35

Boyd, D. B. (1973). *J. med. Chem.*, **16**, 1195–9

Boyd, D. B. (1977). *Proc. natn. Acad. Sci. U.S.A.*, **74**, 5239–43

Boyd, D. B., Hermann, R. B., Presti, D. E. and Marsh, M. M. (1975). *J. med. Chem.*, **18**, 408–17

Brown, A. G., Butterworth, D., Cole, M., Hanscomb, G., Hood, J. D., Reading, C. and Rolinson, G. N. (1976). *J. Antibiot.*, **29**, 668–9

Cartwright, S. J. and Coulson, A. F. W. (1979). *Nature, Lond.*, **278**, 360–1

Charnas, R. L., Fisher, J. and Knowles, J. R. (1978). *Biochemistry*, **17**, 2185–9

Citri, N. (1971). In *The Enzymes*, 3rd ed., Vol. 4, (ed. P. D. Boyer), Academic Press, New York, pp. 23–41

Citri, N. (1973). *Adv. Enzymol.*, **37**, 397–648

Citri, N., Samuni, A. and Zyk, N. (1976). *Proc. natn. Acad. Sci. U.S.A.*, **73**, 1048–52

Corbeth, D. F., Eglington, J. and Howarth, T. T. (1977). *J. chem. Soc.*, **24**, 953–4

Coyette, J., Ghuysen, J. M. and Fontana, R. (1978). *Eur. J. Biochem.*, **88**, 297–305

Csanyi, V., Mile, I., Koczka, I., Badar, E. and Horvath, I. (1970). *Biochim. biophys. Acta*, **220**, 317–24

Davies, R. D., Abraham, E. P. and Melling, J. (1974). *Biochem. J.*, **143**, 115–27

Degelaen, J., Feenay, J., Roberts, G. C. K., Frère, J. M. and Ghuysen, J. M. (1979). *Fedn Eur. biochem. Socs Lett.*, **98**, 53–7

Dideberg, O., Frère, J. M. and Ghuysen, J. M. (1979). *J. molec. Biol.*, **129**, 677–99

Dusart, J., Marquet, A., Ghuysen, J. M., Frère, J. M., Moreno, R., Leyh-Bouille, M., Johnson, K., Lucchi, C., Perkins, H. R. and Nieto, M. (1973). *Antimicrobial Agents Chemother.*, **3**, 181–7

Dusart, J., Leyh-Bouille, M. and Ghuysen, J. M. (1977). *Eur. J. Biochem.*, **81**, 33–44

English, A. R., Retsema, J. A., Geraid, A. E., Lynch, J. E. and Barth, W. E. (1978). *Antimicrobial Agents Chemother.*, **14**, 414–9

Fisher, J., Charnas, R. L., Knowles, J. R. (1978). *Biochemistry*, **17**, 2180–4

Frère, J. M., Geurts, F. and Ghuysen, J. M. (1978*b*). *Biochem. J.*, **175**, 801–5

Frère, J. M., Ghuysen, J. M. and De Graeve, J. (1978*a*). *Fedn Eur. biochem. Socs Lett.*, **88**, 147–50

Frère, J. M., Ghuysen, J. M. and Iwatsubo, M. (1975*a*). *Eur. J. Biochem.*, **57**, 343–51

Frère, J. M., Ghuysen, J. M. and Perkins, H. R. (1975*c*). *Eur. J. Biochem.*, **57**, 353–9

Frère, J. M., Duez, C., Ghuysen, J. M. and Vandekerckhove, J. (1976*b*). *Fedn Eur. biochem. Socs Lett.*, **70**, 257–60

Frère, J. M., Ghuysen, J. M., Perkins, H. R. and Nieto, M. (1973). *Biochem. J.*, **135**, 483–92

Frère, J. M., Leyh-Bouille, M., Ghuysen, J. M. and Perkins, H. R. (1974*a*). *Eur. J. Biochem.*, **50**, 203–14

Frère, J. M., Ghuysen, J. M., Degelaen, J., Loffet, A. and Perkins, H. R. (1975*b*). *Nature, Lond.*, **258**, 168–70

Frère, J. M., Ghuysen, J. M., Reynolds, P. E., Moreno, R. and Perkins, H. R. (1974*b*). *Biochem. J.*, **143**, 241–9

Frère, J. M., Ghuysen, J. M., Vanderhaeghe, H., Adriaens, P., Degelaen, J. and De Graeve, J. (1976*a*). *Nature, Lond.*, **260**, 451–4

Frieden, C. (1970). *J. biol. Chem.*, **245**, 5788–99

Fuad, N., Frère, J. M., Ghuysen, J. M., Duez, C. and Iwatsubo, M (1976). *Biochem. J.*, **155**, 623–9

Furth, A. J. (1975). *Biochim. biophys. Acta,* **377**, 431–43

Georgopapadakou, N., Hammarström, S. and Strominger, J. L. (1977). *Proc. natn. Acad. Sci. U.S.A.*, **74**, 1009–12

Ghuysen, J. M., Reynolds, P. E., Perkins, H. R., Frère, J. M. and Moreno, R. (1974). *Biochemistry,* **13**, 2539–47

Ghuysen, J. M., Frère, J. M., Leyh-Bouille, M., Coyette, J., Dusart, J. and Nguyen-Distèche, M. (1979). *A. Rev. Biochem.*, **48**, 73–101

Ghuysen, J. M., Leyh-Bouille, M., Frère, J. M., Dusart, J., Johnson, K., Nakel, M., Coyette, J., Perkins, H. R. and Nieto, M. (1972). In *Molecular Mechanisms of Antibiotic Action on Protein Biosynthesis and Membranes*, Elsevier, Amsterdam, pp. 406–26

Gorman, M. and Ryan, C. W. (1972). In *Cephalosporins and Penicillins: Chemistry and Biology* (ed. E. H. Flynn), Academic Press, New York, pp. 533–82

Gunda, T. E. and Jászberényi, J. C. (1977). *Progr. med. Chem.*, **14**, 181–248.

Hammarström, S. and Strominger, J. L. (1975). *Proc. natn. Acad. Sci. U.S.A.*, **72**, 3463–7

Hammarström, S. and Strominger, J. L. (1976). *J. biol. Chem.*, **251**, 7947–9

Ho, P., Towner, R., Indelicato, J., Wilham, W., Spitzer, W. and Koppel, G. (1973). *J. Antibiot.*, **26**, 313–4

Izaki, K. and Strominger, J. L. (1968). *J. biol. Chem.*, **243**, 3193–201

Jászberényi, J. C. and Gunda, T. E. (1975). *Progr. med. Chem.*, **12**, 395–477

Jencks, W. P. (1969). *Catalysis in Chemistry and Enzymology*, McGraw-Hill, New York

Johnson, K. J., Dusart, J., Campbell, J. N. and Ghuysen, J. M. (1973). *Antimicrobial Agents Chemother.*, **3**, 289–98

Kiener, P. A. and Waley, S. G. (1977). *Biochem. J.*, **165**, 279–85

Knott-Hunziker, V., Orlek, B. S., Sammes, P. G. and Waley, S. G. (1979*a*). *Biochem. J.*, **177**, 365–7

Knott-Hunziker, V., Waley, S. G., Orlek, B. S. and Sammes, P. G. (1979*b*). *Fedn Eur. biochem. Socs Lett.*, **99**, 59–61

Knox, J. R., Kelly, J. A., Moews, P. C., Murthy, N. S. (1976). *J. molec. Biol.*, **104**, 865–75

Knox, J. R., DeLucia, M. L., Murthy, N. S., Kelly, J. A., Moews, P. C., Frère, J. M. and Ghuysen, J. M. (1979). *J. molec. Biol.*, **127**, 217–8

Kozarich, J. W. and Strominger, J. L. (1978). *J. biol. Chem.*, **253**, 1272–8

Kuwabara, S. and Abraham, E. P. (1967). *Biochem. J.*, **103**, 27c–29c

Lee, B. (1971). *J. molec. Biol.*, **61**, 463–9

Leyh-Bouille, M., Coyette, J., Ghuysen, J. M., Idczak, J., Perkins, H. R. and Nieto, M. (1971). *Biochemistry,* **10**, 2163–70

Leyh-Bouille, M., Ghuysen, J. M., Bonaly, R., Nieto, M., Perkins, H. R., Schleifer, K. H. and Kandler, O. (1970). *Biochemistry,* **9**, 2961–70

Leyh-Bouille, M., Nakel, M., Frère, J. M., Johnson, K., Ghuysen, J. M., Nieto, M. and Perkins, H. R. (1972). *Biochemistry,* **11**, 1290–7

Maeda, K., Takahashi, S., Sezaki, M., Iinuma, K., Nagawana, H., Kondo, S., Ohno, M. and Umezawa, H. (1977). *J. Antibiot.*, **30**, 770–2

Marquet, A., Frère, J. M. and Ghuysen, J. M. (1979). *Biochem. J.*, in press

Meadway, R. J. (1969). PhD thesis, University of Edinburgh

Nayler, J. H. C. (1971). *Proc. R. Soc. B*, **179**, 357–67

Nieto, M., Perkins, H. R., Leyh-Bouille, M., Frère, J. M. and Ghuysen, J. M. (1973). *Biochem. J.*, **131**, 163–71

Nishino, T., Kozarich, J. W. and Strominger, J. L. (1977). *J. biol. Chem.*, **252**, 2934–9

O'Callaghan, C. H., Muggleton, P. and Ross, G. (1968). In *Antimicrobial Agents and Chemotherapy* (ed. G. Hobby), American Society for Microbiology, Washington, DC, pp. 57–63

Ohya, S., Yamazaki, M., Sugarawa, S. and Matsuhashi, M. (1979). *J. Bact.*, **137**, 474–9

Ozer, J. H. and Saz, A. K. (1970). *J. Bact.*, **102**, 64–71

Rando, R. R. (1975). *Biochem. Pharmac.*, **24**, 1153–60

Rasmussen, J. R. and Strominger, J. L. (1978). *Proc. natn. Acad. Sci. U.S.A.*, **75**, 84–8

Richmond, M. H. and Sykes, R. B. (1973). *Adv. microbiol. Physiol.*, **9**, 31–88

Richmond, M. H. (1975). *Meth. Enzymol.*, **43**, 664–72

Sagai, H. and Sato, T. (1973). *J. Antibiot.*, **26**, 315–9

Samuni, A. and Citri, N. (1975). *Biochem. Biophys. Res. Commun.*, **62**, 7–11

Savidge, T. A. and Cole, M. (1975). *Meth. Enzymol.*, **43**, 705–21

Saz, A. K. (1970). *J. cellul. Physiol.*, **76**, 397–400

Schilf, W., Frère, P., Frère, J. M., Martin, H. H., Ghuysen, J. M., Adriaens, P. and Meesschaert, B. (1978). *Eur. J. Biochem.*, **85**, 325–30

Scott, K. G. (1973). *Biochem. Soc. Trans.*, **1**, 159–62

Shepherd, S. T., Chase, H. A. and Reynolds, P. E. (1977). *Eur. J. Biochem.*, **78**, 521–32

Spratt, B. G., Jobanputra, V. and Zimmermann, W. (1977). *Antimicrobial Agents Chemother.*, **12**, 406–9

Spratt, B. G. (1978). *Sci. Progr., Oxford*, **65**, 101–28

Strom, R., Ravagnan, G. and Salfi, V. (1976). *Eur. J. Biochem.*, **62**, 95–101

Sweet, R. M. (1972). In *Cephalosporins and Penicillins: Chemistry and Biology* (ed. E. H. Flynn), Academic Press, New York, pp. 280–309

Tamura, T., Imae, Y. and Strominger, J. L. (1976). *J. biol. Chem.*, **251**, 414–23

Thatcher, D. R. (1975*a*). *Meth. Enzymol.*, **43**, 640–52

Thatcher, D. R. (1975*b*). *Meth. Enzymol.*, **43**, 653–64

Thatcher, D. R. (1975*c*). *Biochem. J.*, **147**, 313–26

Tipper, D. J. and Strominger, J. L. (1965). *Proc. natn. Acad. Sci. U.S.A.*, **54**, 1133–41

Umbreit, J. N. and Strominger, J. L. (1973). *J. biol. Chem.*, **248**, 6767–71

Vanderhaeghe, H. (1975). *Meth. Enzymol.*, **43**, 721–8

Virden, R., Bristow, A. F. and Pain, R. (1978). *Biochem. Biophys. Res. Commun.*, **82**, 951–6

Virudachalam, R. and Rao, V. S. R. (1977). *Int. J. Peptide Protein Res.*, **10**, 51–9

Waley, S. G. (1974). *Biochem. J.*, **139**, 789–90

Yocum, R. R., Blumberg, P. M. and Strominger, J. L. (1974). *J. biol. Chem.*, **249**, 4863–71

Zeiger, A. R., Frère, J. M., Ghuysen, J. M. and Perkins, H. R. (1975). *Fedn Eur. biochem. Socs. Lett.*, **52**, 221–25

13
The inactivation of β-lactamase by mechanism-based reagents

Jed F. Fisher and Jeremy R. Knowles (Department of Chemistry, Harvard University, Cambridge, Massachusetts 02138, USA)

ABSTRACT

The analogies between the behaviour of clavulanic acid and of 6-*des*-aminopenicillin sulphone with the *E. coli* RTEM β-lactamase allow some tentative conclusions to be drawn about the routes for enzyme inactivation taken by these two compounds. These conclusions have led to the synthesis of the sulphones of methicillin and quinacillin, both of which are, as predicted, potent inactivators of the β-lactamase.

The β-lactamases are bacterial enzymes which catalyse the rapid hydrolysis of the β-lactam ring of penicillins and cephalosporins (Abraham, 1977; Fisher and Knowles, 1978). The hydrolytic activity of these enzymes, often coupled with a decreased membrane permeability towards the β-lactam (Sawai *et al.*, 1979; Scudamore *et al.*, 1979), eliminates the bacteriocidal action of many β-lactam antibiotics and makes the organism resistant to these molecules. For this reason, the β-lactamases have long been regarded as promising targets for the action of specific inhibitors which, by their action, would overcome the main defensive mechanism of resistant organisms. Until very recently, however, efforts in this direction were surprisingly unsuccessful.

The first portent of possible benefits from an approach based on β-lactamase inhibition was the discovery of the cephamycins (Nagarajan *et al.*, 1971). These cephalosporin derivatives have a 7α-methoxy substituent, which makes them quite resistant to β-lactamase hydrolysis, and gives them a wide action-spectrum as antibiotics. The first step from these poor enzyme substrates towards potent mechanism-based enzyme inactivators, was taken several years later with the isolation of clavulanic acid (**1**) from *Streptomyces clavigerus* (Howarth *et al.*, 1976). While

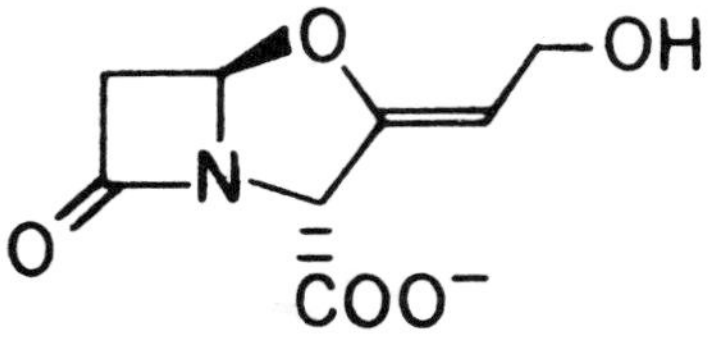

1

itself a poor antibiotic, clavulanic acid acts in synergy with other β-lactam antibiotics to lower dramatically the minimum inhibitory antibiotic concentration towards a large number of resistant bacteria (Reading and Cole, 1977; Dumon *et al.*, 1979). Clavulanic acid did not long remain unique in its ability to act synergistically by protecting the antibiotic from the action of the bacterial β-lactamase. The isolation of clavulanate was quickly followed by the discovery of the 5, 6-*trans*-carba-penem group of compounds (e.g. theinamycin, Kahan *et al.*, 1979; PS-5, Okamura *et al.*, 1978) and the 5, 6-*cis*-carba-penem family (e.g. olivanic acid, Brown *et al.*, 1977; Maeda *et al.*, 1977). It has not as yet been ascertained whether the carba-penem synergy results from enzyme inactivation or enzyme inhibition. In the past year we have also witnessed the appearance of the first truly effacious *synthetic* inactivators, 6β-bromo-*des*-aminopenicillanic acid (Pratt and Loosemore, 1978; Knot-Hunziker *et al.*, 1979) and 6-*des*-amino-penicillanic acid sulphone (English *et al.*, 1978). The field has been transformed from a desert, into a garden of delights for the mechanistic enzymeologist.

Our studies on the inactivation of the *Escherichia coli* RTEM β-lactamase by clavulanic acid exposed a chemical problem of unusual complexity (Charnas *et al.*, 1978*a,b*; Fisher *et al.*, 1978), and subsequent experiments on the interaction of culvanate with β-lactamases from other sources have not provided any simpler a picture (Labia and Peduzzi, 1978; Durkin and Viswanatha, 1978; Cartwright and Coulson, 1979; Reading and Hepburn, 1979). Amongst the unusual aspects of the inactivation of the RTEM enzyme by clavulanate are the heterogeneity of the inactivated protein (three bands of almost equal intensity are detected by isoelectric focusing) and the presence of a chromophore at 280 nm ($\epsilon \sim 20\,000\ \text{M}^{-1}\ \text{cm}^{-1}$) in one of these bands. This chromophore is derived from clavulanate (even though clavulanate itself is transparent at 280 nm) and seems likely to be due to a β-amino-acrylate structure (2). Such structures are not unknown in β-lactam chemistry. One of

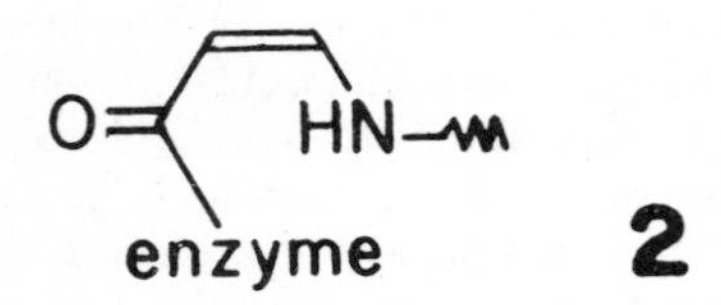

the products of the acid-catalysed decomposition of benzylpenicillin is benzylpenamaldic acid (Degelean *et al.*, 1979), and (in closer analogy to clavulanic acid) Golding and Hall (1973) have observed the fragmentation of **(3)** to **(4)** under very mild conditions. Simpler molecules containing the array of atoms in (2) are known to absorb at the observed wavelength maximum with high extinction coefficients (Ostercamp, 1970; Bell *et al.*, 1970). One of several explanations that could account for the development of this chromophore when clavulanate reacts with the enzyme, is that fragmentations occur from an acyl-enzyme intermediate (*vide infra*). Unfortunately, the existence of a catalytically competent acyl-β-lactamase has been demonstrated for only one substrate, cefoxitin, which is a sluggish 7α-methoxy-cephalosporin (J. Fisher, J. Belasco, S. Khosla, and J. R. Knowles, unpublished work). To propose that an acyl-enzyme forms during clavulanate hydrolysis and enzyme inactivation merely on the basis of the evident existence of an acyl-enzyme during cefoxitin hydrolysis was premature, and we therefore investigated the behaviour of a simpler β-lactam in order to clarify this problem.

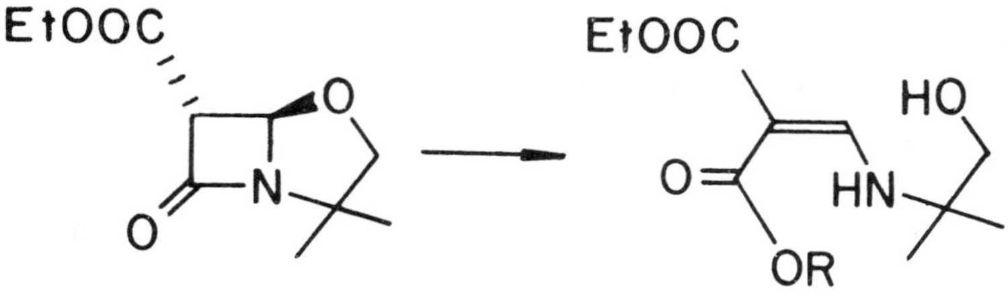

The penicillin sulphone CP-45, 899 (**5**) was recently reported by the Pfizer group (English *et al.*, 1978), and appears to duplicate the antagonistic action of clavulanate towards β-lactamases. This sulphone structurally resembles clavulanate in that the C-6 acylamino side-chain is missing and, as will become clear, there are close functional parallels with clavulanate in terms of its interaction with the enzyme. *In vitro* microbiological experiments have established the synergistic action of (**5**) with ampicillin and with cephalothin, against otherwise resistant bacterial strains (English *et al.*, 1978; Aswapokee and Neu, 1978). Indeed, the sulphone (**5**) and clavulanate possess very similar synergistic action-spectra (Neu and Fu, 1978; Fu and Neu, 1979). These observations led to our expectation that the basis for these similarities was the ability of (**5**) to act, like clavulanate, as a 'suicide' reagent towards the β-lactamase. We have tested this hypothesis, and report here our initial observations. These suggest that penicillin sulphones represent a general class of β-lactamase inhibitors and inactivators.

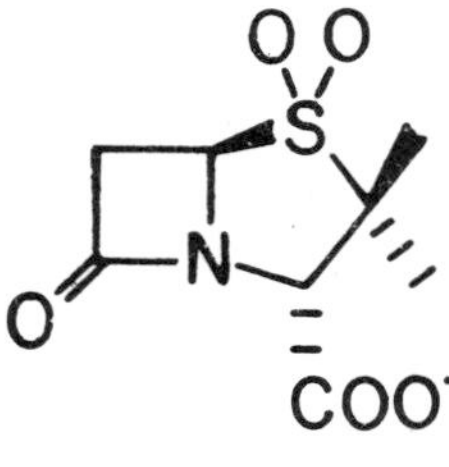

Examination of the sulphone (**5**) established that it is a potent competitive inhibitor ($K_i = 0.9$ μM) of the homogeneous RTEM β-lactamase. Although bound tightly, (**5**) is quite resistant to enzymatic hydrolysis ($k_{cat} = 1.2\ s^{-1}$: compare $k_{cat} = 2000\ s^{-1}$ for benzylpenicillin). Unlike clavulanate, there is no evidence for the formation of a transient complex between (**5**) and the enzyme, and the sulphone is only a lethargic inactivator. The half-time for irreversible inactivation of β-lactamase by (**5**) is 44 min, with an inactivating event occurring approximately every 4500 turnovers. This is much less effective than clavulanic acid, the inactivation half-time of which is 8 min, with 150 turnovers before inactivation. Even though the synergistic activity of (**5**) against *E. coli* RTEM is not as marked as with other resistant bacteria, it seems that the effect of **(5)** is—at least at early times—due to strong competitive inhibition rather than to irreversible inactivation of the enzyme (see also Fu and Neu, 1979).

An examination of the ultraviolet spectrum of the RTEM β-lactamase after irreversible inactivation by **(5)** showed an increase in the absorbance at 280 nm (beyond that of the protein itself), just as is found with inactivated enzyme derived from clavulanate. Since both clavulanate and **(5)** are transparent at this wavelength, we must conclude that both compounds achieve inactivation by a mechanism that involves the formation of a chromophoric fragment attached to the active site. The chromophore is present in about 35 per cent of the clavulanate-derived inactivated enzyme, as evidenced by the reaction of hydroxylamine with this material ($k = 0.10\ \mathrm{M}^{-1}\ \mathrm{s}^{-1}$), which restores native enzyme and concurrently eliminates the chromophore (Charnas *et al.*, 1978*a*). With the sulphone **(5)**, the chromophore appears in about 25 per cent of the inactivated enzyme, and this too reacts with hydroxylamine (resulting in reactivation) at an almost identical rate. The chromophores produced by the interaction of these two inactivators with the enzyme are evidently very similar.

The analogies between clavulanate and **(5)** mesh well with the notion that acyl-enzyme intermediates are initially produced. From these acyl-enzymes, we suggest that a series of fragmentations may occur in competition with the normal deacylation reaction (scheme I). The driving force for these fragmentations is analogous for the two inactivators. With clavulanate it is the weakness of the carbon-5-to-oxygen bond (which results from ring strain and from the nature of the enolate leaving group); with the sulphone it is the weakness of the carbon-5-to-sulphur link (as exemplified by the facility with which sulphones participate in β-elimination processes (Hofmann *et al.*, 1964). The fact that fragmentation *follows* the formation of the acyl-enzyme (rather than the elimination occurring in the intact β-lactam) is consistent with the recent study of Pant and Stoodley (1978), who have established that carbon–sulphur cleavage does not occur during the epimerisation of penicillin sulphones in base. Enzyme inactivation is thus achieved by the generation on the enzyme of a species covalently attached via an ester linkage, which can deacylate only slowly, if at all. This stability must result from some combination of chemical reasons (formation of a stable conjugated carbonyl system, or interception of an enzymic nucleophile by conjugate Michael addition) and steric factors.

We decided to test the validity of the postulates outlined above by investigating several new β-lactam derivatives. We assumed first that acyl-enzymes are ubiquitous intermediates in the β-lactamase-catalysed hydrolysis of penicillin derivatives. Second, we assumed that the rate constants for the inactivating fragmentation reactions remain relatively unperturbed by the presence of a 6β-substituent. Third, we believed that the rate of β-lactam hydrolysis (that is, acylation then deacylation) *is* dependent on the nature of the 6β group. (This is certainly true for hydrolysis of penicillins and cephalosporins by the RTEM enzyme.) These assumptions may be summarised in the following statement: any factor that increases the lifetime of the acyl-enzyme derived from a penicillin sulphone increases the likelihood that a fragmentation (resulting in inactivation) will occur during that turnover. Therefore, to increase the efficiency of the lethal blow dealt by a penicillin sulphone, one should incorporate a sulphone into progressively poorer substrates of the β-lactamase.

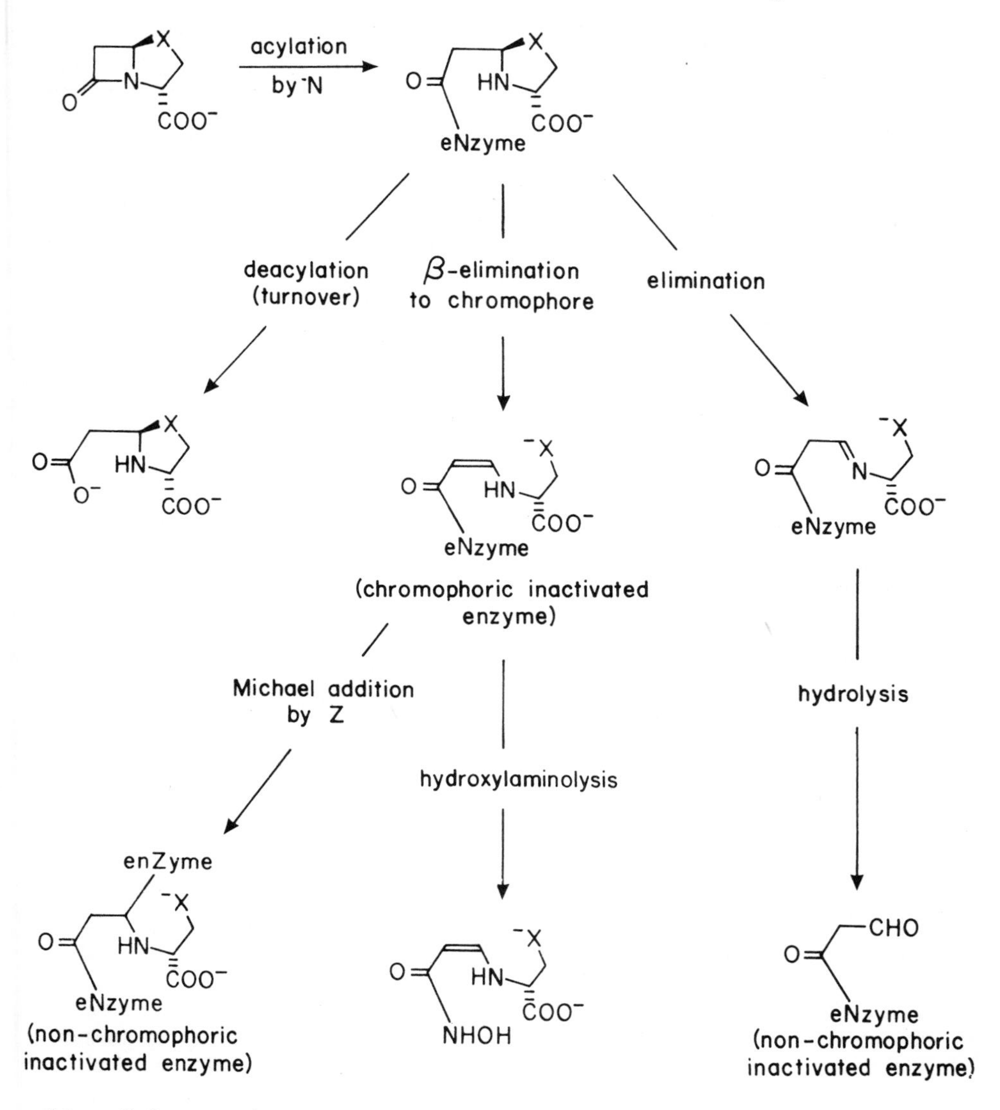

Scheme I Some reaction pathways which may account for the behaviour of β-lactamase inactivators with the *E. coli* enzyme.

Three penicillin sulphones were synthesised using the aqueous permanganate procedure of Johnson *et al.* (1963). The structures of these compounds are shown (with **(5)**) in table 13.1, in descending order of the rate at which the parent (i.e. unoxidised) penicillin is hydrolysed by the RTEM β-lactamase. Of the compounds in table 13.1, one is derived from a penicillin (phenoxymethylpenicillin) that is a substantially better substrate than 6-*des*-aminopenicillin, and two are from parent penicillins that are poorer substrates (methicillin and quinacillin). It may be seen

Table 13.1 Kinetic characteristics of the interaction of various penam derivatives, with the β-lactamase from *E. coli* RTEM

	k_{cat} of parent penam (s^{-1})	$t_{1/2}$ for inactivation (min)	Turnovers before inactivation
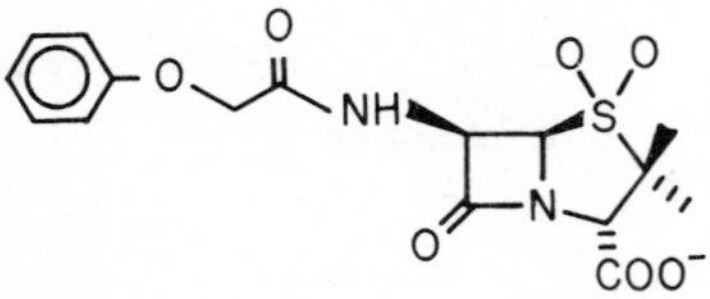	1000	–	> 100 000
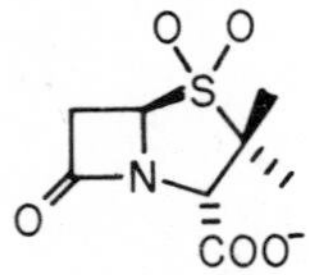	40	44	4500
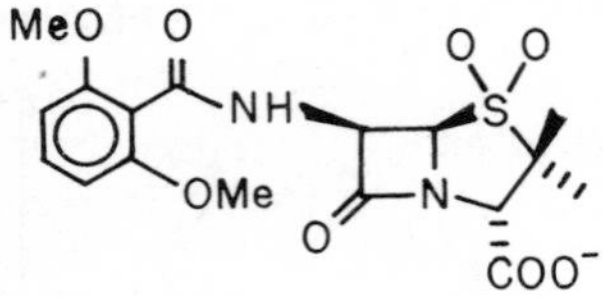	10	~ 1	22 500
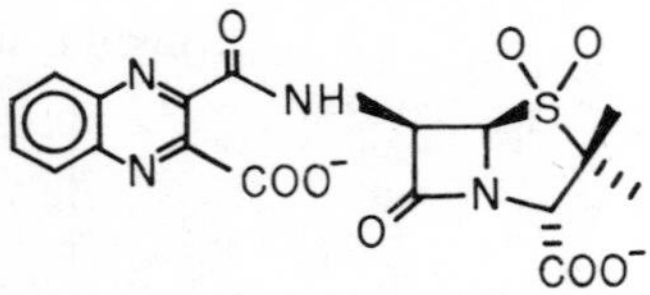	7	~ 1	400

from table 13.1 that these last two sulphones are irreversible inactivators, while the sulphone derived from phenoxymethylpenicillin is not. This progression is as predicted. The correlation is not absolute–methicillin sulphone is a better substrate than quinacillin sulphone despite the similar k_{cat} values of the parent penicillins–but the results lie within the limits of our original assumptions. That analogous inactivation mechanisms are operating is attested to by the appearance of the chromophore around 280 nm in the inactivated enzyme samples prepared from each of the three inactivators. On the basis of these observations we suggest that penicillin sulphones are a general class of β-lactamase inactivator, and that it may be proper to consider acyl-enzyme species as normal intermediates in the hydrolysis of β-lactams by the RTEM β-lactamase.

Of the materials listed in table 13.1, quinacillin sulphone is the most potent inactivator. This compound also contains a 6β substituent that is intensely chromophoric in the ultraviolet (λ_{max} 245 nm, $\epsilon = 32\,300\ M^{-1}\ cm^{-1}$; 327 nm, $\epsilon = 7100\ M^{-1}\ cm^{-1}$; see figure 13.1). The fragmentations illustrated in scheme I predict that the 6β substituent is retained in the inactivated enzyme, and should therefore contribute (with the new 280 nm chromophore discussed above) to the ultraviolet

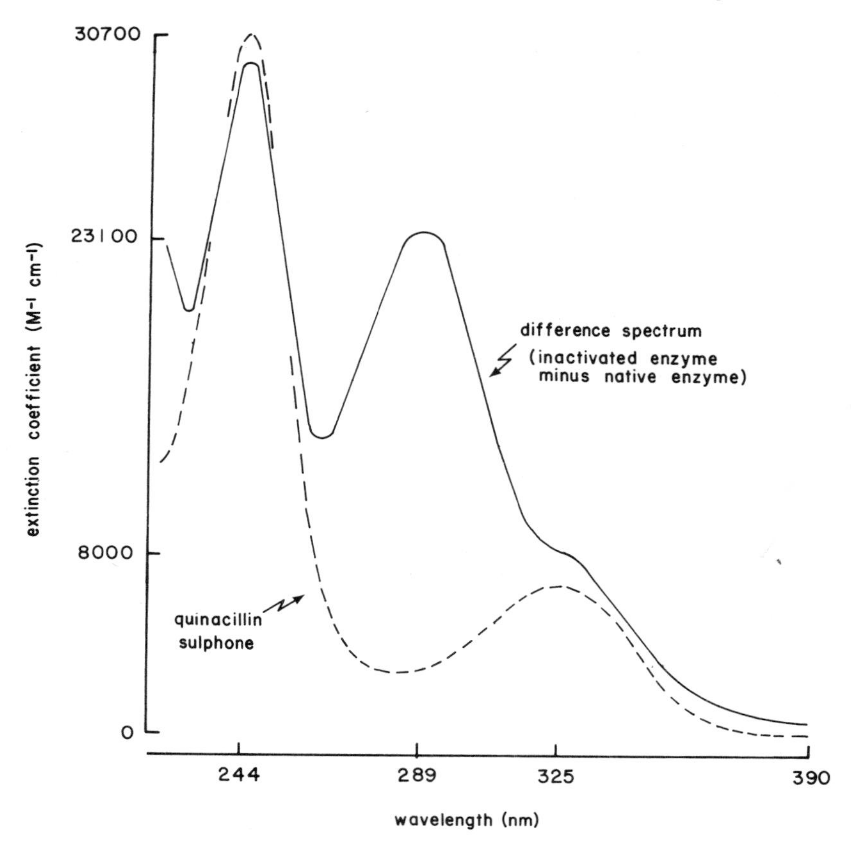

Figure 13.1 Ultraviolet spectrum of quinacillin sulphone (- - -), and of the difference spectrum of the RTEM β-lactamase before and after quinacillin sulphone inactivation (——). The ordinate values refer to the latter spectrum; the values for quinacillin sulphone are λ_{max} 245 (32 300 M^{-1} cm^{-1}), 326 (7100 M^{-1} cm^{-1}).

spectrum of the inactivated enzyme. Treatment of the RTEM β-lactamase with a 600-fold molar excess of quinacillin sulphone for twenty minutes (pH 7.0, 30 °C) yields, after purification, enzyme that has approximately 7 per cent of its initial activity. In a control experiment with quinacillin in place of quinacillin sulphone, no detectable decrease in enzymic activity was observed. The absorption spectrum of the sulphone-inactivated enzyme is shown, with that of native enzyme, in figure 13.2. The covalent attachment of the additional chromophore to the protein was demonstrated by the co-elution of protein and new chromophore from a gel filtration column under denaturing conditions. The nature of the new chromophore is shown in the difference spectrum between inactive and native enzyme (figure 13.1, with intact quinacillin sulphone for comparison). This spectrum confirms the presence of the quinacillin side-chain, as well as the new band centred at 289 nm. We suggest that the structure present at the active site is **(6)**, or something analogous

6

to it. The magnitude of the extinction coefficient at 289 nm for this chromophore (23 100 M^{-1} cm^{-1}) suggests that it is present in all of the inactivated enzyme. We have not, however, been able to confirm this directly, since the quinacillin sulphone-inactivated enzyme is unaffected by hydroxylamine treatment. This contrasts with the inactivated enzymes prepared from either the sulphone **(5)** or clavulanate. The lowered reactivity may, again, result from steric or chemical factors.

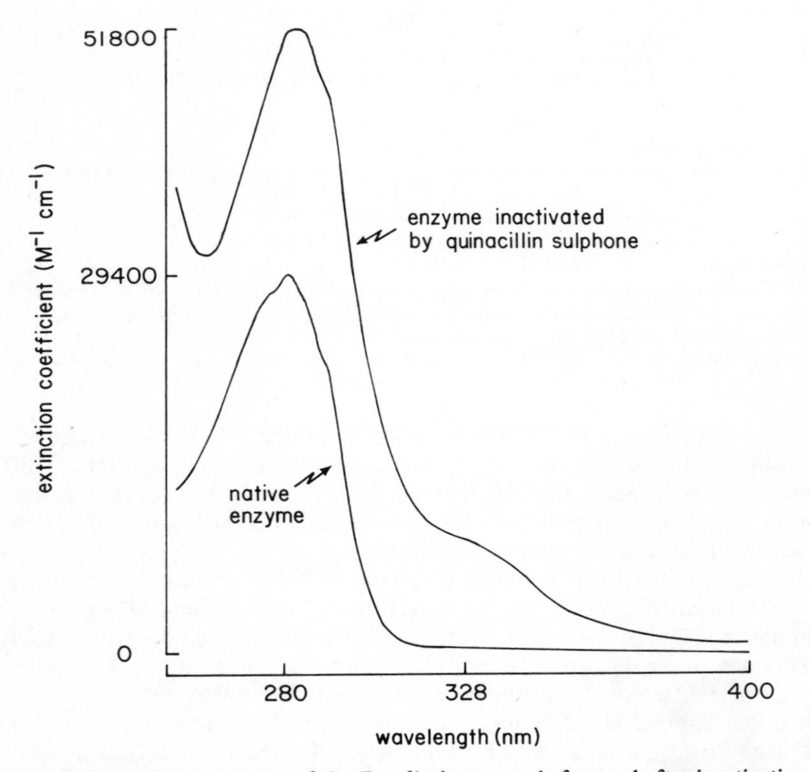

Figure 13.2 Ultraviolet spectrum of the *E. coli* β-lactamase before and after inactivation by quinacillin sulphone.

The ideas of Barth (English *et al.*, 1978) and of Coulson (Cartwright and Coulson, 1979) that culminated in the synthesis of CP-45, 899 (**5**) and 6-chloro-*des*-aminopenicillanic acid sulphone respectively, have offered a splendid opportunity further to delineate the mechanisms of β-lactamase catalysis and inactivation. It is our expectation that the full understanding of β-lactam sulphones will bear on both of these issues.

ACKNOWLEDGEMENTS

We are grateful to Beechams, Ltd, Boots, Ltd, Merck, Sharp and Dohme, and Pfizer, Inc., for generous gifts of β-lactams, and to the National Institutes of Health and Merck, Sharp and Dohme for support.

REFERENCES

Abraham, E. P. (1977). *J. Antibiotics,* **30**, S1-S26
Aswapokee, N. and Neu, H. C. (1978). *J. Antibiotics,* **31**, 1238-44
Bell, M. R., Cleamans, S. D. and Oesterlin, R. (1970). *J. med. Chem.,* **13**, 389-94
Brown, A. G., Corbett, D. F., Eglington, A. J. and Howarth, T. T. (1977). *J. chem. Soc., chem. Commun.,* **1977**, 953-4
Cartwright, S. J. and Coulson, A. F. W. (1979). *Nature, Lond.,* **278**, 360-1
Charnas, R. L., Fisher, J. and Knowles, J. R. (1978*a*). *Biochemistry,* **17**, 2185-9
Charnas, R. L., Fisher, J. and Knowles, J. R. (1978*b*). In *Enzyme-Activated Irreversible Inhibitors* (ed. N. Seiler, M. J. Jung and J. Koch-Weser), Elsevier, Amsterdam, pp. 315-22
Degelaen, J. P., Loukas, S. L., Feeney, J., Roberts, G. C. K. and Burgen, A. S. V. (1979). *J. chem. Soc. Perkin II,* **1979**, 86-90
Dumon, L., Adriaens, P., Anné, J. and Eyssen, H. (1979). *Antimicrobial Agents Chemother.,* **15**, 315-7
Dukin, J. P. and Viswanatha, T. (1978). *J. Antibiotics,* **31**, 1162-9
English, A. R., Retsema, J. A., Girard, A. E., Lynch, J. E. and Barth, W. E. (1978). *Antimicrobial Agents Chemother.,* **14**, 414-9
Fisher, J., Charnas, R. L. and Knowles, J. R. (1978). *Biochemistry,* **17**, 2180-4
Fisher, J. and Knowles, J. R. (1978). *A. Rep. med. Chem.,* **13**, 239-48
Fu, K. P. and Neu, H. C. (1979). *Antimicrobial Agents Chemother.,* **15**, 171-6
Golding, B. T. and Hall, D. R. (1973). *J. chem. Soc., chem. Commun.,* **1973**, 293
Hofmann, J. E., Wallace, T. J. and Schriesheim, A. (1964). *J. Am. chem. Soc.,* **86**, 1561-1563
Howarth, T. T., Brown, A. G. and King, T. J. (1976). *J. Chem. Soc., chem. Commun.,* **1976**, 266-7
Johnson, D. A., Panetta, C. A. and Cooper, D. E. (1963). *J. org. Chem.,* **28**, 1927-8
Kahan, J. S., Kahan, F. M., Goegelman, R., Currie, S. A., Jackson, M., Stapely, E. O., Miller, T. W., Miller, A. K., Hendlin, D., Mochales, S., Hernandez, S., Woodruff, H. B. and Birnbaum, J. (1979). *J. Antibiotics,* **32**, 1-12
Knott-Hunziker, V., Orlek, B. S., Sammes, P. G. and Waley, S. G. (1979). *Biochem. J.,* **177**, 365-7
Labia, R. and Peduzzi, J. (1978). *Biochim. biophys. Acta,* **526**, 572-9
Maeda, K., Takahashi, S., Sezaki, M., Iinuma, K., Naganawa, H., Kondo, S., Ohno, M. and Umezawa, H. (1977). *J. Antibiotics,* **30**, 770-3
Nagarajan, R., Boeck, L. D., Gorman, M., Hamill, R. L., Higgens, C. E., Hoehn, M. M., Stark, W. M. and Whitney, J. G. (1971). *J. Am. chem. Soc.,* **93**, 2308-10
Neu, H. C. and Fu, K. P. (1978). *Antimicrobial Agents Chemother.,* **14**, 650-5
Okamura, K., Hirata, S., Okumura, Y., Fukagawa, Y., Shimauchi, Y., Kouno, K., Ishikura, T. and Lein, J. (1978). *J. Antibiotics,* **31**, 480-2
Ostercamp, D. L. (1970). *J. org. Chem.,* **35**, 1632-41

Pant, C. M. and Stoodley, R. J. (1978). *J. chem. Soc. Perkin I,* **1978**, 1366–9
Pratt, R. F. and Loosemore, M. J. (1978). *Proc. natn. Acad. Sci. U.S.A.,* **75**, 4145–9
Reading, C. and Cole, M. (1977). *Antimicrobial Agents Chemother.,* **11**, 852–7
Reading, C. and Hepburn, P. (1979). *Biochem. J.,* **179**, 67–76
Sawai, T., Matsuba, K., Tamura, A. and Yamagishi, S. (1979). *J. Antibiotics,* **32**, 59–64
Scudamore, R. A., Beveridge, T. J. and Goldner, M. (1979). *Antimicrobial Agents Chemother.,* **15**, 182–9

14
Proteinase inhibitors: potential drugs?

Alan J. Barrett (Biochemistry Department, Strangeways Laboratory, Cambridge CB1 4RN, UK)

INTRODUCTION

There are at present few, if any, drugs of proven value that undoubtedly act by virtue of their capacity to inhibit proteinases, that is to say, endopeptidases. There are, however, clear indications that proteinases mediate a variety of processes that one would like to see under therapeutic control, and there are some good guidelines as to the way in which inhibitors for these enzymes may eventually be developed as drugs.

THE MAJOR CLASSES OF PROTEINASE

Many important properties of a proteinase, including the types of inhibitor to which it is likely to be susceptible, depend upon the chemical nature of the groups responsible for its catalytic activity. Four classes of proteinase are recognised on this basis, the serine, thiol, carboxyl and metallo-proteinases (Barrett, 1977). Some characteristics of these groups are summarised in table 14.1.

SOME PHYSIOLOGICAL FUNCTIONS OF PROTEINASES

Other contributors to this symposium deal fully with the roles of proteolytic enzymes in blood coagulation and regulation of blood pressure, so we shall be concerned here with a variety of other functions of proteinases, particularly in cells and tissues.

Infection

It seems that a number of pathogenic organisms make use of proteinases in entering the body or living there. A few diverse examples from the many one could cite include the parasitic worm *Schistosoma mansonii* which seems to pass through the skin by use of a chymotrypsin-like proteinase (Dresden and Asch, 1972) and the ringworm organism, *Trichophyton mentagraphytes*, which uses a keratin-degrading enzyme as it grows through the skin (Yu *et al.*, 1972).

Table 14.1 A summary of the four recognised classes of proteinase, or endopeptidase

	Serine (EC 3.4.21)	Thiol* (EC 3.4.22)	Carboxyl (EC 3.4.23)	Metallo (EC 3.4.24)
Examples	Trypsin, chymotrypsin, elastases, cathepsin G, chymase, plasmin, thrombin	(Papain), cathepsin B, cathepsin H, cathepsin L, Ca^{2+}-proteinase	Pepsin, cathepsin D, chymosin, renin, cathepsin D	Collagenase, gelatinase, brush-border proteinase
pH optimum	pH 7–9	pH 3–7	pH 2–7	pH 7–9
Inhibitors	Dip-F, Pms-F (also numerous selective synthetic and natural inhibitors)	IAA, *p*CMB, Ma1NEt_2, leupeptin	Pepstatin, N_2 Ac-Nle-OMe + Cu^{2+}	EDTA, 1, 10-phenanthroline, DTT. (phosphoramidon)

Abbreviations: Dip-F, diisopropylfluorophosphate; Pms-F, phenylmethane sulphonyl fluoride; IAA, iodoacetate; *p*CMB, *p*-chloromercuribenzoate; Ma1NEt_2 ; *N*-ethylmaleimide; N_2 Ac-Nle-OMe, diazoacetylnorleucine methyl ester; DTT, dithiothreitol. Leupeptin, pepstatin and phosphoramidon are microbial proteinase inhibitors described by Umezawa and Aoyagi (1977).

*Note added in proof: It has recently been proposed that the 'thiol proteinases' be renamed 'cysteine proteinases' and the 'carboxyl proteinases' be called 'aspartic proteinases' (Barrett, A. J., Chairman's introduction, in *Proteolysis in Health and Disease: Ciba Foundation Symposium No. 75* (ed. D. Evered and J. Whelan), Excerpta Medica, Amsterdam, 1980).

Clostridium histolyticum secretes a collagenase as it liquefies and invades the tissues. The malarial parasite, *Plasmodium*, seems to use an acid proteinase to digest and metabolise haemoglobin during its life within the red cells of the host, and its growth may be retarded by an inhibitor of carboxyl proteinases (Levy *et al.*, 1974; Levy and Chou, 1975). The powerful elastases produced by some strains of *Pseudomonas aeruginosa* and *Staphylococcus* can do serious damage to tissues including those of the lung, vasculature, skin and eyes (Wretlind and Wadstrom, 1977; Hartman and Murphy, 1977).

A particularly interesting way in which pathogenic bacteria make use of proteinases is exemplified by the 'IgA-proteases' of *Streptococcus sanguis, Neisseria gonorrhoeae* and *Neisseria meningiditis*, which inactivate the host's IgA1 and apparently thus resist the immune response (Plaut *et al.*, 1975; Blake and Swanson, 1978), and the complement-inactivating proteinase of *Pseudomonas aeruginosa*, which may have a similar effect (Schultz and Miller, 1974).

Proteins of animal viruses are commonly synthesised within mammalian cells as 'polyprotein' precursors which require processing by proteolytic enzymes before they become functional (Korant, 1975). Some of the specific proteinases responsible for the segmentation of the precursors are themselves coded for by the viruses. The important rhinoviruses responsible for the common cold, influenza and poliomyelitis certainly employ proteolytic processing of their precursor proteins. The proteinases involved may be trypsin-like serine proteinases, or thiol proteinases (Korant, 1975; Shapiro and August, 1976; Dittmar and Moelling, 1978).

It is clear that specific inhibition of proteinases used by pathogens in any of these ways could be of help in the prevention or treatment of the diseases they give rise to. The prospects seem particularly good as the proteinases are different from those of the host, and might therefore be susceptible to highly specific inhibition by well-designed reagents.

Fertility

There is evidence for a role of proteinase at several key stages in reproductive physiology (Morton, 1977) and it has been suggested that proteinase inhibitors might contribute to contraception (Zaneveld *et al.*, 1970; Suominen *et al.*, 1973). Attention has been focused on acrosin, the proteinase present in the acrosome covering the head of the spermatozoan, and revealed in sperm capacitation. The location of this proteinase, and some more direct evidence, suggests that it may play a role in penetration of the *zona pellucida* around the egg prior to fertilisation. Acrosin is a serine proteinase related to trypsin, but it is not known to have other functions in the body, so that if a specific inhibitor were to be designed for it, this might possibly be administered without unwanted effects. Inhibitors of serine proteinases have also been found to interfere with implantation (Denker *et al.*, 1978).

Processing of peptides and proteins

There are now numerous examples of sequences of amino acids, ranging from two or three to hundreds in length, which acquire their biological activity as a result of the limited proteolysis of higher molecular weight precursors (Neurath and Walsh, 1977). Sometimes the inactive precursor is stored, so that it is reasonable to think that the proteolytic step is rate-limiting, and therefore a potential point of external control. One of the simplest examples of proteolytic generation of a hormone is the liberation of the thyroxines from thyroglobulin, after iodination of tyrosine residues in the whole protein. The iodinated protein is taken up by colloid-containing follicle cells, digested, probably by lysosomal proteinases, and the products released into the circulation. Normal control by thyroid-stimulating hormone seems to be exerted by increasing endocytosis by the cells, rather than by any direct action on proteolysis, however (Wollman, 1969).

Many polypeptide hormones and mediators are liberated from their precursors by cleavage of peptide bonds on the carboxyl side of pairs of arginine or lysine residues (reviewed by Steiner *et al.*, 1974; Geisow, 1978). In the pancreatic islets, both insulin and glucagon are generated from precursors in this way, and the same is true for parathyrin, in the parathyroid gland (Hamilton *et al.*, 1974). A peculiarly interesting situation is seen in the pituitary, where a glycoprotein of about 31 000 molecular weight is the precursor for ACTH and β-lipotropin, and therefore also for the various fragments of these, α-MSH, CLIP, α-lipotropin, β-MSH, endorphins and methionine-enkephalin. Again, most of the cleavage points for liberation of the active peptides are marked by pairs of basic residues, and a major new development in this field has been the elucidation of the nucleotide sequence of the DNA for this bovine corticotropin-β-lipotropin precursor (Nakanishi *et al.*, 1979).

It is presumed that the various cell types concerned in processing the precursor proteins contain one or more proteinases specific for cleavage at pairs of basic residues in appropriate secondary conformation. Sometimes, such cleavages may be rate-limiting, although both insulin and parathyrin are stored in the active, processed form.

Lysosomal cathepsin B has a marked selectivity for synthetic substrates with pairs of arginine residues, and may be responsible at least for the conversion of proalbumin to albumin (Quinn and Judah, 1978), although our own recent work has shown that cathepsin B can also cause the inactivation of hormones and enzymes (Aronson and Barrett, 1978; Bond and Barrett, 1979).

Since the pituitary cells do not secrete all of the peptides derived from the large precursor in equimolar amounts, there must be selective generation or subsequent destruction of the peptides. Perhaps the proteinases responsible for these steps, or the subsequent conversion and inactivation of the peptides in the blood and tissues, will one day be controlled by proteinase inhibitors.

Malignancy and invasion

Much has been written in the last few years about the fact that tumour cells often secrete large amounts of plasminogen activator, and that the production of this enzyme tends to be activated by transformation of normal cells with a tumour virus (Rifkin *et al.*, 1974). Opinions on the significance of this phenomenon have varied, the extreme view being that the plasminogen activator, and the plasmin it generates, are directly responsible for the changed growth pattern of the cells, perhaps through the degradation of a cell surface protein with recognition or growth-controlling properties. If this were true, an inhibitor of the activator might well restore the cells to a normal growth pattern, but when such results have been reported, they have often been attributable to cytotoxic effects of the inhibitor (Schnebli, 1975). Nevertheless, there are indications that some aspects of the changes in cell behaviour, notably migration and adhesion, are plasminogen-dependent, and probably effected by plasmin.

It has often been suspected that proteolytic activity facilitates the advance of the invading front of a tumour, and it is notable that the rabbit V8 carcinoma is one of the few tissues from which collagenase is directly extractable (McCroskery *et al.*, 1975). Proteinases may also be responsible for the dissociation of a tumour which leads to metastasis (Sylvén, 1974). Ascites fluid accumulation may be due to the action of a cathepsin D-like carboxyl proteinase (Greenbaum *et al.*, 1978; Esumi *et al.*, 1978).

All these possibilities could well justify an interest in the inhibition of proteinases in the field of cancer research.

Inflammation and tissue damage

The fundamental processes underlining local chronic inflammation, such as that in rheumatoid arthritis, remain mysterious. Various hypothetical mechanisms involving proteinases have been proposed, and have received limited support from the finding that a number of compounds possess both anti-inflammatory and proteinase-inhibitory activities (e.g. sodium aurothiomalate, penicillamine). Hayashi (1975) reported that thiol proteinases play an important part in the passive subcutaneous Arthus reaction in the rabbit, and that this is naturally limited by the appearance of a proteinase inhibitor at the site. Even when some inflammatory models are affected by rather selective inhibitors such as leupeptin or pepstatin, however, one cannot be certain that the activity is due to proteinase inhibition. The significance of the anti-inflammatory activity of several proteinase inhibitors has recently been considered by Hyman and Vischer (1978).

In contrast to the situation for inflammation itself, there is very little doubt that the tissue damage so often associated with inflammation is due to proteinases. Only the action of proteolytic enzymes seems able to explain the destruction of articular cartilage in rheumatoid arthritis, for example, and the important questions are the cellular origin of the enzymes and their identity (Barrett and Saklatvala, 1980). Cathepsin D, a lysosomal carboxyl proteinase, is secreted by cells at the time and place of cartilage degradation, and certainly is responsible for most of the autolytic degradation of the cartilage matrix which occurs when the tissue is incubated in a slightly acidic buffer (Dingle *et al.*, 1972; Poole *et al.*, 1974). Nevertheless, experiments with living cartilage in organ culture, in which attempts were made to block the degradation of the matrix by inhibition of cathepsin D with specific antibodies or high concentrations of pepstatin, gave no clear confirmation of the idea that cathepsin D plays a major role in the breakdown (R. M. Hembry, C. G. Knight, A. J. Barrett and J. T. Dingle, in preparation). The evidence of release of cathepsin D at the sites of cartilage degradation can be taken as an indication of release of the other lysosomal proteinases too, and there is also evidence of secretion of collagenase, a non-lysosomal metalloproteinase, at the invasion site (Woolley *et al.*, 1977). Experiments with inhibitors of the other classes of proteinase were as inconclusive as those with pepstatin in our organ culture experiments, however, so that the identity of the proteinases presumed to act extracellularly in the cartilage matrix remains obscure.

Organ culture systems for study of the degradation of cartilage matrix are, of course, devoid of leucocytes, and clearly show that degradation can occur in the absence of these cells. Nevertheless, neutrophil granulocytes are abundant in the joint affected by rheumatoid arthritis, and there is a distinct possibility that their proteinases contribute to the tissue damage. The azurophil granules of the neutrophils contain large amounts of two serine proteinases, elastase and cathepsin G (a chymotrypsin-like enzyme). Both of these enzymes have the capacity not only to degrade cartilage proteoglycan, but also to attack molecules of cartilage collagen in the non-helical cross-link region, eliminating the stabilising cross-links, so that the stability of the collagen fibres against mechanical stress, thermal denaturation of monomers, and attack by collagenase and proteinases, is lost (Starkey *et al.*, 1977). The evidence for and against a role of neutrophil proteinases in damage to joint tissues has been further considered by Barrett (1978) and Barrett and Saklatvala (1980).

Starkey (1977) has discussed the possibility of the neutrophil serine proteinases contributing to damage to collagenous structures in a variety of other situations, including the inflamed cornea. The idea that the destruction of collagen may be mediated by serine proteinases rather than the specific collagenases, metallo-proteinases, is relatively new, and it has been usual to attribute the damage in the ulcerated or alkali-burned cornea to collagenase (reviewed by Brown 1975; Berman, 1975). Limited success has been claimed for the use of penicillamine (a drug the activities of which include moderate inhibition of collagenase) in the management of corneal ulcers (Francois *et al.*, 1972–3; Berman, 1975).

Pulmonary emphysema

The possibility that leucocyte elastase is responsible for serious destruction of elastin and collagen in the lung, in some forms of emphysema has now become widely accepted. Certainly, the enzyme has the capacity to degrade the macromolecules involved, *in vitro*, but the evidence is otherwise somewhat indirect. It has been noted that people with an inherited abnormality in the structure of α_1-proteinase inhibitor (α_1-antitrypsin) are prone to develop a form of pulmonary emphysema which has some resemblance to that which can be induced in experimental animals by introducing elastinolytic enzymes into their lungs (Janoff *et al.*, 1977). Since α_1-proteinase inhibitor is the major inhibitor of leucocyte elastase in the plasma, it was natural to suggest that the deficiency of the inhibitor led directly to excess activity of elastase which had been released by leucocytes in the lung.

The risk of emphysema in association with deficiency of α_1-proteinase inhibitor is known to be increased by smoking, and this can be explained in terms of the above theory through the finding that α_1-proteinase inhibitor is inactivated by a component of tobacco smoke (Janoff, 1978), apparently through oxidation of an essential methionyl side chain (Johnson and Travis, 1978).

The hypothesis that leucocyte elastase is an important causative factor in pulmonary emphysema is not without its difficulties. One consideration is that neutrophil leucocytes are not normally so numerous in the lung, except perhaps during infection, as the macrophages which are also capable of producing an elastase. Far less is known about the elastase of macrophages than that of neutrophils, but it seems likely to be a metallo-proteinase, and as such would probably not be affected by α_1-proteinase inhibitor (Werb and Gordon, 1975). A second point requiring consideration is the finding that although the elastase-inhibiting activity of *plasma* is very much dependent on its content of α_1-proteinase inhibitor, the fluid bathing the surface of the lung contains another powerful inhibitor of leucocyte elastase which is apparently produced by the lung itself, and is not affected by deficiency of α_1-proteinase inhibitor (Ohlsson and Tegner, 1976).

Nevertheless, the idea that emphysema can be due to deficiency of a natural inhibitor of leucocyte elastase clearly leads to the thought that exogenous inhibitors of the enzyme might be of therapeutic value, and this has been the stimulus for some excellent work on the discovery or design of selective inhibitors for the leucocyte proteinases (Dorn *et al.*, 1977; Powers *et al.*, 1977; Schiessler *et al.*, 1976).

Skin diseases

The normal distribution and activities of proteinases in the skin are known to be disturbed in a number of skin diseases (Hopsu-Havu *et al.*, 1977), and in some the proteinases may be important in disrupting the structure of the tissue. This has been suggested recently by Farb *et al.* (1978) for pemphigus, on the basis of an *in-vitro* model system in which detachment of epidermal cells was produced by pemphigus serum, but could be blocked by soya bean trypsin inhibitor or α_2-macroglobulin. This finding might be taken to suggest that topical application of proteinase inhibitors, perhaps in dimethylsulphoxide, would be worth testing.

The possibility that thiol proteinases play an important part in the passive subcutaneous Arthus reaction in the rabbit has been examined in detail by Hayashi and his colleagues (reviewed by Hayashi, 1975), as was mentioned earlier.

Glomerulonephritis

Basement membranes often appear to be remarkably stable structures: the type IV collagen which many of them contain is unaffected by the specific collagenases, the membranes often resist dissolution in the face of tumour invasion, and they are solubilised chemically only under the most aggressive conditions. Nevertheless, there are important disease situations in which degradation of basement membranes does occur, notable glomerulonephritis. In animal studies of glomerulonephritis, there is evidence that antibodies against glomerular basement membrane are fixed to the membrane. They then activate and bind complement, which leads to the chemotactic attraction and degranulation of neutrophil leucocytes (Cochrane, 1977). Although antibodies to glomerular basement membrane are not commonly detected in patients, other routes can be imagined which would lead to a similar involvement of leucocytes in the ultimate tissue damage.

Davies *et al.* (1978) have shown that both of the neutrophil serine proteinases, elastase and cathepsin G, have the power to digest human glomerular basement membrane, solubilising the type IV collagen. This observation supports the theory that enzymes from neutrophils are important in causing the crucial damage to the basement membrane, at least in some forms of glomerulonephritis. Further support came from the detection of an enzyme resembling leucocyte elastase in the urine of patients with rapidly progressive glomerulonephritis.

Demyelinating diseases

Degradation of the insulating myelin sheath leads to serious impairment of nerve function, and there are at least two possible connections of tissue proteinases with this pathological process.

Chemically, myelin is a complex material containing both protein and lipid components (reviewed by Rumsby and Crang, 1977). One of the proteins, the myelin basic protein, is of particular interest because it is very susceptible to proteolysis yielding fragments some of which produce allergic encephalomyelitis in experimental animals. The condition inadvertently induced in human patients by use of anti-rabies serum containing rabbit myelin was probably due to the same effect. Allergic encephalomyelitis shows some resemblances to natural demyelinating disease, and is used as an experimental model, but differs from multiple sclerosis, say, in lacking the typical relapsing course. Cathepsin D is one of the enzymes which can liberate allergenic fragments from the myelin basic protein (Brostoff *et al.*, 1974; Benuck *et al.*, 1975).

Whether or not proteolytic fragments of myelin trigger human demyelinating disease (presumably through an autoimmune reaction), proteinases may well be involved in the actual demyelination process. Cammer *et al.* (1978) have reported the release from stimulated macrophages of a proteinase which degrades myelin basic protein. The multiple sclerotic plaques formed at sites of demyelination are likely to contain cells such as these.

Acute pancreatitis

Although the primary cause of acute pancreatitis is unknown, there is reasonably strong evidence that the severe damage to the pancreas itself, and the generalised shock reaction with its potentially fatal outcome, are associated with conversion of latent precursors of kallikrein, trypsin and other pancreatic proteinases to their active forms within the pancreas, and their release into the circulation and peritoneal fluid. If these enzymes are not promptly inhibited, they are likely to produce large quantities of kallidin and other pharmacologically active peptides, and to damage the blood vessel walls.

Aprotinin, commercially produced as Trasylol, is a polypeptide inhibitor of kallikrein and trypsin extractable from various bovine organs. It has been advocated for use in acute pancreatitis, but its effectiveness in decreasing mortality seems not to have been decisively proved. Since human pancreatic chymotrypsin and elastase are not well inhibited by aprotinin there may be scope for the use of other proteinase-inhibiting drugs in this disease.

Amyloidosis

Amyloidosis is a rare but serious disorder in which an abnormal protein material accumulates in intercellular deposits, and disrupts normal tissue function. There is no satisfactory treatment, and the progressive build up of amyloid usually leads to death. The amyloid deposits consist primarily of fibrils which can be assembled from either of two proteins referred to as AL and AA. AL is a protein homologous with the N-terminal (variable) region of immunoglobulin light chains, whereas protein AA seems to be derived from a normal plasma protein. There is much evidence to suggest that both AL and AA are formed by partial proteolysis of circulating precursor molecules, abnormal light chains in primary or myeloma-associated amyloid, and serum amyloid A in secondary amyloid.

There are indications that the amyloid components may be formed in the lysosomes of mononuclear phagocytes in the affected tissues (Shirahama and Cohen, 1975), but the recent results of Lavie *et al.* (1978) favour the conversion of serum amyloid A to AA proper by a serine proteinase on the outer membrane of mononuclear cells. It is not clear, however, what sort of abnormality of the normal proteolytic processes amyloidosis represents, i.e. whether formation of the fibril components occurs as a result of partial proteolysis which would normally either not have occurred at all or would have been complete. Clearly, this is a question that has to be answered before any serious consideration can be given to the use of proteinase inhibitors in the management of amyloidosis.

Muscular dystrophy

The wasting of muscle in the various forms of muscular dystrophy is obviously the consequence of breakdown occurring at a rate exceeding that of resynthesis. A natural approach to the control of this situation is to try to inhibit the proteinases responsible for the excessive degradation.

Breakdown of the myofibrils is, of course, an intracellular process, and muscle cells are like most other tissue cells in containing predominantly the lysosomal thiol and carboxyl proteinases. Schwartz and Bird (1977) have shown that both cathepsin B and cathepsin D are capable of degrading myosin and actin, and Bird *et al.* (1978) have suggested that the modified lysosomal system of skeletal muscle cells is responsible for much of the normal and pathological proteolysis. In agreement with this view, Libby and Goldberg (1978), and Stracher *et al.* (1978) have been able to inhibit the protein degradation in muscle cells in culture, and the muscle wasting in dystrophic chickens, by use of pepstatin and leupeptin, potent inhibitors of cathepsins D and B, respectively. Many further questions remain to be answered of course, a crucial one being that of whether the protein that is spared is in the form of functional myofibrils, so that muscle strength as well as bulk is preserved.

In conclusion, then, it seems clear that there are many diseases in which proteinases may play an important part, and on that basis, proteinase inhibitors do potentially form a useful class of drugs. Before their potential can be realised, however, various kinds of problems will have to be solved. One difficulty is that an enzyme may often be performing a necessary function at one location in the body, while being harmful elsewhere. A second consideration is that an undesired degradative process may be catalysed by several proteinases acting together (several of the lysosomal proteinases, for example), so that selective inhibition of any one will be ineffective. These difficulties are serious, but are not unique to proteinase inhibitors, and the design of proteinase inhibitors as drugs remains a worthwhile and stimulating objective.

REFERENCES

Aronson, N. N., Jr and Barrett, A. J. (1978). *Biochem. J.*, **171**, 759–65

Barrett, A. J. (1977). In *Proteinases in Mammalian Cells and Tissues* (ed. A. J. Barrett), North Holland, Amsterdam, pp. 1–55

Barrett, A. J. (1978). *Agents Actions*, **8**, 11–18

Barrett, A. J. and Saklatvala, J. (1980). In *Textbook of Rheumatology* (W. N. Kelley, E. D. Harris, Jr, S. Ruddy and C. B. Sledge), Saunders, Philadelphia, in press

Benuck, M., Marks, N. and Hashim, G. (1975). *Eur. J. Biochem.*, **52**, 615–21

Berman, M. (1975). In *International Ophthalmology Clinics*, Vol 15, No. 4 (ed. D. Pavan-Langston), Little, Brown, Boston, pp. 49–66

Bird, J. W. C., Spanier, A. M. and Schwartz, W. N. (1978). In *Protein Turnover and Lysosome Function* (ed. H. Segal and D. Doyle), Academic Press, New York, pp. 589–604
Blake, M. S. and Swanson, J. (1978). *Infect. Immunol.*, **22**, 350–8
Bond, J. S. and Barrett, A. J. (1979). *Fedn Proc. Fedn Am. Socs exp. Biol.*, **39**, 946
Brostoff, S. W., Reuter, W., Hichens, M. and Eylar, E. H. (1974). *J. biol. Chem.*, **249**, 559–67
Brown, S. I. (1975). *Arch. Ophthalmol., Paris*, **35**, 91–4
Cammer, W., Bloom, B. R., Norton, W. T. and Gordon, S. (1978). *Proc. natn. Acad. Sci. U.S.A.*, **75**, 1554–8
Cochrane, C. G. (1977). *Inflammation*, **2**, 319–33
Davies, M., Barrett, A. J., Travis, J., Sanders, J. and Coles, G. A. (1978). *Clin. Sci. molec. Med.*, **54**, 233
Denker, H.-W., Eng, L. A. and Hamner, C. E. (1978). *Anat. Embryol.*, **154**, 39–54
Dingle, J. T., Barrett, A. J., Poole, A. R. and Stovin, P. (1972). *Biochem. J.*, **127**, 443–4
Dittmar, K. J. and Moelling, K. (1978). *J. Virol.*, **28**, 106–18
Dorn, C. P., Zimmerman, M., Yang, S. S., Yurewicz, E. C., Ashe, B. M., Frankshun, R. and Jones, H. (1977). *J. med. Chem.*, **20**, 1464–8
Dresden, M. H. and Asch, H. L. (1972). *Biochim. biophys. Acta*, **289**, 378–84
Esumi, H., Sato, S., Sugimara, T. and Okasaki, N. (1978). *Biochim. biophys. Acta*, **523**, 191–7
Farb, R. M., Dykes, R. and Lazarus, G. S. (1978). *Proc. natn. Acad. Sci. U.S.A.*, **75**, 459–63
Francois, J., Cambie, E., Feher, J. and van den Eeckout, E. (1972–3). *Ophthalmic Res.*, **4**, 223–36
Geisow, M. J. (1978). *Fedn Eur. biochem. Socs Lett.*, **87**, 111–4
Greenbaum, L. M., Semente, G., Prakash, A. and Roffman, S. (1978). *Agents Actions*, **8**, 80–4
Hamilton, J. W., Niall, H. D., Jacobs, J. W., Keutmann, H. T., Potts, J. T. and Cohn, D. V. (1974). *Proc. natn. Acad. Sci. U.S.A.*, **71**, 653–6
Hartman, D. P. and Murphy, R. A. (1977). *Infect. Immunol.*, **15**, 59–65
Hayashi, H. (1975). *Int. Rev. Cytol.*, **40**, 101–51
Hopsu-Havu, V. K., Fräki, J. E. and Järvinen, M. (1977). In *Proteinases in Mammalian Cells and Tissues* (ed. A. J. Barrett), North Holland, Amsterdam, pp. 545–81
Hyman, Y. G. and Vischer, T. L. (1978). *Agents Actions*, **8**, 532–5
Janoff, A., Sloan, B., Weinbaum, G., Damiano, V., Sandhaus, R., Elias, J. and Kimbel, P. (1977). *Am. Rev. resp. Dis.*, **115**, 461–78
Janoff, A. (1978). In *Neutral Proteases of Human Polymorphonuclear Leukocytes* (ed. K. Havermann and A. Janoff), Urban and Schwarzenberg, Baltimore, pp. 390–417
Johnson, D. and Travis, J. (1978). *J. biol. Chem.*, **253**, 7142–4
Korant, B. (1975). In *Proteases and Biological Control* (ed. E. Reich, D. B. Rifkin and E. Shaw), Cold Spring Harbor Laboratory, Cold Spring Harbor, N.Y., pp. 621–44
Lavie, G., Zucker-Franklin, D. and Franklin, E. C. (1978). *J. exp. Med.*, **148**, 1020–31
Levy, M. R. and Chou, S. C. (1975). *Experientia*, **31**, 52–4
Levy, M. R., Siddiqui, W. A. and Chou, S. C. (1974). *Nature, Lond.*, **247**, 546–9
Libby, P. and Goldberg, A. L. (1978). *Science, N.Y.*, **199**, 534–6
McCroskery, P. A., Richard, J. F. and Harris, E. D., Jr (1975). *Biochem. J.*, **152**, 131–42
Morton, D. B. (1977). In *Proteinases in Mammalian Cells and Tissues* (ed. A. J. Barrett), North Holland, Amsterdam, pp. 445–500
Nakanishi, S., Inoue, A., Kita, T., Nakamura, M., Chang, A. C. Y., Cohen, S. N. and Numa, S. (1979). *Nature, Lond.*, **278**, 423–7
Neurath, H. and Walsh, K. A. (1977). *Proc. natn. Acad. Sci. U.S.A.*, **73**, 3825–32
Ohlsson, K. and Tegner, H. (1976). *Scand. J. clin. Lab. Invest.*, **36**, 437–45
Plaut, A. G., Gilbert, J. V., Artenstein, M. S. and Capra, J. D. (1975). *Science, N.Y.*, **190**, 1103–5
Poole, A. R., Hembry, R. M. and Dingle, J. T. (1974). *J. cellul. Sci.*, **14**, 139–61
Powers, J. C., Gupton, B. F., Harley, A. D., Nishino, N. and Whitley, R. J. (1977). *Biochim. biophys. Acta*, **485**, 156–66
Quinn, P. S. and Judah, J. D. (1978). *Biochem. J.*, **172**, 301–9
Rifkin, D. B., Leob, J. N., Moore, G. and Reich, E. (1974). *J. exp. Med.*, **139**, 1317–28
Rumsby, M. G. and Crang, A. J. (1977). In *The Synthesis, Assembly and Turnover of Cell Surface Components* (ed. G. Poste and G. L. Nicholson) Elsevier, Amsterdam, pp. 247-362

Schiessler, H., Arnold, M., Ohlsson, K. and Fritz, H. (1976). *Hoppe-Seyler's Z. physiol. Chem.*, **357**, 1251–60
Schnebli, H. P. (1975). In *Proteases and Biological Control* (ed. E. Reich, D. B. Rifkin and E. Shaw), Cold Spring Harbor Laboratory, Cold Spring Harbor, N.Y., pp. 785–94
Schultz, D. R. and Miller, K. D. (1974). *Infect. Immunol.*, **10**, 128–135
Schwartz, W. N. and Bird, J. W. C. (1977). *Biochem. J.*, **167**, 811–20
Shapiro, S. Z. and August, J. T. (1976). *Biochem. biophys. Acta*, **458**, 375–96
Shirahama, T. and Cohen, A. S. (1975). *Am. J. Path.*, **81**, 101–9
Starkey, P. M. (1977). *Acta biol. med. germ.*, **36**, 1549–54
Starkey, P. M., Barrett, A. J. and Burleigh, M. C. (1977). *Biochem. biophys. Acta*, **483**, 386–97
Steiner, D. F., Kemmler, W., Tager, H. S. and Peterson, J. D. (1974). *Fedn Proc. Fedn Am. Socs exp. Biol.*, **33**, 2105–15
Stracher, A., McGowan, E. B. and Shafiq, S. A. (1978). *Science, N.Y.*, **200**, 50–1
Suominen, J., Kaufman, M. H. and Setchell, B. P. (1973). *J. Reprod. Fertil.*, **34**, 385–8
Sylvén, B. (1974). *Schweiz. med. Wschr.*, **104**, 258–61
Umezawa, H. and Aoyagi, T. (1977). In *Proteinases in Mammalian Cells and Tissues* (ed. A. J. Barrett), North Holland, Amsterdam, pp. 637–62
Werb, Z. and Gordon, S. (1975). *J. exp. Med.*, **142**, 361–77
Woolley, D. E., Crossley, M. J. and Evanson, J. M. (1977). *Arthritis Rheum.*, **20**, 1213–39
Wollman, S. H. (1969). In *Lysosomes in Biology and Pathology*, Vol 2 (ed. J. T. Dingle and H. B. Fell), North Holland, Amsterdam, pp. 483–512
Wretlind, B. and Wadstrom, T. (1977). *J. gen. Microbiol.*, **103**, 319–27
Yu, R. J., Grapple, F. and Blank, F. (1972). *Experientia*, **28**, 886
Zanefeld, L. J. D., Robertson, R. T. and Williams, W. L. (1970). *Fedn Eur. biochem. Socs Lett.*, **11**, 345–7

15
Angiotensin-converting enzyme inhibitors

D. W. Cushman, M. A. Ondetti, H. S. Cheung, E. F. Sabo,
M. J. Antonaccio and B. Rubin (The Squibb Institute for Medical Research,
Princeton, New Jersey 08540, USA)

INTRODUCTION

A fundamental but elusive goal of molecular pharmacology is the chemical characterisation of drug receptors and the design of drugs with optimal structures for specific receptor binding. Most drug receptors have not been isolated or purified, and existing knowledge offers few if any clues as to their chemical structure. Great progress has been made, however, in delineation of the chemical structures of the active sites of enzymes, biological catalysts which may serve as receptor sites for drug action. The development of inhibitors of biologically-important enzymes thus represents one of the most fruitful areas for attempts at rational drug design. An example of a drug designed for optimal binding to the active site of an enzyme is captopril (SQ 14 225), an orally active inhibitor of angiotensin-converting enzyme which has been found to be a remarkably safe and effective antihypertensive drug.

ANGIOTENSIN-CONVERTING ENZYME

Angiotensin-converting enzyme (peptidyldipeptide carboxy hydrolase, EC 3.4.15.1) owes its trivial name to its ability to convert the inactive decapeptide angiotensin I (AI) to the vasoconstrictor octapeptide angiotensin II (AII) (Skeggs *et al.*, 1956). It is a key enzymic component of the renin-angiotensin-aldosterone system which participates in maintenance of blood pressure and tissue blood flow following haemorrhage, sodium restriction, or other events serving to lower effective blood volume (Peach, 1977). Angiotensin-converting enzyme is a unique kind of exopeptidase which cleaves dipeptides from the C-terminal end of polypeptide substrates (Erdös, 1977). As shown diagrammatically in figure 15.1, two such dipeptide-releasing reactions catalysed by the converting enzyme may play important roles in the regulation of blood pressure and blood volume: one results in formation of the vasoconstrictor, antidiuretic and antinatriuretic octapeptide AII, the other in inactivation of the vasodilator, diuretic and natriuretic nonapeptide bradykinin (BK).

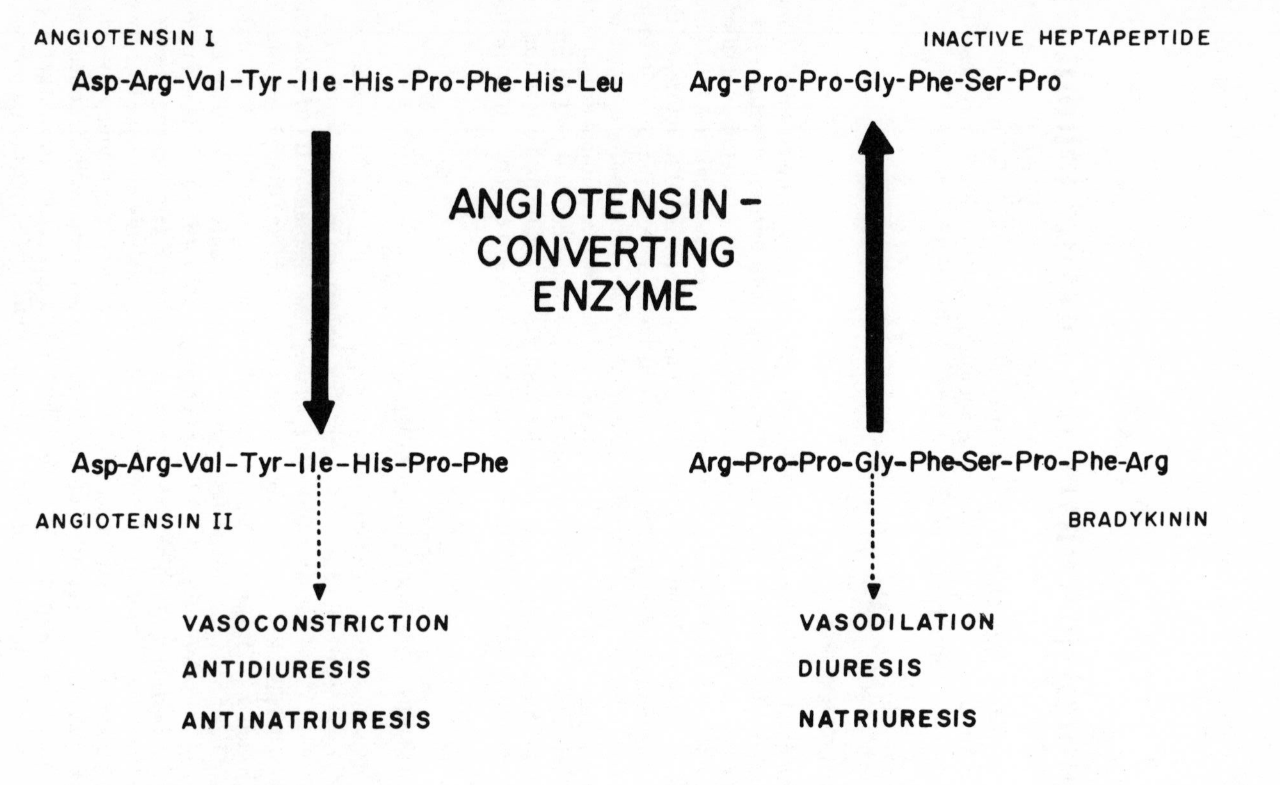

Figure 15.1 Biologically important reactions of angiotensin-converting enzyme (kininase II).

Table 15.1 Affinity of angiotensin-converting enzyme for peptide substrates and competitive inhibitors

	Peptide	Substrate K_m (μM)	Inhibitor I_{50} (μM)	Reference
1 (AI)	Asp–Arg–Val–Tyr–Ile–His–Pro–Phe–His–Leu	50		Cheung and Cushman, 1973
2	Bz–**Gly**–His–Leu	2200		
3	Z–Phe–His–Leu	40		Piquilloud *et al.*, 1970
4 (AII)	Asp–Arg–Val–Tyr–Ile–His–**Pro**–Phe	NS		Skeggs *et al.*, 1956
5	Z–Gly–**Pro**–Gly	NS		
6	Bz–Gly–Phe–**Glu**	NS		Elisseeva *et al.*, 1971
7	Bz–Gly–Phe–**Phe–NH$_2$**	NS		
8 (BPP$_{5a}$)	< Glu–Lys–Trp–Ala–Pro		0.07	
9	< Glu–Lys–Phe–Ala–Pro		0.09	
10	< Glu–Lys–Phe–Ala–**Pyn**		> 350	
11	< Glu–Lys–Phe–Ala–**Glu**		3.3	
12	< Glu–Lys–Phe–**Pro**–Pro		5.5	
13	< Glu–Lys–**Ile**–Ala–Pro		3.3	Cushman *et al.*, 1973
14 (SQ 20, 881)	< Glu–Trp–Pro–Arg–Pro–Gln–Ile–Pro–Pro		1.0	
15	< Glu–Trp–Pro–Arg–Pro–Gln–Ile–Pro–**Pyn**		> 180	
16	< Glu–Trp–Pro–Arg–Pro–Gln–Ile–**Ala**–Pro		0.37	
17	< Glu–Trp–Pro–Arg–Pro–Gln–**Phe**–Pro–Pro		0.25	
18	< Glu–Trp–Pro–Arg–Pro–**Lys–Phe–Ala**–Pro		0.05	

NS = not a substrate; < Glu = pyrrolidone-5-carbonyl; Pyn= pyrrolidine

In several of its structural and enzymic properties, angiotensin-converting enzyme is similar to the well-characterised carboxypeptidase A of bovine pancreas (Quiocho and Lipscomb, 1971). Like carboxypeptidase A, the converting enzyme is a zinc metalloprotein (Das and Soffer, 1975). Its substrate specificity also has many similarities to that of carboxypeptidase A, even though the converting enzyme releases dipeptide products rather than amino acids (table 15.1). Like carboxypeptidase A, converting enzyme will not hydrolyse a peptide which lacks a free C-terminal carboxyl group (**7** table 15.1), one with proline donating the amino group for the susceptible peptide bond (**4, 5**) or one with a C-terminal dicarboxylic amino acid (**6**). Angiotensin-converting enzyme also appears to have a high affinity for peptide substrates with an aromatic amino acid in the antepenultimate position (compare **1–3**, table 15.1).

SNAKE VENOM PEPTIDE INHIBITORS

The first inhibitors of angiotensin-converting enzyme with the requisite potency and specificity for use as therapeutic agents were polypeptides obtained from snake venoms. Use of such peptides as pharmacological tools provided the first evidence for the great potential of converting enzyme inhibitors as antihypertensive drugs.

History

Ferreira (1965) isolated a non-toxic 'bradykinin-potentiating factor' from the venom of the Brazilian arrowhead viper *Bothrops jararaca*. Three years later, this mixture of polypeptides was shown to inhibit angiotensin-converting enzyme (Bakhle, 1968), which was not at that time known to be an important bradykinin-degrading enzyme. Two groups of investigators independently undertook the isolation of peptides from the venom. Ferreira *et al.* (1970) purified nine bradykinin-potentiating peptides, but determined the sequence of only one, a pentapeptide < Glu-Lys-Trp-Ala-Pro (BPP_{5a}), later shown to be a converting enzyme inhibitor and short-acting antihypertensive agent (Greene *et al.*, 1972). Ondetti and co-workers (1971) purified and determined sequences for six nona- to tridecapeptide inhibitors of angiotensin-converting enzyme from *Bothrops jararaca* venom, relying on a specific spectrophotometric assay (Cushman and Cheung, 1971) for determination of converting enzyme inhibitory actions of various fractions. These six peptides were shown to be both converting enzyme inhibitors and bradykinin-potentiating peptides *in vitro* and *in vivo* (Cheung and Cushman, 1973; Engel *et al.*, 1972). The most potent of these peptidic inhibitors, and the one with the longest duration of inhibitory and antihypertensive activity *in vivo*, was a nonapeptide SQ 20 881 (< Glu-Trp-Pro-Arg-Pro-Gln-Ile-Pro-Pro). Five other bradykinin-potentiating and angiotensin-converting enzyme inhibitory peptides, designated A–E, have been isolated from the venom of a Japanese pit viper *Agkistrodon halys blomhoffii*; sequences of four of these peptides have been determined (Kato and Suzuki, 1977).

Mechanism of action

The peptidic inhibitors from *Bothrops jararaca* venom compete with substrates for binding to the active-site of angiotensin-converting enzyme (Cheung and Cushman,

1973). A substrate for this enzyme must, at the very least, bind to the active site via its C-terminal dipeptide leaving group and its antepenultimate amino acid residue, the groups linked by the peptide bond cleaved by the enzyme. Competitive peptide inhibitors would be expected to bind to the active site of the enzyme in an analogous manner, and their C-terminal tripeptide residues would be expected to share with the terminal tripeptides of substrates certain structural requirements for binding (table 15.1). Analogues of the venom peptides lacking free C-terminal carboxyl groups (**10** and **15**) are virtually without inhibitory activity. Inhibitor analogues with penultimate proline residues (**12**, **14**) or with C-terminal dicarboxylic amino acid residues (**11**) are bound relatively weakly to the enzyme, as reflected by their inhibitory activities, and inhibitors with antepenultimate aromatic amino acids have the greatest inhibitory action (**8** or **9** versus **13**; **17** versus **14**, table 15.1). In studies with 57 analogues of the venom peptide inhibitors (Cushman *et al.*, 1973), the C-terminal tripeptide sequence Trp-Ala-Pro (or Phe-Ala-Pro) was found to confer the greatest inhibitory activity (**8**, **9**, **18**, table 15.1). However, the penultimate proline residue of SQ 20 881 (**14**), although unfavourable for binding, helps to prevent cleavage of this peptide by converting enzyme and other peptidases.

Although we have postulated a similar mode of active-site binding for the C-terminal tripeptide residues of substrates and venom peptide inhibitors, the venom peptides bind much more tightly than substrates to the active site of the enzyme (table 15.1): K_i values (enzyme–inhibitor dissociation constants) for BPP_{5a} (**8**) and SQ 20 881 (**14**) are 0.06 μM and 0.10 μM, respectively (Cheung and Cushman, 1973). The extra binding affinity of the venom peptides is due to enzyme-binding interactions of their amino acid residues beyond the competitive C-terminal tripeptide residue.

Antihypertensive activity

The pentapeptide BPP_{5a} lowers blood pressure in the two-kidney, one-clip model of renal hypertension in the rat (both kidneys intact, with one renal artery constricted), but not in the one-kidney, one-clip model (Greene *et al.*, 1972). The short duration of antihypertensive activity of the pentapeptide, however, precludes its use as an antihypertensive drug. The nonapeptide SQ 20 881, administered intravenously, lowers blood pressure for several hours in two-kidney, one-clip renal hypertensive rats (Engel *et al.*, 1973). SQ 20 881 has also been shown to prevent development of malignant hypertension in one-kidney, one-clip hypertensive rabbits (Muirhead *et al.*, 1974), and to prevent or delay the onset of one-kidney, one-clip hypertension in dogs (Miller *et al.*, 1975; Watkins *et al.*, 1978*a*, *b*). The antihypertensive actions of SQ 20 881 are associated with an increase in plasma renin activity which appears to be due to release from the feedback inhibition of renin release normally exerted by AII.

The nonapeptide SQ 20 881, under the generic name teprotide, has been studied clinically as a parenterally active antihypertensive drug (Gavras *et al.*, 1974; Case *et al.*, 1977; Johnson *et al.*, 1975). It was found to lower blood pressure in renovascular and essential hypertensive patients for up to 16 h when administered at intravenous doses of 1–4 mg kg^{-1}. Its antihypertensive effect was associated with increased plasma renin activity and decreased plasma aldosterone concentration. It was a safe and effective antihypertensive drug for most patients with renovascular or malignant hypertension and for a high percentage of essential hypertensive patients with high or 'normal' levels of plasma renin. The antihypertensive action of the drug was enhanced by moderate sodium depletion induced by diet or diuretics. The only major limitation for use of teprotide in chronic therapy of human hypertension has been its lack of oral activity.

SPECIFIC NON-PEPTIDIC INHIBITORS

Active-site model

The substrate specificity of angiotensin-converting enzyme and the structural specificity of analogues of the competitive snake venom peptide inhibitors, as discussed above, suggest that this enzyme is a novel carboxypeptidase, similar in many respects to the well-characterised carboxypeptidase A. We thus felt that it was likely that the active site of the converting enzyme might be similar to that of carboxypeptidase A, although adapted for cleavage of carboxyl-terminal dipeptides rather than carboxyl-terminal amino acids.

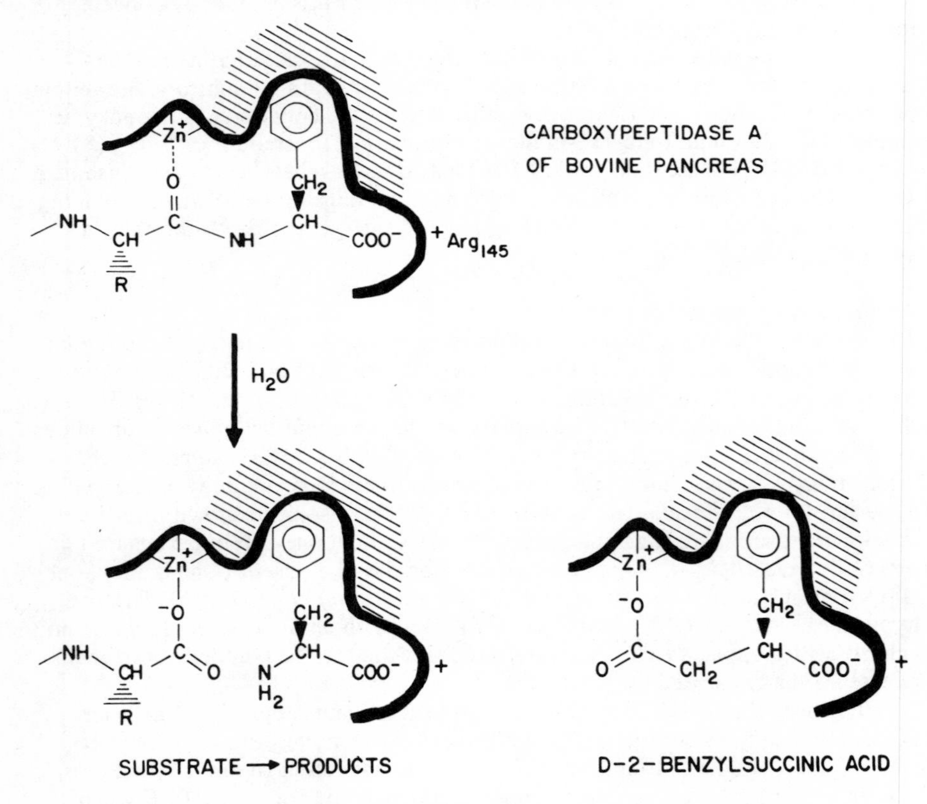

Figure 15.2 Diagrammatic model of the binding of peptide substrate, products, and competitive inhibitor to the active site of bovine pancreatic carboxypeptidase A. The cross-hatched area represents the 'hydrophobic pocket' of the enzyme.

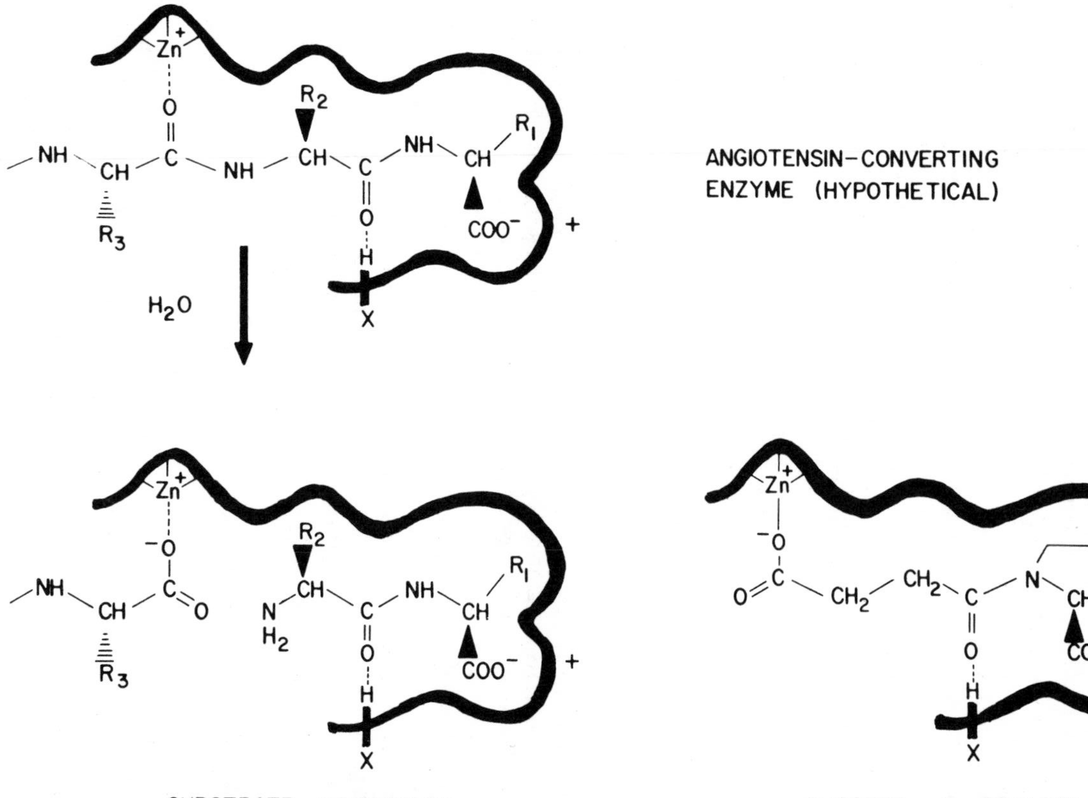

Figure 15.3 Diagrammatic representation of a hypothetical model for binding of peptide substrate, products and competitive inhibitor to the active site of angiotensin-converting enzyme. Amino acid side-chains R_1 and R_2 are proposed to interact with 'pockets' of unspecified nature; X–H represents a group at the surface of the enzyme proposed to form a hydrogen bond with the carbonyl of the terminal peptide bond of the substrate and with that of the dipeptide product.

The exact chemical structure of the active site of carboxypeptidase A is known from X-ray crystallographic studies (Quiocho and Lipscomb, 1971). The basic features are shown diagrammatically in a simple two-dimensional form in figure 15.2. The enzyme is a carboxypeptidase by virture of its ability to bind ionically the free carboxyl group of the peptide substrate with its positively-charged arginine residue number 145. An adjacent 'hydrophobic pocket' further anchors the side-chain of the C-terminal amino acid, thus positioning the terminal peptide bond of the substrate in the vicinity of the zinc ion of the enzyme participating in its hydrolysis. The peptide substrate is cleaved as shown in figure 15.2 to yield an amino acid product. Figure 15.2 also shows the structure and postulated mode of binding of a potent competitive inhibitor of carboxypeptidase A, D-2-benzylsuccinic acid. Byers and Wolfenden (1972) suggested that benzylsuccinic acid is a 'biproduct analogue' binding to the active site of carboxypeptidase A in a manner which combines binding interactions of each of the two products of the enzyme's action, although they did not specifically indicate that the succinyl carboxyl group is bound to the zinc ion of the enzyme as represented in figure 15.2.

If, indeed, the active site of converting enzyme is similar to that of carboxypeptidase A, we hypothesised that it must look something like the model depicted in figure 15.3 (Ondetti *et al.*, 1977*a,b*; Cushman *et al.*, 1977, 1978). In this model, the zinc ion and the carboxyl binding group are separated by a greater distance than in carboxypeptidase A, so that ionic binding of the carboxyl group of the substrate brings the second peptide bond from its C-terminal end into alignment with the zinc ion of the enzyme, allowing the resultant peptide bond cleavage to liberate a dipeptide product. Amino acid side-chains R_1 and R_2 of peptide substrates are visualised as interacting with 'pockets' which may or may not be hydrophobic as in carboxypeptidase A. It also seemed reasonable that some mechanism would have evolved for binding the terminal (nonscissile) peptide bond of a converting enzyme substrate; as shown in figure 15.3, we postulated that a group X–H on the surface of the enzyme would form a hydrogen bond with the carbonyl of this peptide bond. If this reasonable guess as to the nature of the active site of angiotensin-converting enzyme is basically correct, it seems possible that a succinyl amino acid might serve as a competitive 'biproduct analogue' inhibitor of angiotensin-converting enzyme, binding to the active site of the enzyme as shown in figure 15.3.

Design of specific active-site binding agents

The first succinyl amino acid synthesised as a possible 'biproduct analogue' inhibitor of angiotensin-converting enzyme was succinyl-L-proline. Figure 15.3 shows conjecturally how this compound might combine binding interactions of two products of converting enzyme action, although it has no side-chain analogous to R_2 of the dipeptide product. Its amino acid moiety, proline, the terminal amino acid of all the snake venom peptide inhibitors, was expected to bind to the enzyme as well as or better than other terminal amino acid residues. As shown in figure 15.4, succinyl-L-proline (**1**) was a fairly weak inhibitor of angiotensin-converting enzyme of rabbit lung. However, its probable specificity as a converting enzyme inhibitor was indicated by its spectrum of activities versus various agonists contracting guinea pig ileum strips: it inhibited the contractile action of AI

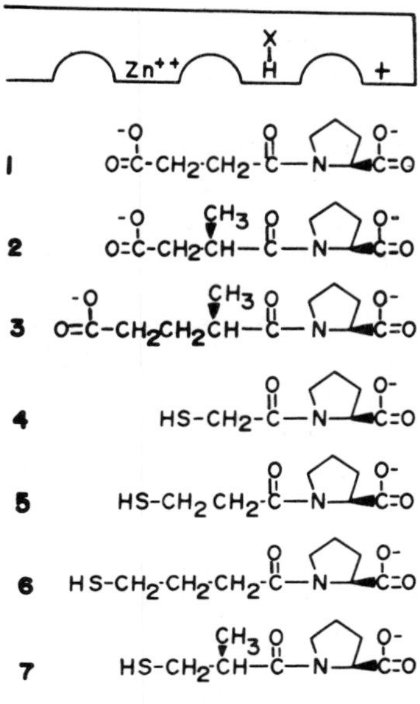

		RABBIT LUNG CONVERTING ENZYME		GUINEA PIG ILEUM	
				AI	BK
	Zn^{++} … X–H … +	I_{50} (μM)	K_I (μM)	I_{50} (μM)	A_{50} (μM)
1	$^{-}$O–C(=O)–CH$_2$·CH$_2$–C(=O)—N(proline)–C(=O)O^{-}	330	-	440	37
2	$^{-}$O–C(=O)–CH$_2$CH(CH$_3$)—C(=O)—N(proline)–C(=O)O^{-}	22	2.5	18	0.87
3	$^{-}$O–C(=O)–CH$_2$CH$_2$CH(CH$_3$)—C(=O)—N(proline)–C(=O)O^{-}	4.9	0.8	23	4.9
4	HS–CH$_2$–C(=O)—N(proline)–C(=O)O^{-}	1.1	-	0.19	0.037
5	HS–CH$_2$CH$_2$–C(=O)—N(proline)–C(=O)O^{-}	0.20	0.012	0.30	0.025
6	HS–CH$_2$·CH$_2$·CH$_2$–C(=O)—N(proline)–C(=O)O^{-}	9.7	-	12	5.1
7	HS–CH$_2$·CH(CH$_3$)—C(=O)—N(proline)–C(=O)O^{-}	0.023	0.0017	0.023	0.0032

SQ 14,225 (CAPTOPRIL)

Figure 15.4 Activities *in vitro* of carboxyalkanoyl and mercaptoalkanoyl amino acid inhibitors of angiotensin-converting enzyme. I_{50} is the concentration of a compound producing a 50 per cent inhibition of the activity of rabbit lung converting enzyme, or a 50 per cent inhibition of the contractile activity of angiotensin I on guinea pig ileum strip; A_{50} is the concentration producing a 50 per cent augmentation of the contractile action of bradykinin.

and augmented that of BK, but had no effect on muscle tone or the contractile actions of various other agonists.

The apparently specific inhibitory activity of succinyl-L-proline prompted the synthesis of additional compounds such as those shown in figure 15.4, in an attempt to design analogues which would interact more strongly with the active site of the enzyme, as visualised in the hypothetical model (Ondetti *et al.*, 1977*a,b*; Cushman *et al.*, 1977, 1978). The model (figure 15.3) predicts that the proper 2-succinyl substituent, analogous in position and stereochemistry to amino acid side-chain R_2 of the dipeptide product, can enhance binding of the inhibitor by its specific interaction with a pocket at the active site of the enzyme. In confirmation of this prediction, it was found that substituents in the 2-position and with the D configuration (isosteric with L-amino acid side-chains of substrates and products) enhance the inhibitory activity of succinyl amino acid derivatives. The most favourable substituent of several tested was the methyl group. D-2-Methylsuccinyl-L-proline (**2**, figure 15.4) is 15–20 times more potent than succinyl-L-proline as a converting enzyme inhibitor; its structure is analogous to the most favourable terminal sequence of the snake venom peptides, Ala-Pro. This compound was the first shown to be a specific inhibitor of angiotensin conversion after oral administration to rats (Ondetti *et al.*, 1977*a*).

Our model (figure 15.3) would suggest that the succinyl chain length should be approximately optimal for interaction of succinyl carboxyl and carbonyl groups with the zinc ion and hydrogen bonding function of the enzyme. However, studies with carboxyalkanoyl groups of different chain lengths indicate that glutaryl derivatives such as compound **3** (figure 15.4) are slightly more potent than the succinyl derivatives but, as expected, derivatives with longer or shorter acyl chain lengths than glutaryl or succinyl are much less inhibitory.

The most important structural modification made in this new class of nonpeptidic converting enzyme inhibitors came when attempts were made to replace the putative zinc-binding carboxyl group by other functional groups which might have a greater zinc-binding affinity. Mercaptoalkanoyl amino acids (**4–7**, figure 15.4) proved to be about 1000 times more inhibitory than the corresponding carboxyalkanoyl amino acids. In this class of compounds, the optimal chain length was that of the mercaptopropanoyl derivative, although mercaptoacetyl derivatives were often nearly as active (compare **4–6**, figure 15.4). The most potent mercaptoalkanoyl amino acid was D-3-mercapto-2-methylpropanoyl-L-proline (7, figure 15.4). This compound (SQ 14 225 or captopril) is a tightly-binding competitive inhibitor of angiotensin-converting enzyme, with a K_i value of 1.7×10^{-9} M. It did not inhibit most other peptidases tested until added at a concentration of 10^{-3} M, and had no effect at very high concentrations *in vitro* or *in vivo* on agonists other than AI or BK (Cushman *et al.*, 1977, 1978; Rubin *et al.*, 1978*b,c*; Murthy *et al.*, 1977, 1978).

Verification of the hypothetical model

In addition to the compounds shown in figure 15.4, a great number of structure–activity correlations, including many not yet published, have consistently supported the basic validity of the model of the active site of angiotensin-converting enzyme shown in figure 15.3 (Ondetti *et al.*, 1977*b*, 1978; Cushman *et al.*, 1977, 1978; Petrillo *et al.*, 1978).

The importance of the amino acid carboxyl group for binding of mercaptoalkanoyl amino acid inhibitors to the active site of angiotensin-converting enzyme is best illustrated by compound **8** of figure 15.5, which differs from captopril (**7**) only in its lack of this carboxyl group, and is 10 000 times less inhibitory than captopril. The amino acid side-chain (R_1 in figure 15.3) is proposed to interact with a pocket at the active site of the converting enzyme. In support of this proposal, the proline analogue (**5**) is seven times more inhibitory than glycine analogue **9**, which has no amino acid side-chain, and 350 times more inhibitory than an analogue (not shown) which has the undesirable carboxylate side-chain of aspartic acid (Cushman *et al.*, 1977). The requirement for an amide carbonyl for hydrogen bonding to the unspecified active-site residue X–H (Ondetti *et al.*, 1978) is nicely demonstrated by compound **10**, which lacks such a group and is 1000 times less active than the corresponding amide analogue (**5**). The proposed interaction of the acyl side-chain (R_2, figure 15.3) with a pocket on the enzyme has been verified by numerous analogues (for example 7 versus **5**, figure 15.5). Finally, compound **11** demonstrates better than most analogues the great importance of the zinc-binding function for the potent inhibitory activity of mercaptoalkanoyl amino acids. The alcohol function of compound **11** is similar in many respects to the sulphydryl group of compound **5**, but it has little if any affinity for zinc ion, and **11** is more than 10 000 times less inhibitory than **5**.

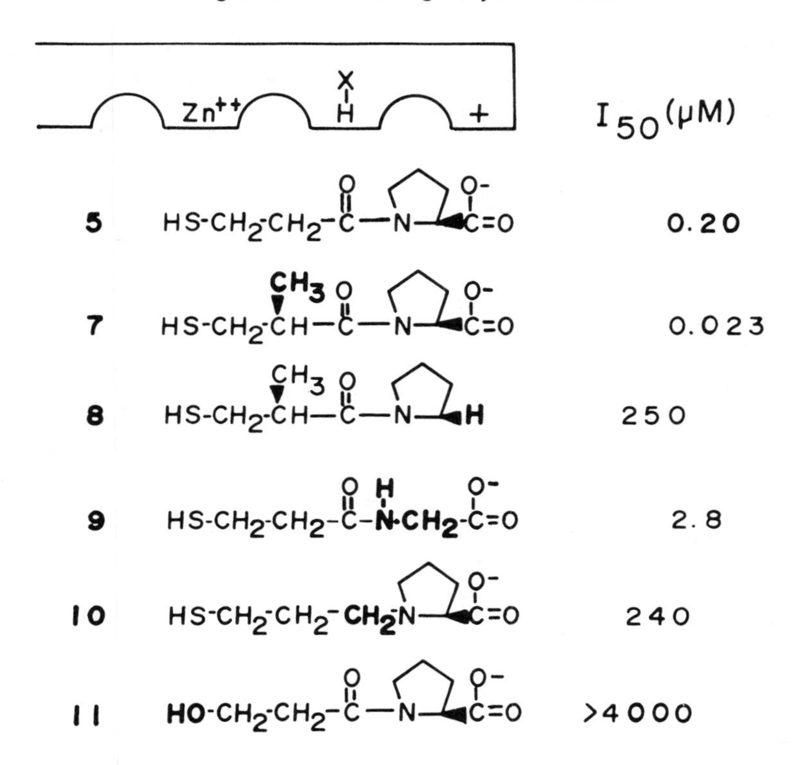

Figure 15.5 Effect of structural alterations on the angiotensin-converting enzyme inhibitory activities of mercaptopropanoyl amino acids. I_{50} is the concentration of a compound producing a 50 per cent inhibition of the activity of angiotensin-converting enzyme of rabbit lung.

CAPTOPRIL: A SPECIFIC ORALLY-ACTIVE ANTIHYPERTENSIVE DRUG

Captopril or SQ 14 225 (7, figure 15.4), although quite specific as an inhibitor of angiotensin-converting enzyme, may lower blood pressure by inhibiting formation of AII or by inhibiting destruction of BK (figure 15.1). This converting enzyme inhibitor is active by the oral route of administration, but it apparently does not readily pass the blood–brain barrier (Vollmer and Boccagno, 1977). It is a remarkably safe compound, showing no acute toxic effects until doses are reached that are at least 1000 times those required for maximal inhibition of converting-enzyme activity *in vivo* (Sibley *et al.*, 1978); for instance, in rats, no overt toxicological effects were seen until the oral dose of captopril exceeded 4000 mg kg^{-1}. The effective oral antihypertensive dose of captopril in various species is usually between 3 and 30 mg kg^{-1}.

Normotensive animals

Inhibition of angiotensin-converting enzyme by captopril has only a slight effect on the normal blood pressure of most animal species, including man, in the conscious state (Laffan *et al.*, 1978; Rubin *et al.*, 1978*c*; Murthy *et al.*, 1977; Bengis *et al.*, 1978; Ferguson *et al.*, 1977). However, a moderate decrease in blood pressure did occur in conscious dogs which were not sodium restricted (Harris *et al.*, 1978). Normotensive rats or dogs which had been moderately sodium depleted by a low sodium diet or by administration of diuretics showed a marked hypotensive response to captopril (Bengis *et al.*, 1978; McCaa *et al.*, 1978). In sodium-depleted rats or dogs, the elevated level of plasma aldosterone was reduced by captopril (McCaa *et al.*, 1978; Aguilera and Catt, 1978). In sodium-depleted dogs, in addition to the usual reflex increase in plasma renin activity, a doubling of blood kinin concentration was observed (McCaa *et al.*, 1978).

One-kidney, one-clip renal hypertensive animals

The renal hypertensive animal model based on constriction of one renal artery and removal of the contralateral kidney is supposed to involve active mediation of the renin–angiotensin system for only the first few days, as judged by measurements of plasma renin activities. One-kidney, one-clip renal hypertensive rats studied 4–6 weeks after renal artery constriction did not respond to acute administration of captopril (Laffan *et al.*, 1978; Bengis *et al.*, 1978). However, a progressive lowering of blood pressure was obtained in such rats when the drug was administered for several days, and this was accompanied by increases in urine output and urinary sodium excretion (Bengis *et al.*, 1978; Rubin *et al.*, 1978*b*). Captopril administered during the period of onset of renal hypertension in one-kidney, one-clip hypertensive rats and dogs was shown to delay, but not to prevent, the onset of hypertension (Watkins *et al.*, 1978*a,b*; Freeman *et al.*, 1979).

Two-kidney, one-clip renal hypertensive animals

The two-kidney, one-clip model of renal hypertension (no nephrectomy) is thought to be mediated by the renin-angiotensin system during the first few weeks following renal artery clipping. In this model in rats, acute oral administration of captopril several weeks after clipping results in a rapid and marked lowering of blood pressure (Ondetti *et al.*, 1977*a*; Laffan *et al.*, 1978; Bengis *et al.*, 1978; Rubin *et al.*, 1978*b*). At high doses of captopril, the total duration of this antihypertensive action exceeded 24 h. Captopril prolonged the renin-dependent phase of two-kidney, one-clip hypertension in rats (Weed *et al.*, 1979), and prevented the onset of hypertension for at least 12 days (Freeman *et al.*, 1979). In two-kidney, one-wrapped, renal hypertension in dogs, captopril prevented the development of elevated blood pressure during the 12 weeks of its administration, with a typical development of hypertension occurring after cessation of drug treatment (Rubin *et al.*, 1978*b*).

Bengis and coworkers (1978) showed that captopril maintains its antihypertensive action in two-kidney, one-clip hypertension in rats when administered for 1 week, with a progressively greater lowering of blood pressure during the first 4 days of drug administration. Rubin and his colleagues (1978*a,b*) have reported results of long-term treatment of such rats with captopril, and have compared its effects with those of other antihypertensive regimes. As shown in figure 15.6, captopril given orally at 30 mg kg^{-1} maintained systolic blood pressure 25–30 mm Hg lower than initial hypertensive levels for a period of at least 10 months. Similar groups of rats given only water showed blood pressures which were slightly increased during the first few months, followed by an apparent decrease in systolic

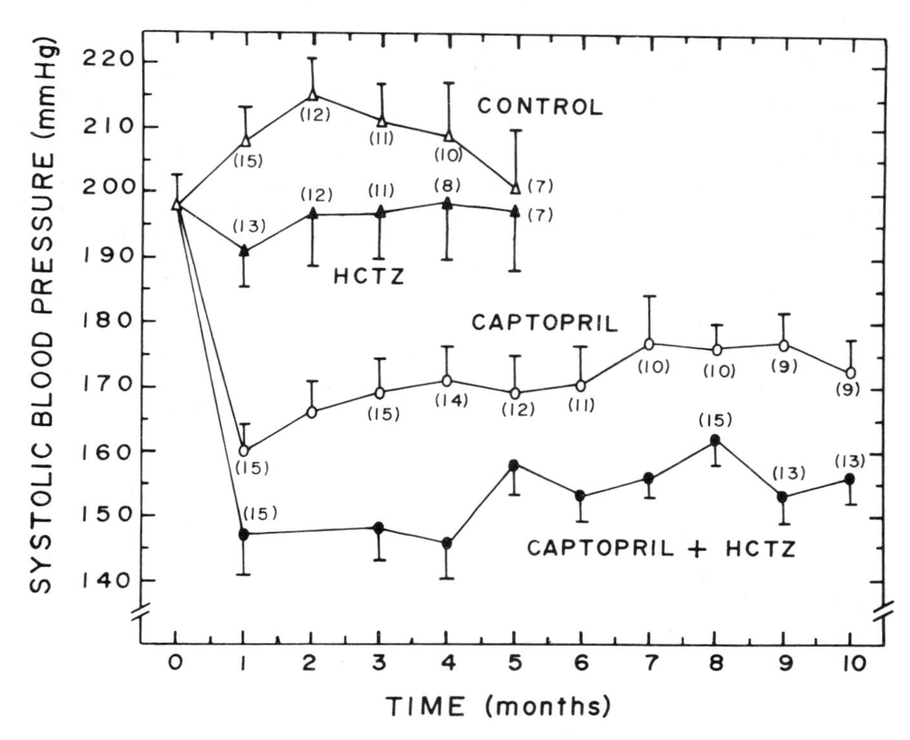

Figure 15.6 Effect of chronic oral administration of captopril (30 mg kg^{-1}), hydrochlorothiazide (6 mg kg^{-1}), or their combination on the systolic blood pressures of two-kidney, one-clip hypertensive rats. Groups of 15 rats had renal arterial clips placed 6 weeks before the initiation of drug therapy. Systolic blood pressures were determined indirectly and are reported as mean ± standard error for a maximum of 15 rats. The number of rats surviving in each group are listed in parenthesis.

blood pressure which may have been related to the poor survival of the rats in this group with the highest blood pressures. An effective diuretic dose of hydrochlorothiazide (6 mg kg^{-1}) had little effect on blood pressure over the 10 month period, but the combination of captopril plus diuretic maintained systolic blood pressure at a level 45-50 mm Hg below the initial hypertensive level. Hydralazine, not shown in figure 15.6, produced an initial dramatic lowering of blood pressure, followed by a rapidly developing tolerance during the first month of administration. As shown in figure 15.6, captopril alone dramatically increased survival of two-kidney, one-clip renal hypertensive rats, and captopril in combination with hydrochlorothiazide so increased survival that after 8 months all 15 rats in this group remained alive, as opposed to only three rats in the untreated control group. Captopril also reversed the cardiac hypertrophy observed with the untreated renal hypertensive rats. Withdrawal of the drug at various stages of treatment did not result in rebound hypertension, but in a gradual return of blood pressure to the hypertensive level over a period of several days.

Spontaneously hypertensive rats

Spontaneously developing hypertension in rats of the Wistar-Kyoto Okamoto-Aoki strain is not associated with increased plasma renin activity. Nevertheless, orally administered captopril produces a significant dose-related antihypertensive effect in these animals (Laffan *et al*., 1978; Rubin *et al*., 1978*b*; Muirhead *et al*., 1978), achieved entirely by a decrease in peripheral vascular resistance. Chronic administration of captopril to spontaneously hypertensive rats (Antonaccio *et al*., 1979; Rubin *et al*., 1978*b*) produced a complete normalisation of blood pressure after 6-12 months of treatment (figure 15.7), and a reversal of cardiac hypertrophy. Hydralazine, in this model, resulted in an antihypertensive effect which was maintained over the 12 month period, but it did not reduce blood pressure to the level of the normotensive controls, and only partially reversed the cardiac hypertrophy.

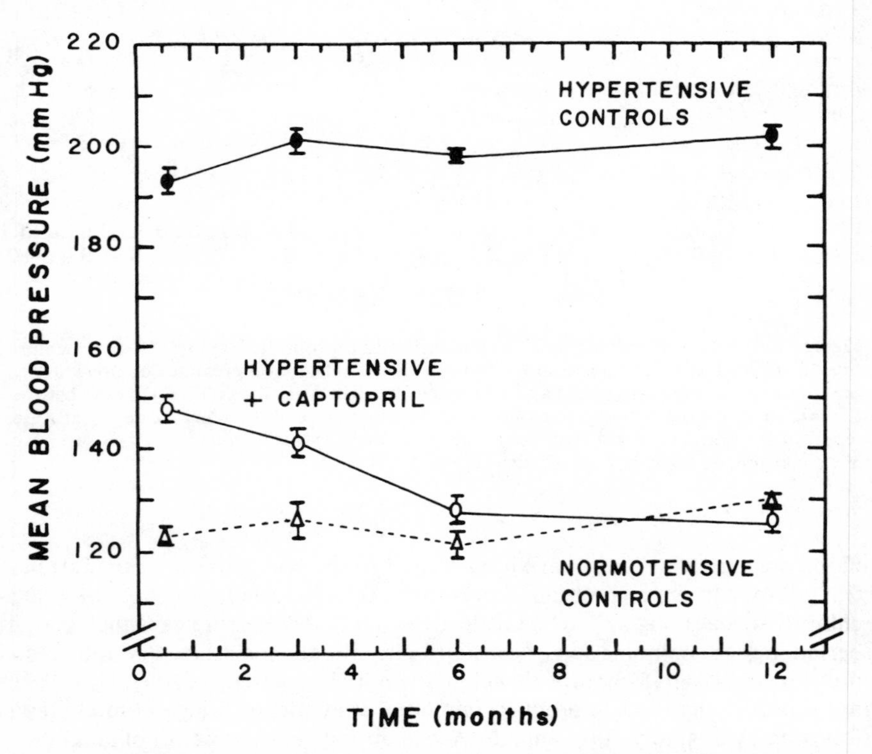

Figure 15.7 Effect of chronic oral administration of captopril (100 mg kg^{-1}) on the blood pressure of spontaneously hypertensive rats. Mean blood pressures were determined directly at the indicated times with groups of 10-12 rats from each treatment group and are reported as mean blood pressure ± standard error.

Therapeutic action in human hypertension

Captopril has now been studied for acute and chronic treatment of human renovascular and essential hypertension, and the hypertension of chronic renal failure

(Gavras *et al.*, 1978; Case *et al.*, 1978; Cody *et al.*, 1978; Brunner *et al.*, 1978*a,b*, 1979; Bravo and Tarazi, 1979). In all of these studies, it was shown to be effective in a high percentage of patients including, in some of the studies, patients previously found to be unresponsive to or unable to tolerate various combination drug therapies. A correlation has not always been observed between lowering of blood pressure and initial plasma renin activity, although this may be an inadequate index of the subtle involvement of the renin–angiotensin system in hypertensive disease.

Results obtained with captopril in hypertensive patients have, for the most part, been quite similar to those obtained in animal models of hypertension. Captopril lowered blood pressure in most patients with high or 'normal' levels of plasma renin, and in some low renin patients. Patients with incompletely normalised blood pressures usually showed a greater antihypertensive response to the drug after sodium depletion achieved through diet or diuretics. No rebound hypertension has been observed following cessation of captopril treatment, and orthostatic hypotension has not been observed. In those studies where the drug was administered for several months (Case *et al.*, 1978; Brunner *et al.*, 1978*a*, 1979), no tolerance was observed to the antihypertensive action of the drug; in fact, the antihypertensive effect was often observed to become progressively greater with time. The antihypertensive action of captopril was accompanied by relatively large increases in plasma renin activity and moderate decreases in plasma aldosterone concentration. Captopril did not, however, cause any changes in plasma levels of cortisol or noradrenaline, and the heart rate, cardiac output, and baroreceptor reflexes were not significantly altered (Bravo and Tarazi, 1979; Cody *et al.*, 1978).

Captopril is an exceptionally non-toxic drug. It has none of the untoward side-effects often obtained with conventional antihypertensive drugs. It does not normally lower blood pressure below the normotensive level, but can produce a marked hypotensive response in severely volume-depleted patients. Use of captopril at relatively high doses has been associated with a moderate incidence of rash which was readily reversible, apparently non-allergic, and usually not severe.

CONCLUSIONS

Captopril is an antihypertensive drug initially designed for specific and maximally effective binding to the active site of angiotensin-converting enzyme. Such a potent and specific inhibition of a key enzyme involved in hypertensive disease has been found to translate clinically into marked efficacy, large therapeutic ratio, and relative freedom from side effects. Despite some ambiguity with regard to captopril's exact mechanism of action (angiotensin II inhibition versus bradykinin augmentation; antivasoconstrictor versus volume lowering effects), there is general agreement as to its great potential for use as an antihypertensive drug.

REFERENCES

Aguilera, G. and Catt, K. J. (1978). *Proc. natn. Acad. Sci. U.S.A.*, **75**, 4057

Antonaccio, M. J., Rubin, B., Horovitz, Z. P., Laffan, R. J., Goldberg, M. E., High, J. P., Harris, D. N. and Zaidi, I. (1979). *Japan. J. Pharmac.*, **29**, 275

Bakhle, Y. S. (1968). *Nature, Lond.*, **220**, 919
Bengis, R. G., Coleman, T. G., Young, D. B. and McCaa, R. E. (1978). *Circulation Res.*, **43**, Suppl. I, I–45
Bravo, E. L. and Tarazi, R. C. (1979). *Hypertension*, **1**, 39
Brunner, H. R., Gavras, H., Turini, G. A., Waeber, B., Chappuis, P. and McKinstry, D. N. (1978*a*). *Clin. Sci. molec. Med.*, **55**, 293s
Brunner, H. R., Waeber, B., Wauters, J. P., Turini, G., McKinstry, D. and Gavras, H. (1978*b*). *Lancet*, **ii**, 704
Brunner, H. R., Gavras, H., Waeber, B., Kershaw, G. R., Turini, G. A., Vukovich, R. A., McKinstry, D. N. and Gavras, I. (1979). *Ann. intern. Med.*, **90**, 19
Byers, L. D. and Wolfenden, R. (1972). *J. biol. Chem.*, **247**, 606
Case, D. B., Atlas, S. A., Laragh, J. H., Sealey, J. E., Sullivan, P. A. and McKinstry, D. N. (1978). *Progr. cardiovasc. Dis.*, **21**, 195
Case, D. B., Wallace, J. M., Keim, H. J., Weber, M. A., Sealey, J. E. and Laragh, J. H. (1977). *New Eng. J. Med.*, **296**, 641
Cheung, H. S. and Cushman, D. W. (1973). *Biochim. biophys. Acta*, **293**, 451
Cody, R. J., Jr, Tarazi, R. C., Bravo, E. L. and Fouad, F. M. (1978). *Clin. Sci. molec. Med.*, **55**, 453
Cushman, D. W. and Cheung, H. S. (1971). *Biochem. Pharmac.*, **20**, 1637
Cushman, D. W., Cheung, H. S., Sabo, E. F. and Ondetti, M. A. (1977). *Biochemistry*, **16**, 5484
Cushman, D. W., Cheung, H. S., Sabo, E. F. and Ondetti, M. A. (1978). *Progr. cardiovasc. Dis.*, **21**, 176
Cushman, D. W., Pluščec, J., Williams, N. J., Weaver, E. R., Sabo, E. F., Kocy, O., Cheung, H. S. and Ondetti, M. A. (1973). *Experientia*, **29**, 1032
Das, M. and Soffer, R. L. (1975). *J. biol. Chem.*, **250**, 6762
Elisseeva, Y. E., Orekhovich, V. N., Pavlikhina, L. V. and Alexeenko, L. P. (1971). *Clin. chim. Acta*, **31**, 413
Engel, S. L., Schaeffer, T. R., Gold, B. I. and Rubin, B. (1972). *Proc. Soc. exp. Biol. Med.*, **140**, 240
Engel, S. L., Schaeffer, T. R., Waugh, M. H. and Rubin, B. (1973). *Proc. Soc. exp. Biol. Med.*, **143**, 483
Erdös, E. G. (1977). *Fedn Proc. Fedn Am. Socs exp. Biol.*, **36**, 1760
Ferreira, S. H. (1965). *Br. J. Pharmac.*, **24**, 163
Ferreira, S. H., Bartelt, D. C. and Greene, L. S. (1970). *Biochemistry*, **9**, 2583
Ferguson, R. K., Turini, G. A., Brunner, H. R., Gavras, H. and McKinstry, D. N. (1977). *Lancet*, **i**, 775
Freeman, R. H., Davis, J. O., Watkins, B. E., Stephens, G. A. and DeForrest, J. M. (1979). *Am. J. Physiol.*, **236**, F21
Gavras, H., Brunner, H. R., Laragh, J. H., Sealey, J. E., Gavras, I. and Vukovich, R. A. (1974). *New Eng. J. Med.*, **291**, 817
Gavras, H., Brunner, H. R., Turini, G. A., Kershaw, G. R., Tifft, C. P., Cuttelod, S., Gavras, I., Vukovich, R. A. and McKinstry, D. N. (1978). *New Eng. J. Med.*, **298**, 991
Greene, L. J., Camargo, A. C. M., Krieger, E. M., Stewart, J. M. and Ferreira, S. H. (1972). *Circulation Res.*, **30**, **31**, Suppl. II, II–62
Harris, D. N., Heran, C. L., Goldenberg, H. J., High, J. P., Laffan, R. J., Rubin, B., Antonaccio, M. J. and Goldberg, M. E. (1978). *Eur. J. Pharmac.*, **51**, 345
Johnson, J. G., Black, W. D., Vukovich, R. A., Hatch, F. E., Jr., Friedman, B. I., Blackwell, C. F., Shenouda, A. N., Share, L., Shade, R. E., Acchiardo, S. R. and Muirhead, E. E. (1975). *Clin. Sci. molec. Med.*, **48**, 53s
Kato, H. and Suzuki, T. (1977). In *Chemistry and Biology of the Kallikrein-Kinin System in Health and Disease*, (ed. J. J. Pisano and K. F. Austen), US Government Printing Office, Washington, DC, p. 299
Laffan, R. J., Goldberg, M. E., High, J. P., Schaeffer, T. R., Waugh, M. H. and Rubin, B. (1978). *J. Pharmac. exp. Ther.*, **204**, 281
McCaa, R. E., Hall, J. E. and McCaa, C. S. (1978). *Circulation Res.*, **43**, Suppl. I, I–32
Miller, E. D., Jr, Samuels, A. I., Haber, E. and Barger, A. C. (1975). *Am. J. Physiol.*, **228**, 448
Muirhead, E. E., Brooks, B. and Arora, K. K. (1974). *Lab. Invest.*, **30**, 129

Muirhead, E. E., Prewitt, R. L., Brooks, B. and Brosius, W. L., Jr (1978). *Circulation Res.*, **43**, Suppl. I, I–53
Murthy, V. S., Waldron, T. L. and Goldberg, M. E. (1978). *Proc. Soc. exp. Biol. Med.*, **157**, 121
Murthy, V. S., Waldron, T. L., Goldberg, M. E. and Vollmer, R. R. (1977). *Eur. J. Pharmac.*, **46**, 207
Ondetti, M. A., Williams, N. J., Sabo, E. F., Pluščec, J., Weaver, E. R. and Kocy, O. (1971). *Biochemistry*, **10**, 4033
Ondetti, M. A., Rubin, B. and Cushman, D. W. (1977*a*). *Science, N.Y.*, **196**, 441
Ondetti, M. A., Sabo, E. F., Losee, K. A., Cheung, H. S., Cushman, D. W. and Rubin, B. (1977*b*). In *Proceedings of the Fifth American Peptide Symposium* (ed. M. Goodman and J. Meienhofer), John Wiley, New York, p. 576
Ondetti, M. A., Petrillo, E. W., Condon, M. E., Cushman, D. W., Puar, M., Heikes, J. E., Sabo, E. F. and Reid, J. (1978). *Fedn Proc. Fedn Am. Socs exp. Biol.*, **37**, 1386
Peach, M. J. (1977). *Physiol. Rev.*, **57**, 313
Petrillo, E. W., Jr, Ondetti, M. A., Cushman, D. W., Weaver, E. R. and Heikes, J. E. (1978). *Abstracts 176th National Meeting of the American Chemical Society*, MEDI-27
Piquilloud, Y., Reinharz, A. and Roth, M. (1970). *Biochim. biophys. Acta*, **206**, 136
Quiocho, F. A. and Lipscomb, W. N. (1971). *Adv. Protein Chem.*, **25**, 1
Rubin, B., Antonaccio, M. J., Goldberg, M. E., Harris, D. N., Itkin, A. G., Horovitz, Z. P., Panasevich, R. E. and Laffan, R. J. (1978*a*). *Eur. J. Pharmac.*, **51**, 377
Rubin, B., Antonaccio, M. J. and Horovitz, Z. P. (1978*b*). *Progr. cardiovasc. Dis.*, **21**, 183
Rubin, B., Laffan, R. J., Kotler, D. G., O'Keefe, E. H., DeMaio, D. A. and Goldberg, M. E. (1978*c*). *J. Pharmac. exp. Ther.*, **204**, 271
Sibley, P. L., Keim, G. R., Keysser, C. H., Kulesza, J. S., Miller, M. M. and Zaidi, I. H. (1978). *Toxicol. appl. Pharmac.*, **45**, 315
Skeggs, L. T., Jr, Kahn, J. R. and Shumway, N. P. (1956). *J. exp. Med.*, **103**, 295
Vollmer, R. R. and Boccagno, J. A. (1977). *Eur. J. Pharmac.*, **45**, 117
Watkins, B. E., Davis, J. O., Freeman, R. H., DeForrest, J. M. and Stephens, G. A. (1978*a*). *Circulation Res.*, **42**, 813
Watkins, B. E., Davis, J. O., Freeman, R. H., Stephens, G. A. and DeForrest, J. M. (1978*b*). *Proc. Soc. exp. Biol. Med.*, **157**, 245
Weed, W. C., Vaughan, E. D., Jr and Peach, M. J. (1979). *Hypertension*, **1**, 8

16

Inhibitors of arachidonic acid metabolism

S. Moncada and J. R. Vane (Wellcome Research Laboratories,
Langley Court, Beckenham, Kent, BR3 3BS, UK)

INTRODUCTION

Aspirin was first used in medicine some 80 years ago and was certainly not designed as an enzyme inhibitor. Neither were any of the numerous other anti-inflammatory acids which came to the market mainly between 1950 and 1970. However, in 1971, we made the discovery that aspirin inhibits prostaglandin biosynthesis (Vane, 1971; Ferreira *et al.*, 1971; Smith and Willis, 1971). Since then, there has been abundant confirmation that non-steroid anti-inflammatory drugs (aspirin-like drugs) block prostaglandin synthetase in numerous species (including man) and in many biological preparations, ranging from cell-free microsomal preparations of the synthetase itself through to whole organs or organisms. Elsewhere, we have argued (Vane , 1973; Flower *et al.*, 1972; Ferreira and Vane, 1974; Moncada and Vane, 1979) that the concentration of anti-inflammatory drugs achieved after therapeutic dosage (despite plasma protein binding) is adequate to inhibit prostaglandin biosynthesis and this view is substantiated by measurements in man. Smith and Willis (1971) showed that oral aspirin (600 mg) almost completely prevents prostaglandin generation by platelets subsequently isolated and incubated with thrombin. Hamberg (1972) and Samuelsson (1974) calculated daily prostaglandin turnover in man from the amounts of metabolites in the urine. Therapeutic doses of indomethacin, aspirin, or salicylate gave a 77–98 per cent inhibition of prostaglandin turnover. We have also shown (Higgs *et al.*, 1974) that therapeutic doses of aspirin-like drugs substantially reduce the prostaglandin content of synovial fluid taken from arthritic patients. The evidence, therefore, is overwhelming that aspirin-like drugs inhibit the biosynthesis of prostaglandins in all animal species so far tested and that in man this effect is achieved with therapeutic dosage.

THE ARACHIDONIC ACID CASCADE

Arachidonic acid is released from cell membranes, probably by the action of the enzyme phospholipase A_2. From this precursor, a 'cyclo-oxygenase' generates cyclic endoperoxides, isolated and characterized by Nugteren and Hazelhof (1973), Hamberg and Samuelsson (1973, 1974*a,b*), Hamberg *et al.* (1974*a*) and Willis (1974). These intermediates have been variously named but Samuelsson's nomenclature of PGG_2 and PGH_2 is now generally accepted. They are relatively unstable in watery solution, decomposing with a half life of about 5 min.

The enzyme complex 'prostaglandin synthetase' also contains an isomerase which generates PGE_2 from PGH_2. The presence of a reductase which forms $PGF_{2\alpha}$ is now questioned. Hamberg *et al.* (1976) have characterised a non-prostaglandin metabolite of PGG_2 or PGH_2 which they have called thromboxane A_2 (TXA_2). This has a half life of 30 s and it is now evident (Moncada and Vane, 1977) that the activity of rabbit aorta contracting substance (RCS) described by Piper and Vane (1969) was mainly due to TXA_2, along with some endoperoxides (Svensson *et al.*, 1975). We have isolated an enzyme from platelets which generates TXA_2 from the endoperoxides (Needleman *et al.*, 1976). Thromboxane A_2 (or RCS) can also be formed by lungs, spleen and white cells (Hamberg and Samuelsson, 1974*b*; Nijkamp *et al.*, 1977; Higgs *et al.*, 1976).

Metabolism of arachidonic acid to prostacyclin is the dominant pathway in vessel walls. This has been shown for arteries and veins from pig, rabbit and man (Moncada *et al.*, 1976*a,b*, 1977*a*; Gryglewski *et al.*, 1976). The enzyme which metabolises prostaglandin endoperoxides to prostacyclin (prostacyclin synthetase) is most highly concentrated in the intimal surface and progressively decreases towards the adventitial surface (Moncada *et al.*, 1977*b*). Prostacyclin formation followed by chemical breakdown to 6-oxo-$PGF_{1\alpha}$ is also the major route of endoperoxide metabolism in rat stomach (Bunting *et al.*, 1976; Pace-Asciak *et al.*, 1976). Microsomal fractions from other tissues such as rabbit lung and ram seminal vesicle (Cottee *et al.*, 1977) generate smaller amounts of prostacyclin along with other prostaglandin products. However, conditions for prostacyclin formation have been well defined and it is evident that addition of external cofactors such as glutathione will divert arachidonic acid metabolism towards PGE_2 and $PGF_{2\alpha}$ (Cottee *et al.*, 1977). The reports of formation of 6-oxo-$PGF_{1\alpha}$ by lungs (Dawson *et al.*, 1976) and by uterine tissue (Fenwick *et al.*, 1977; Jones *et al.*, 1977) suggest that this route of endoperoxide metabolism will become of increasing importance.

Another enzyme transforms arachidonic acid, and this is not inhibited by concentrations of aspirin-like drugs which affect the cyclo-oxygenase. It is a lipoxygenase, isolated from human platelets and guinea pig lung (Hamberg *et al.*, 1974*b*; Hamberg and Samuelsson, 1974*b*) which generates a 12-hydroperoxide derivative of arachidonic acid (HPETE) and 12-L-hydroxy-5,8,10,14-eicosatetraenoic acid (HETE). The importance of HPETE and HETE in inflammatory processes is as yet unexplored, but there is one report that HETE is chemotactic for polymorphonuclear (PMN) leukocytes (see later).

EFFECTS OF ASPIRIN IN MAN

The effects of the aspirin-like drugs in man include the therapeutic actions of reducing fever, pain, and inflammation and, to a greater or lesser extent, several other effects such as gastrointestinal irritation, renal toxicity, inhibition of platelet aggregation, and delayed and prolonged parturition. When a whole group of chemically dissimilar compounds—and there must be several hundred of them—share the same therapeutic effects, this is strongly indicative of intervention at a single biochemical pathway. When those same compounds have common side effects, there is the further implication that the side effects are also induced by interference with the same biochemical pathway. The proposition is, therefore, that the shared therapeutic effects of the aspirin-like drugs are accounted for by reduction in that biosynthesis of prostaglandins (or other products of endoperoxides) which accompanies pathological processes. As the same time, any prostaglandin biosynthesis necessary for physiological processes will also be reduced, and it is this reduction which leads to the unwanted effects of aspirin-like drugs.

Certainly, with so many dissimilar compounds sharing this common enzyme inhibition, there are bound to be some exhibiting extra effects which may or may not contribute to the anti-inflammatory action. For example, the fenamates also antagonise receptors for prostaglandins (Collier and Sweatman, 1968); this action could add to the anti-inflammatory activity induced by prostaglandin synthetase inhibition. Indomethacin, at rather higher concentrations than are needed to inhibit prostaglandin production, will also inhibit phosphodiesterase (Flores and Sharp, 1972). However, such other activities are not shared by the whole group of aspirin-like compounds and do not in any way detract from the main theory.

What, then, is the evidence in support of the theory? First, prostaglandins or other endoperoxide derivatives, are released when cell membranes are distorted or damaged. Interestingly, cells do not store prostaglandins, so prostaglandin release depends on fresh prostaglandin biosynthesis (Piper and Vane, 1971). Continued release will thus depend on continued synthesis. Second, all cell types so far studied have enzymes associated with the microsomal fraction which can synthesise prostaglandins or other endoperoxide products. Third, the aspirin-like drugs inhibit the synthesis and release of prostaglandins but, in general, do not reduce the inflammatory effects of injected prostaglandins (Ferreira and Vane, 1974).

PROSTAGLANDIN RELEASE IN INFLAMMATION

Many mediators are liberated locally in tissues during inflammatory reactions. Among these are histamine, 5-hydroxytryptamine (5-HT), slow-reacting substance in anaphylaxis (SRS-A), various chemotactic factors, bradykinin, thromboxane A_2, prostaglandins of the E and F series, and lysosomal enzymes. Each substance, as its involvement is proposed or demonstrated, has been studied intensively and over the years attempts have been made to link the actions of anti-inflammatory drugs to an antagonism of the release or actions of each known mediator.

In fact, anti-inflammatory substances have little or no antagonistic activity toward the direct effects of histamine, 5-HT, or bradykinin. Similarly, known potent antagonists, for example of histamine, have very limited use in inflammation. As a consequence of this lack of antagonism, the importance of the long-standing mediators in the maintenance of inflammation must be questioned. At the same time, the finding that aspirin prevents prostaglandin biosynthesis, together with the emergence of prostaglandins as putative mediators of inflammation, increases the importance of the contribution of prostaglandins to this process.

In considering the relative importance of different mediators, we should also take into account differences in tissue sensitivity to them. The fact that some tissues, such as lung, spleen, and platelets, preferentially generate thromboxane A_2 also has obvious implications. In addition, the sequence of mediator release may be important. For instance, in anaphylactic shock there is an explosive and virtually simultaneous release of histamine, SRS-A, TXA_2, PGE_2 and $PGF_{2\alpha}$ (Piper and Vane, 1971). However, in the inflammatory response to subcutaneous injection of carrageenin in the rat, there is a sequential release first of histamine, then bradykinin, and then, much later, prostaglandins with a further increase in histamine levels (Willis, 1969). The breakdown product of prostacyclin, 6-oxo-$PGF_{1\alpha}$ can also be generated in chronic granulomas (Çhang *et al.*, 1976).

Prostaglandin E_2 generation has been shown in many forms of damage to the skin, in both animals and man. These damaging agents include contact dermatitis (Greaves *et al.*, 1971), inflammation due to ultraviolet light (Greaves and Søndergaard, 1970) and scalding (Änggård and Jonsson, 1971).

The invasion of the inflamed area by polymorphonucleocytes may also be important for the maintenance of prostaglandin generation (and thereby inflammation). Phagocytosis substantially increases prostaglandin release from polymorphs (Higgs *et al.*, 1975). In view of the fact that HETE is highly chemotactic (Turner *et al.*, 1975; Goetzl *et al.*, 1977), it will be interesting to determine whether PMN leukocytes can also generate this substance.

THE INFLAMMATORY EFFECTS OF PROSTAGLANDINS

Erythema

Prostaglandins of the E and F series (the ones likely to be generated in inflammation) cause erythema, and PGE_1 is effective at doses as low as 1 ng; for $PGF_{2\alpha}$, 1 μg is needed (Solomon *et al.*, 1968). There are, however, two features of the vascular effects of prostaglandins not shared by other putative mediators of inflammation. The first is a sustained action and the second is the ability to counteract the vasoconstriction caused by substances such as noradrenaline and angiotensin. Intradermal injections or short subdermal infusions of PGEs produce erythema lasting for up to 10 h (Ferreira, 1972). The long-lasting action is an important property of some prostaglandins, in that the appearance and the magnitude of their effects depend not only on the actual concentration but also on the duration of their release or infusion (Ferreira, 1972). Prostacyclin and its degradation product 6-oxo-$PGF_{1\alpha}$ induce erythema when injected into the rabbit skin but are less active and of shorter duration than prostaglandin E_2 (Peck and Williams, 1978).

Oedema

Prostaglandins, like bradykinin, histamine and 5-hydroxytryptamine, cause increased vascular leakage at the postcapillary and collecting venules (Kaley and Weiner, 1971). Unlike erythema, which is caused by local pooling of blood due to a relaxation of the smooth muscles of the walls of the arterioles and venules, increased vascular permeability is thought to result from the contraction of the venular endothelial cells (Majno *et al.*, 1972). In fact, prostaglandins produce vasodilatation more effectively than oedema.

Prostacyclin, PGE_1 or PGE_2 potentiate carageenin-induced paw oedema in the rat (Higgs *et al.*, 1978*a*; Komoriya *et al.*, 1978), increase vascular permeability (Murota and Morita, 1978) and enhance the vascular permeability induced by other agents, such as bradykinin, histamine and 5-hydroxytryptamine (Peck and Williams, 1978; Komoriya *et al.*, 1978). Thus, prostaglandin release can sensitise (as happens with pain receptors, see later) blood vessels to the permeability effects of other mediators, and the actions of anti-inflammatory drugs on oedema can be explained by removal of this sensitisation. In other words, the contribution which prostaglandins make to oedema of inflammation is by increasing the effects of the other known mediators. This potentiating effect has been shown for vascular leakage caused by histamine, bradykinin, carrageenin, or various immune responses (Smith and Willis, 1971; Moncada *et al.*, 1973; Williams and Morley, 1977).

Williams and Peck (1977) have proposed that there is a dichotomy between those inflammatory mediators which are predominantly vasodilators (such as prostaglandins) and those more important in increasing vascular permeability (e.g. histamine and bradykinin). They have shown that arachidonic acid, the precursor of PGE_2 and PGI_2, increases vasodilatation without increasing exudation in rabbit skin. Furthermore, indomethacin inhibits the production of the vasodilator but not the permeability-increasing mediator in an inflammatory response. It is probable, therefore, that prostacyclin contributes to the formation of oedema through its vasodilator effects.

Pain

In man, prostaglandins cause headache and pain along the veins into which they are infused (Bergstrom *et al.*, 1959). When given intradermally (Karim, 1971) or intramuscularly (Juhlin and Michaelsson, 1969) in concentrations higher than those occurring in inflammation, PGE_1 causes long-lasting pain. However, induction of hyperalgesia (i.e. a state in which pain can be elicited by normally painless mechanical or chemical stimulation) seems to be a typical effect of low concentrations of prostaglandins. Ferreira's (1972) subdermal infusion experiments showed that the hyperalgesic effect of PGE_1 is cumulative, because it depends not only on concentration, but on duration of the infusions. During separate subdermal infusions of PGE_1, bradykinin or histamine (or a mixture of bradykinin and histamine), there was no overt pain; but when PGE_1 was infused before bradykinin or histamine (or a mixture of both), strong pain occurred (Ferreira, 1972). This sensitising action of prostaglandins to pain induced by bradykinin occurs also in the dog's spleen (Ferreira *et al.*, 1973), a preparation used by Guzman *et al.* (1964) and Lim and colleagues (1964) to show that aspirin-like drugs act peripherally as analgesics. Moncada *et al.* (1972), using either PGE_1 or PGE_2, have confirmed these results in a preparation of the dog knee joint. Juan and Lembeck (1974) have also shown, in rabbit ear preparations, that PGE_1 enhances the nociceptive actions of bradykinin, substance P, 5-hydroxytryptamine and histamine.

Staszewska-Barczak *et al.* (1976) measured the reflex rise in blood pressure induced by topical application of bradykinin to the exposed left ventricle of the anaesthetised dog. This, too, is decreased by indomethacin and increased again by PGE_1 or PGE_2; prostaglandin release may, therefore, contribute to the pain associated with cardiac ischaemia. All these results, taken together, suggest that sensitisation by prostaglandins of the sensory nerves to chemical stimulation by other substances is a general phenomenon.

Another importance observation concerns pruritus (Ferreira, 1972). Neither histamine, nor bradykinin, nor PGE_1 infusions by themselves cause itch. However, when PGE_1 was infused with histamine, itching was always recorded. This role of prostaglandins in exacerbating the itch produced by histamine was confirmed by Greaves and McDonald-Gibson (1973) who also found that PGE_1 infused with bradykinin causes pain rather than itch. Prostacyclin, but not 6-oxo-$PGF_{1\alpha}$, induces strong hyperalgesia when injected in the paw of rats and in this respect PGI_2 was more active than PGE_2 (Higgs *et al.*, 1978*a*). Moreover, unlike PGE_2, prostacyclin induces a short-lasting hyperalgesia (Ferreira *et al.*, 1978).

The analgesic action of morphine occurs centrally, whereas aspirin has a peripheral effect (Guzman *et al.*, 1964; Lim *et al.*, 1964). By preventing prostaglandin release in inflammation, aspirin prevents the sensitisation of the pain receptors to mechanical stimulation or to the other chemicals. This hypothesis also explains why aspirin is ineffective as an analgesic in uninflamed tissues, as shown by the Randall and Selitto (1957) test. Presumably, aspirin is only effective as an analgesic in tissues in which increased prostaglandin formation is taking place.

Fatty acid hydroperoxides can also cause pain when injected intradermally in man (Ferreira, 1972). Thus, lipoperoxides formed during arachidonic acid metabolism may also be important as pain-producing substances. It should be remembered too that TXA_2 has strong pharmacological activity in that it contracts rabbit aorta and many other arterial muscle strips (Palmer *et al.*, 1973), as do the lipid peroxides mentioned above. These observations take on a new importance now that it is known that lipid peroxides can be generated enzymically (Hamberg *et al.*, 1974*b*; Hamberg and Samuelsson, 1974*b*). This pathway, together with the formation of TXA_2, accounts for the major part of the metabolism of archidonic acid in platelets, lung and white cells. These observations, together with the high activity of the endoperoxides and to TXA_2, all support the suggestion (Vane, 1972) that each intermediate generated in the cascade of arachidonic acid metabolism might have its own contribution to make to the inflammatory process. In this context, it is interesting that Kuehl *et al.* (1977) have recently questioned the role of PGE_2 in inflammation. They suggested that the unstable and possibly non-prostaglandin metabolites of PGG_2 are more likely to be directly relevant to inflammation than either PGG_2 itself or PGE_2. Candidates they suggested as primary effectors included TXA_2 and a destructive moiety (not singlet oxygen) tentatively designated as a free radical, formed when the hydroperoxide PGG_2 is transformed to the hydroxy-containing PGH_2. A similar destructive moiety could be formed in the transformation of HPETE to HETE.

Fever

Fever is often associated with an inflammatory process. Prostaglandin E_1 is the most powerful pyretic agent known, when injected either into cerebral ventricles or directly into the anterior hypothalamus (Feldberg and Saxena, 1971). The hyperthermic effect is dose dependent, almost immediate, and lasts for about 3 h. Arachidonic acid, the precursor of the prostaglandins, also induces fever when injected intraventricularly in rats but, in contrast to the stable prostaglandins, its effect appears after some latency and can be blocked by aspirin (Splawinski *et al.*, 1974). Prostaglandin E_1 and E_2 causes fever by an action on the same region as that on which monoamines and pyrogens act to affect temperature. Fever also occurs during induction of human abortion with PGE_2 or $PGF_{2\alpha}$ (Hendricks *et al.*, 1971); elevation of temperature shows better correlation with the infusion rate than with the time course of the abortion.

As in peripheral inflammatory responses, there is a generation of PGE-like substance in the central nervous system of cats during fever (Feldberg and Gupta, 1973), and the concentrations in the cerebrospinal fluid rise after intravenous pyrogen by 2.5–4.0 times, sometimes to as much as 35 ng ml^{-1}. Thus, the generation of a prostaglandin in some areas of the central nervous system or its presence in the general circulation may induce fever in animals, including man. Whether prostacyclin or TXA_2 are involved in the effect remains to be investigated. Feldberg and Milton (1978) have developed the concept that the behavioural changes accompanying high fever may also be due to the effects of prostaglandins in the central nervous system.

Aspirin-like drugs do not abolish either the formation of endogenous pyrogen by leukocytes or the pyretic action of prostaglandins injected into the third ventricle of cats (Clark and Moyer, 1972). However, they inhibit both the generation of prostaglandins in the central nervous system, and the fever caused by pyrogens or 5-hydroxytryptamine given into the cerebral ventricles (Milton, 1973).

So far there are no reports of the effect of TXA_2 on inflammation and its role will be difficult to assess in view of its very short half-life ($\simeq$ 30 s). However, phagocytosing white cells (Higgs *et al.*, 1976; Goldstein *et al.*, 1977) and aggregating platelets (Svensson *et al.*, 1975; Needleman *et al.*, 1976) produce TXA_2, and thromboxane B_2 (TBX_2), the stable end product of TXA_2, has been reported to be chemotactic (Boot *et al.*, 1976). Moreover, fluid obtained from carrageenin-induced granuloma converts arachidonic acid into TBX_2 *in vitro* (Chang *et al.*, 1977). In view of this finding, selective inhibition of TXA_2 formation could modify the inflammatory process.

Selective inhibition of thromboxane formation has been attempted using benzydamine (Moncada *et al.*, 1976*c*), a phenyl phosponate derivative of phloretin phosphate designated as N-0164 (Kulkarni and Eakins, 1976) and imidazole or 1-methylimidazole (Moncada *et al.*, 1977*c*; Blackwell *et al.*, 1978) and some endoperoxide analogues (Gorman *et al.*, 1977). Imidazole has little activity on cyclo-oxygenase ($IC_{50} > 800$ μg ml^{-1}) but does have anti-inflammatory activity (Puig Muset *et al.*, 1972). Whether this effect is a result of inhibition of thromboxane synthetase or any other action remains to be clarified. A recent report suggests that *in vivo* in an acute experimental model of arthritis in the pigeon, imidazole acts more like a cyclo-oxygenase inhibitor (Peskar *et al.*, 1978). Interestingly, the main prostaglandin formed was PGD_2 (Anhut *et al.*, 1979).

The contribution of prostacyclin as opposed to other products of arachidonic acid in inflammation is not yet clear. The effect of lipid peroxides, which are selective inhibitors of PGI_2 formation (Salmon *et al.*, 1978), or the use of a recently described antiserum which inactivates prostacyclin (Bunting *et al.*, 1978) will help the elucidation of this question.

ANTI-INFLAMMATORY STEROIDS

Initially, there seemed little evidence that the mechanism of action of the anti-inflammatory steroids was connected with the prostaglandin system, but evidence has now accumulated. Vane (1971) could not demonstrate a direct inhibitory effect of steroids on a cell-free preparation of prostaglandin synthetase and in perfused spleens there was also no effect (Ferreira *et al.*, 1971). However, Greaves and McDonald-Gibson (1972) reported inhibition of prostaglandin biosynthesis by unseparated homogenates of skin, albeit using high doses of fluocinolone or hydrocortisone. Lewis and Piper (1975) suggested that steroids inhibit prostaglandin release from the cell, rather than synthesis. Gryglewski *et al.* (1975) found that either indomethacin or dexamethasone inhibit prostaglandin release induced by noradrenaline in the rabbit mesenteric bed. However, the dexamethasone inhibition is easily reversed by infusions of arachidonic acid, whereas that of indomethacin is not. They concluded that steroids may work by impairing the availability of the substrate for prostaglandin generation. Kantrowitz and his colleagues (1975) have also found that steroids reduce prostaglandin production by isolated synovial membrane cells in culture. Nijkamp *et al.* (1976) demonstrated that stimulation by RCS-releasing factor of phospholipase A_2 activity in guinea pig lungs, leading to RCS formation, is inhibited by steroids in proportion to their anti-inflammatory activity. Thus steroids could be affecting the prostaglandin synthetase system by limiting the substrate availability, possibly through 'membrane stabilisation' or by inhibition of phospholipase A_2. Such an activity would also reduce substrate availability for other enzymes which metabolise arachidonic acid, such as the lipoxygenase (which is not inhibited by aspirin). If this were so, it might be the basis of the improved performance of the steroids over the non-steroids against inflammation and allergic conditions. Recent exciting work by Flower and Blackwell (1979) has demonstrated that anti-inflammatory steroids induce the generation and release of a peptide which inhibits phospholipase A_2.

THE LIPOXYGENASE PATHWAY

In 1974 an alternative pathway of arachidonic acid oxygenation was described (Hamberg and Samuelsson, 1974*a*, Nugteren, 1975) with the discovery of a platelet lipoxygenase. The hydroxy acid product of arachidonate lipoxygenase, 12-L-hydroxyeicosatetraenoic acid (HETE), is a potent chemotactic agent (Turner *et al.*, 1975; Goetzl *et al.*, 1977) and a lipid function derived from *Escherichia coli* has the same chemotactic properties and chromatographic mobility as HETE (Tainer *et al.*, 1975). Leukocyte lipoxygenases generate a number of different mono- and dihydroxy acids (Borgeat *et al.*, 1976; Samuelsson, 1979) which may also be involved in chemotaxis.

At concentrations which are effective against cyclo-oxygenase, the aspirin-like drugs do not inhibit lipoxygenase (Hamberg and Samuelsson, 1974*a*) and this may explain why these drugs do not abolish leukocyte accumulation *in vivo*. BW 755C (3-amino-l-[*m*-(trifluoromethyl)phenyl]-2-pyrazoline) inhibits both pathways of arachidonate metabolism *in vitro*, causes a dose-dependent reduction in carrageenin-induced oedema in the rat paw and has a significantly greater effect on leukocyte migration *in vivo* than indomethacin (Higgs *et al.*, 1978*b*, 1979). BW 755C, which reduces prostaglandin concentrations in experimental inflammation, may inhibit leukocyte migration by preventing the generation of chemotactic hydroxy acids.

BW 755C has a similar effect to dexamethasone on leukocyte migration and prostaglandin production *in vivo* (Higgs *et al*., 1979) and it is possible that the simultaneous inhibition of both pathways of arachidonic acid oxygenation may lead to more selective anti-inflammatory actions, without the complicating side-effects of the corticosteroids. Such an anti-inflammatory mechanism would also have advantages over the aspirin-like drugs in the treatment of chronic inflammation by having a greater effect on leukocyte migration (Higgs *et al*., 1979).

Recent work by Adcock *et al*. (1978*a,b*) has also highlighted the possible importance of lipid peroxides on lung pathophysiology. They have demonstrated that, unlike indomethacin, a compound which inhibits both pathways does not increase mediator release (SRS-A and histamine) during anaphylactic reaction in isolated guinea pig lungs. This, coupled with the fact that exogenously added lipid peroxides greatly potentiate mediator release, suggests that endogenously released products of the lipoxygenase pathway in some way act by potentiating the anaphylactic reaction.

SIDE EFFECTS OF ASPIRIN-LIKE DRUGS

Some of the toxic effects of large doses of aspirin-like drugs may be unrelated to inhibition of prostaglandin formation. However, those which are shared may well be due to inhibition of prostaglandin biosynthesis leading to other unwanted side effects in organs depending on prostaglandins for normal physiological function.

The aspirin-like drugs all induce varying degrees of gastro-intestinal irritation, which may lead to ulceration. The prostacyclin breakdown product, 6-oxo-$PGF_{1\alpha}$, is the major product of endoperoxide metabolism in homogenates of rat stomach (Pace-Asciak, 1976). We have now shown that prostacyclin is the major prostaglandin product of the gastric mucosa of several species (Moncada *et al*., 1977*d*) and, furthermore, that it is a potent vasodilator of rat stomach mucosa *in vivo* (Whittle *et al*., 1978). It also reduces acid secretion induced by pentagastrin (Whittle *et al*., 1978). Thus, prostacyclin release may be involved both in functional hyperaemia of the mucosa during acid secretion and in acting as a natural brake on the secretion. Removal or reduction of prostacyclin formation in the stomach could, therefore, lead to increased acid secretion and relative hypoxia, both conditions that could promote mucosal damage.

Another possibility is that inhibition of prostaglandin biosynthesis in the stomach diverts arachidonic acid metabolism into another pathway, such as that which produces HPETE and HETE. One or both of these substances (or the free radical formation) could cause damage.

The anti-inflammatory acids also cause varying degrees of nephrotoxicity, with some incidence of papillary necrosis. Some, such as phenylbutazone, lead to retention of sodium chloride and water. The prostaglandins are natriuretic (McGiff *et al*., 1970) and are found in renal medulla (Daniels *et al*., 1967), together with prostaglandin synthetase (Crowshaw, 1971), which is located in cells forming the collecting tubules (Janszen and Nugteren, 1971). Thus, some of the renal side effects of anti-inflammatory drugs may depend on their interaction with prostaglandin synthetase in the kidney.

The major concern in current research on new drugs is to minimise side effects. To minimise those commonly associated with broad-spectrum, non-specific, anti-inflammatory drugs, Shen (1972) suggested a cocktail mixture consisting of several narrow-spectrum agents, each acting specifically at one of the many facets of complex inflammatory reactions, such as lysosomal enzymes, inflammatory mediators, and so on. Considering the differential sensitivity of the prostaglandin synthetases, it may also be possible to develop specific inhibitors for the synthetase of each tissue or group of tissues. If such a search could include the requirement that the synthetase of stomach and kidney were *not* inhibited, the common side effects may be eliminated.

NEW CLINICAL USES FOR ASPIRIN-LIKE DRUGS

The discovery that aspirin-like drugs inhibit prostaglandin synthesis has suggested some new clinical uses. These include the treatment of Bartter's syndrome. (Norby *et al.*, 1976), treatment of the refractory diarrhoea accompanying some cases of medullary carcinoma of the thyroid and Crohn's disease (Shafran *et al.*, 1977), and closure of patent ductus arteriosus (Heymann *et al.*, 1976; Nadas, 1976). Moreover, as prostaglandins might play a role in the pathogenesis of malignant osteolytic metastases, aspirin-like drugs have been suggested for their treatment (Anon., 1976). They have also been proposed to prevent premature labour and abortion (Zuckerman *et al.*, 1974) and certain forms of renal disease (Romero *et al.*, 1976).

CONCLUSIONS

Aspirin-like drugs inhibit prostaglandin biosynthesis in concentrations similar to those in body fluids during therapy. All the evidence, together with the actions of prostaglandins, overwhelmingly supports the theory that this anti-enzyme effect is the mechanism of action of aspirin-like drugs. The aspirin-like drugs interfere with biosynthesis at the cyclo-oxygenase stage, so that the generation of prostaglandin endoperoxides (PGG_2 and PGH_2) and all of their products is also reduced or abolished. The metabolite of arachidonic acid most strongly established as a mediator of inflammation is PGE_2. Other products, such as prostacyclin, thromboxane A_2 and the 'free radical' liberated by conversion of PGG_2 to PGH_2 may also play a role, but their involvement has yet to be clearly defined.

The fact that one group of substances—the aspirin-like drugs—exhibit the otherwise unrelated and diverse properties of preventing pain, fever, and inflammation simply reflects the involvement of prostaglandins in all of these pathological processes. Similarly, the shared side effects of aspirin-like drugs may also reflect the involvement of prostaglandins in physiological processes.

Despite the general acceptance of this unifying and scientifically satisfactory theory, it has been challenged by Smith (1975, 1978) mainly on the grounds that salicylic acid has no inhibitory effect on prostaglandin synthetase *in vitro* but is as powerful as aspirin as an anti-inflammatory agent *in vivo*. Other arguments are based on relative dosages needed to inhibit synthetases from different tissues. There is clear evidence that different synthetases have different sensitivities to aspirin-like drugs (Flower and Vane, 1974), and this has been confirmed recently in that aspirin is far more effective against platelet synthetase than that from cells obtained from vessel walls (Baenziger *et al.*, 1977). The obvious hypothesis that salicylic acid is converted to a synthetase inhibitor *in vivo* (Flower and Vane, 1974) may not stand the test of time but this does not detract from the strength of the overall theory linking the anti-inflammatory actions of aspirin-like drugs to their biochemical inhibition of prostaglandin biosynthesis. Indeed, it can be confidently predicted that any compound which strongly inhibits prostaglandin synthetase *in vitro* will (if it is absorbed and not rapidly metabolised) have an anti-inflammatory effect. This is the common experience of many pharmaceutical companies which screen several thousands of compounds. This is not to say that all anti-inflammatory drugs (e.g. gold and penicillamine) have to work by this mechanism, for inflammation is a multimediated response. There will surely be other types of drugs acting on other systems and perhaps, one day, we shall see a potent and relatively non-toxic anti-rheumatic drug which is seen to reverse the course of the disease. Until that day comes, we shall have to be satisfied with ameliorating the signs and symptoms with drugs such as prostaglandin synthetase inhibitors.

REFERENCES

Adcock, J. J., Garland, L. G., Moncada, S. and Salmon, J. A. (1978*a*). *Prostaglandins*, **16**, 163

Adcock, J. J., Garland, L. G., Moncada, S. and Salmon, J. A. (1978*b*). *Prostaglandins*, **16**, 179

Änggård, E. and Jonsson, C. E. (1971). *Acta physiol. scand.*, **81**, 440

Anhut, H., Brune, K., Frölich, J. C. and Peskar, B. A. (1979). *Br. J. Pharmac.*, **65**, 357

Anon. (1976). *Lancet*, **ii**, 1063

Baenziger, N. L., Dillender, M. J. and Majerus, P. W. (1977). *Biochem. Biophys. Res. Commun.*, **78**, 294

Bergstrom, S., Duner, H., von Euler, U. S., Pernow, B. and Sjovall, J. (1959). *Acta physiol. scand.*, **45**, 145

Blackwell, G. J., Flower, R. J., Russell-Smith, N., Salmon, J. A., Thorogood, P. B. and Vane, J. R. (1978). *Br. J. Pharmac.*, **64**, 436P

Boot, J. R., Dawson, W. and Kitchen, G. A. (1976). *J. Physiol., Lond.*, **257**, 47P

Borgeat, P., Hamberg, M. and Samuelsson, B. (1976). *J. biol. Chem.*, **251**, 7816

Bunting, S., Gryglewski, R., Moncada, S. and Vane, J. R. (1976). *Prostaglandins*, **12**, 897

Bunting, S., Moncada, S., Reed, P., Salmon, J. A. and Vane, J. R. (1978). *Prostaglandins*, **15**, 565

Chang, W.-C., Murota, S.-I. and Tsurufuji, S. (1977). *Prostaglandins*, **13**, 17

Chang, W.-C., Murota, S.-I., Matsuo, M. and Tsurufuji, S. (1976). *Biochem. Biophys. Res. Commun.*, 72, 1259

Clark, W. G. and Moyer, S. G. (1972). *J. Pharmac. exp. Ther.*, **181**, 183

Collier, H. O. J. and Sweatman, W. J. F. (1968). *Nature, Lond.*, **219**, 864

Cottee, F., Flower, R. J., Moncada, S., Salmon, J. and Vane, J. R. (1977). *Prostaglandins*, **14**, 413

Crowshaw, K. (1971). *Nature new Biol.*, **231**, 240

Daniels, E. G., Hinman, J. W., Leach, B. E. and Muirhead, E. E. (1967). *Nature, Lond.*, **215**, 1298

Dawson, W., Boot, J. R., Cockerill, A. F., Mallen, D. N. B. and Osborne, D. J. (1976). *Nature, Lond.*, **262**, 699

Feldberg, W. and Gupta, K. P. (1973). *J. Physiol., Lond.*, **228**, 41

Feldberg, W. and Milton, A. S. (1978). In *Inflammatory and Anti-inflammatory Drugs* (ed. J. R. Vane and S. H. Ferreira), Springer-Verlag, Berlin, p. 617

Feldberg, W. and Saxena, P. N. (1971). *J. Physiol., Lond.*, **217**, 546

Fenwick, L., Jones, R. L., Naylor, B., Poyser, N. L. and Wilson, M. H. (1977). *Br. J. Pharmac.*, **59**, 191

Ferreira, S. H. (1972). *Nature, Lond.*, **240**, 200
Ferreira, S. H. and Vane, J. R. (1974). *A. Rev. Pharmac.*, **14**, 57
Ferreira, S. H., Moncada, S. and Vane, J. R. (1971). *Nature new Biol.*, **231**, 237
Ferreira, S. H., Moncada, S. and Vane, J. R. (1973). *Br. J. Pharmac.*, **49**, 86
Ferreira, S. H., Nakamura, M. and Abreu Castro, M. S. (1978). *Prostaglandins*, **16**, 31
Flores, A. G. A. and Sharp, G. W. G. (1972). *Am. J. Physiol.*, **223**, 1392
Flower, R. J. and Blackwell, G. J. (1979). *Nature, Lond.*, **278**, 456
Flower, R. J. and Vane, J. R. (1974). *Prostaglandin Synthetase Inhibitors* (ed. H. J. Robinson and J. R. Vane), Raven Press, New York, p. 9
Flower, R. J., Gryglewski, R., Herbaczynska-Cedro, K. and Vane, J. R. (1972). *Nature new Biol.*, **238**, 104
Goetzl, E. J., Woods, J. M. and Gorman, R. R. (1977). *J. clin. Invest.*, **59**, 179
Goldstein, I. M., Malmsten, C. L., Kaplan, H. B., Kindahl, H., Samuelsson, B. and Weissman, G. (1977). *Clin. Res.*, **25**, 518A
Gorman, R. R., Bundy, G. L., Peterson, D. C., Sun, F. F., Miller, O. V. C. and Fitzpatrick, F. A. (1977). *Proc. natn. Acad. Sci. U.S.A.*, **74**, 4007
Greaves, M. W. and McDonald-Gibson, W. (1972). (1972). *Br. med. J.*, **ii**, 83
Greaves, M. W. and McDonald-Gibson, W. (1973). *Br. med. J.*, **iii**, 608
Greaves, M. W. and Søndergaard, J. (1970). *J. invest. Dermatol.*, **54**, 365
Greaves, M. W., Søndergaard, J. and McDonald-Gibson, W. (1971). *Br. med. J.*, **iii**, 258
Gryglewski, R., Bunting, S., Moncada, S., Flower, R. J. and Vane, J. R. (1976). *Prostaglandins*, **12**, 658–713
Gryglewski, R., Panczeko, B., Korbut, R., Grodzinska, L. and Ocetkiewicz, A. (1975). *Prostaglandins*, **10**, 343
Guzman, F., Braun, C., Lim, R. K. S., Potter, G. D. and Rodgers, D. W. (1964). *Arch. int. Pharmacodyn. Ther.*, **149**, 571
Hamberg, M. (1972). *Biochem. Biophys. Res. Commun.*, **49**, 720
Hamberg, M. and Samuelsson, B. (1973). *Proc. natn. Acad. Sci. U.S.A.*, **70**, 899–903
Hamberg, M. and Samuelsson, B. (1974*a*). *Proc. natn. Acad. Sci. U.S.A.*, **71**, 3400
Hamberg, M. and Samuelsson, B. (1974*b*). *Biochim. biophys. Acta.*, **61**, 942
Hamberg, M., Svensson, J. and Samuelsson, B. (1974*b*). *Proc. natn. Acad. Sci. U.S.A.*, **71**, 3824
Hamberg, M., Svensson, J. and Samuelsson, B. (1976). *Adv. Prostaglandin Thromboxane Res.*, **1**, 19
Hamberg, M., Svensson, J., Wakabayashi, T. and Samuelsson, B. (1974*a*). *Proc. natn. Acad. Sci. U.S.A.*, **71**, 345
Hendricks, C. H., Brenner, W. E., Ekbladh, L., Brotanek, V. and Fishburne, J. I. (1971). *Am. J. Obstet. Gynec.*, **3**, 564
Heymann, M. A., Rudolph, A. M. and Silverman, N. H. (1976). *New Engl. J. Med.*, **295**, 530
Higgs, E. A., Moncada, S. and Vane, J. R. (1978*a*). *Prostaglandins*, **16**, 153
Higgs, G. A., Bunting, S., Moncada, S. and Vane, J. R. (1976). *Prostaglandins*, **12**, 749
Higgs, G. A., Copp, F. C., Denyer, C. V., Flower, R. J., Tateson, J. E., Vane, J. R. and Walker, J. M. G. (1978*b*). *Proc. VII int. Cong. Pharmac. Paris*
Higgs, G. A., Flower, R. J. and Vane, J. R. (1979). *Biochem. Pharmac.*, **28**, 1959
Higgs, G. A., McCall, E. and Youlten, L. J. F. (1975). *Br. J. Pharmac.*, **53**, 539
Higgs, G. A., Vane, J. R., Hart, F. D. and Wojtulewski, J. A. (1974). *Prostaglandin Synthetase Inhibitors* (ed. H. J. Robinson and J. R. Vane), Raven Press, New York, p. 165
Janszen, F. H. A. and Nugteren, D. H. (1971). *Histochemie*, **27**, 159
Jones, R. L., Poyser, N. L. and Wilson, N. H. (1977). *Br. J. Pharmac.*, **59**, 436
Juan, H. and Lembeck, F. (1974). *Naunyn Schmiedeberg's Arch. Pharmac.*, **283**, 151
Juhlin, S. and Michaelsson, G. (1969). *Acta derm. venereol., Stockholm*, **49**, 251
Kaley, G. and Weiner, R. (1971). *Ann. N.Y. Acad. Sci.*, **180**, 339
Kantrowitz, F., Robinson, D. W., McGuire, M. B. and Levine, L. (1975). *Nature, Lond.*, **258**, 737
Karim, S. M. M. (1971). *Ann. N.Y. Acad. Sci.*, **180**, 483
Komoriya, K., Ohmori, H., Azuma, A., Kurozumi, S., Hashimoto, Y., Nicolaou, K. C., Barnette, W. G. and Magolda, R. L. (1978). *Prostaglandins*, **15**, 557

Kuehl, F. A., Humes, J. L., Egan, R. W., Ham, E. A., Beveridge, G. C. and Van Arman, C. G. (1977). *Nature, Lond.*, **265**, 170
Kulkarni, P. S. and Eakins, K. E (1976). *Prostaglandins,* **12**, 465
Lewis, G. P. and Piper, P. J. (1975). *Nature, Lond.*, **254**, 308
Lim, R. K. S., Guzman, F., Rodgers, D. W., Goto, K., Braun, C., Dickerson, G. D. and Engle, R. J. (1964). *Arch. int. Pharmacodyn. Ther.*, **152**, 25
Majno, G., Ryan, G. B., Gabbiani, G., Horshel, B. J., Irle, C. and Joris, I. (1972). In *Inflammation, Mechanism and Control* (ed. I. A. Lepow and P. A. Ward), Academic Press, New York, p. 13
McGiff, J. C., Crowshaw, K., Terragno, N. A., Lonigro, A. J., Strand, J. C., Williamson, M. A., Lee, J. B. and Ng, K. K. F. (1970). *Circulation Res.*, **27**, 765
Milton, A. S. (1973). In *Advances in the Biosciences (International Conference on Prostaglandins)* (ed. S. Bergstrom and S. Bernhard), Pergamon Press, Braunschweig, p. 495
Moncada, S. and Vane, J. R. (1977). In *Biochemical Aspects of Prostaglandins and Thromboxanes* (ed. N. Karasch and J. Fried), Academic Press, New York, p. 155
Moncada, S. and Vane, J. R. (1979). *Adv. intern. Med.*, **24**, 1–22
Moncada, S., Ferreira, S. H. and Vane, J. R. (1972). In *Biochemical and Pharmacological Aspects of Imidazole* (*Fifth Congress on Pharmacology, San Francisco*) (ed. P. Puig Muset, P. Puig Parellada and J. Martin Esteve), JIMS, Barcelona, p. 160 (abstract)
Moncada, S., Ferreira, S. H. and Vane, J. R. (1973). *Nature, Lond.*, **246**, 217
Moncada, S., Higgs, E. A. and Vane, J. R. (1977*a*). *Lancet*, **i**, 18
Moncada, S., Gryglewski, R., Bunting, S. and Vane, J. R. (1976*a*). *Nature, Lond.*, **263**, 663
Moncada, S., Gryglewski, R., Bunting, S. and Vane, J. R. (1976*b*). *Prostaglandins,* **12**, 715
Moncada, S., Herman, A. G., Higgs, G. A. and Vane, J. R. (1977*b*). *Thrombosis Res.*, **11**, 323
Moncada, S., Needleman, P., Bunting, S. and Vane, J. R. (1976*c*). *Prostaglandins,* **12**, 323
Moncada, S., Salmon, J. A., Vane, J. R. and Whittle, B. J. R. (1977*d*). *J. Physiol., Lond.*, **275**, 4P
Moncada, S., Bunting, S., Mullane, K. M., Thorogood, P., Vane, J. R., Raz, A. and Needleman, P. (1977*c*). *Prostaglandins,* **13**, 611
Murota, S.-I. and Morita, I. (1978). *Prostaglandins,* **15**, 297
Nadas, A. S. (1976). *New Engl. J. Med.*, **295**, 563
Needleman, P., Moncada, S., Bunting, S., Vane, J. R., Hamberg, M. and Samuelsson, B. (1976). *Nature, Lond.*, **261**, 559
Nijkamp, F. P., Flower, R. J., Moncada, S. and Vane, J. R. (1976). *Nature, Lond.*, **263**, 479
Nijkamp, F. P., Moncada, S., White, H. L. and Vane, J. R. (1977). *Eur. J. Pharmac.*, **44**, 179
Norby, L., Flamenbaum, W., Lentz, R. and Ramwell, P. (1976). *Lancet,* **ii**, 604
Nugteren, D. A. (1975). *Biochim. biophys. Acta,* **380**, 299
Nugteren, D. and Hazelhof, E. (1973). *Biochim. biophys. Acta,* **326**, 448
Pace-Asciak, C. (1976). *Experientia,* **32**, 291
Pace-Asciak, C., Nashat, M. and Menon, N. K. (1976). *Biochim. biophys. Acta,* **424**, 323
Palmer, M. A., Piper, P. J. and Vane, J. R. (1973). *Br. J. Pharmac.*, **49**, 226
Peck, M. J. and Williams, T. J. (1978). *Br. J. Pharmac.*, **62**, 464P
Peskar, B. A., Glatt, M., Anhut, H. and Brune, K. (1978). *Eur. J. Pharmac.*, **50**, 437
Piper, P. J. and Vane, J. R. (1969). *Nature, Lond.*, **223**, 29
Piper, P. J. and Vane, J. R. (1971). *Ann. N.Y. Acad. Sci.*, **180**, 363
Puig Muset, P., Puig Parallada, P. and Martin Esteve, J. (eds) (1972). *Biochemical and Pharmacological Aspects of Imidazole (5th International Congress on Pharmacology, San Francisco, July* 1972), JIMS, Barcelona
Randall, L. O. and Selitto, J. J. (1957). *Arch. int. Pharmacodyn. Ther.*, **111**, 409
Romero, J. C., Duncap, C. L. and Strong, C. G. (1976). *J. clin. Invest.*, **58**, 282
Salmon, J. A., Smith, D. R., Flower, R. J., Moncada, S. and Vane, J. R. (1978). *Biochim. biophys. Acta,* **523**, 250
Samuelsson, B. (1974). *Prostaglandin Synthetase Inhibitors* (ed. H. J. Robinson and J. R. Vane) Raven Press, New York, p. 99
Samuelsson, B. (1979). In *Proceedings of the International Congress on Inflammation, Bologna, 1978*, Raven Press, New York, in press
Shafran, I., Maurer, W. and Thomas, F. B. (1977). *New Engl. J. Med.*, **296**, 694
Shen, T. Y. (1972). *Angew. Chem. int. Edn,* **11**, 460

Smith, J. B. and Willis, A. L. (1971). *Nature new Biol.*, **231**, 235
Smith, M. J. H. (1975). *Agents Actions*, **5**, 315
Smith, M. J. H. (1978). *Agents Actions*, **8**, 427
Solomon, L. M., Juhlin, L. and Kirschenbaum, M. B. (1968). *J. invest. Dermatol.*, **51**, 280
Splawinski, J. A., Reichenberg, K., Vetulani, J., Marchaj, J. and Kaluza, J. (1974). *Pol. J. Pharmac. Pharm.*, **26**, 101
Staszewska-Barczak, J., Ferreira, S. H. and Vane, J. R. (1976). *Cardiovasc. Res.*, **10**, 314
Svensson, J., Hamberg, M. and Samuelsson, B. (1975). *Acta physiol. scand.*, **94**, 222
Tainer, J. A., Turner, S. R. and Lynn, W. S. (1975). *Am. J. Path.*, **81**, 401
Turner, S. R., Tainer, J. A. and Lynn, W. S. (1975). *Nature, Lond.*, **257**, 680
Vane, J. R. (1971). *Nature new Biol.*, **231**, 232
Vane, J. R. (1972). *Hospital Practice*, **7**, 61
Vane, J. R. (1973). In *Advances in the Biosciences (International Conference on Prostaglandins)* (ed. S. Bergstrom and S. Bernhard), Pergamon Press, Braunschweig, p. 395
Whittle, B. J. R., Boughton-Smith, N. K., Moncada, S. and Vane, J. R. (1978). *Prostaglandins*, **15**, 955
Williams, T. J. and Morley, J. (1977). *Nature, Lond.*, **246**, 215
Williams, T. J. and Peck, M. J. (1977). *Nature, Lond.*, **270**, 530
Willis, A. L. (1969). In *Prostaglandins, Peptides and Amines* (ed. P. Mantegazza and E. W. Horton), Academic Press, New York, p. 33
Willis, A. L. (1974). *Prostaglandins*, **5**, 1
Zuckerman, H., Reiss, U. and Rubinstein, I. (1974). *Obstet. Gynecol.*, **44**, 787

17

Dihydrofolate reductases as targets for selective inhibitors

George H. Hitchings and Barbara Roth (The Wellcome Research Laboratories, Research Triangle Park, N.C. 27709, USA)

INTRODUCTION

This presentation is a sequel to one delivered 17 years ago in this same theatre under the same auspices (Hitchings, 1962). That paper described examples of selective inhibitors of dihydrofolate reductase (DHFR) and presented the interpretation that their selectivities reflect variations from species to species in the fine structures of the isofunctional enzymes (5,6,7,8-tetrahydrofolate:NADP+ oxidoreductase (EC 1.5.1.3)). It examined the role of DHFR in cellular metabolism, described host–parasite differences in both biosynthetic and transport capabilities and pointed out how these differences can be exploited in the interest of antimicrobial chemotherapy. It went on to indicate that practical applications were beginning to appear, and to emphasise the potentiation obtainable through the combined use of reductase inhibitors and antagonists of *p*-aminobenzoic acid.

Since then, cotrimoxazole, a combination of trimethoprim and sulphamethoxazole (Garrod *et al.*, 1969; Garrod, 1971; Bernstein and Salter, 1973; Finland and Kass, 1973; Anderson, 1975) has been accepted widely as a useful antibacterial agent with a range of indications of considerable breadth (except in the nation of its origin where the Federal Drug Administration, enmeshed in red tape of its own weave, has been unable to take positive action with respect to many important uses). Moreover, at long last, pyrimethamine is being used in combination with various sulphonamides and sulphones, primarily in the treatment of drug-resistant malarias (Hitchings, 1978). In addition, diaveridine, mentioned in the earlier paper, makes periodic reappearances as a coccidiostat.

The period between 1962 and 1979, therefore, has been a time of fruitful exploitation of the biochemical differences between hosts and parasites identified in the 1962 paper. It has also been a period of documentation, with very substantial expansion and refinement, of the inferences which had been drawn with respect to the 'many subtle variations among species in the geometry and charge of the enzyme surfaces and cell receptors which lend themselves to exploitation by the chemotherapist' (Hitchings, 1962). These advances have included inhibitor analysis of a spectrum of partially purified enzymes, the preparation of pure enzymes, sequence determinations, crystallisations of binary and ternary complexes of enzymes with substrates or inhibitors and, most recently, X-ray conformational studies of the crystals. Much of this work is still in progress and this status report, therefore, may well be out of date before it appears in print.

There is no intention of presenting an exhaustive treatment of these subjects in this paper. It is hoped, however, that representative samples of these endeavours will convey an appreciation of work in progress.

It may be appropriate to begin with a brief summary of the status as of 1962 and to attempt to bring the field up to date as rapidly as is feasible.

BIOCHEMICAL BACKGROUND TO COMBINATIONS

A discussion of the mechanism of action of combinations of reductase inhibitors and sulphonamides (sulphones) will provide a point of departure for an understanding of the whole field (see Hitchings and Burchall, 1965). Each of these depends ultimately on the selectivity of the particular DHFR inhibitor employed; in turn these selectivities depend on differences among the enzymes with respect to primary structure and conformation. Given an inhibitor more active on the parasite's than on the host's enzyme, there are additional biochemical considerations which are important to understand (see figure 17.1).

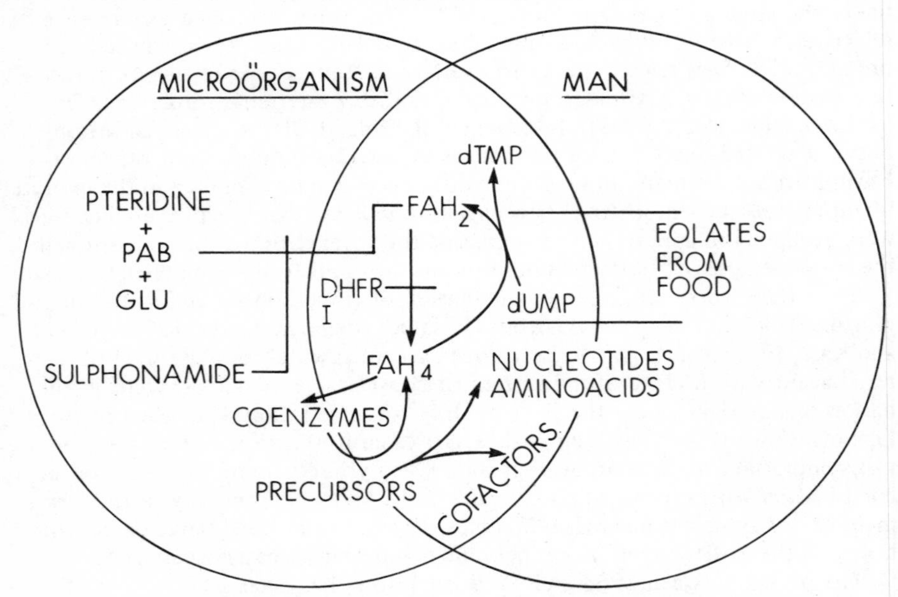

Figure 17.1 Diagrammatic representation of the biosynthesis and utilisation of folates by microorganisms and man. Within the overlapping areas are activities carried out by all species; activities outside this region are unique to the organism indicated. See text for full explanation. Abbreviations used: PAB, *p*-aminobenzoic acid; GLU, glutamate; FAH_2, dihydrofolate; FAH_4, tetrahydrofolate; dUMP, deoxyuridylate; dTMP, deoxythymidylate.

It is clear that most pathogenic microorganisms are unable to utilise exogenous pre-formed folates. Had uptake and utilisation been possible, sulphonamide inhibition could not have occurred, for the block in synthesis would have been bypassed by incorporation of the folates of blood and body fluids. On the other hand, higher organisms are unable to carry out the synthesis *de novo* and are dependent on exogenous (food) sources of the vitamin. Obviously, the acquisition of an uptake mechanism must have preceded the deletion of the biosynthetic route during evolution.

Uptake mechanisms for end-products exist in both hosts and parasites. These consist of kinases (to convert nucleosides to nucleotides) and phosphoribosyl transferases (to convert bases to nucleotides). The success in chemotherapy of agents restricting the function of folic acid testify to the paucity of end-products ordinarily available to pathogens. The other side of the coin is seen when extensive tissue destruction interferes with the effectiveness of sulphonamides, and the failure of sensitivity testing when media containing thymidine are employed (Ferone *et al.*, 1975).

The biosynthetic pathways which employ tetrahydrofolate-containing cofactors, with one exception, are shuttle mechanisms in which the tetrahydrofolate (FH_4) alternately accepts and then donates a one-carbon fragment (figure 17.2).

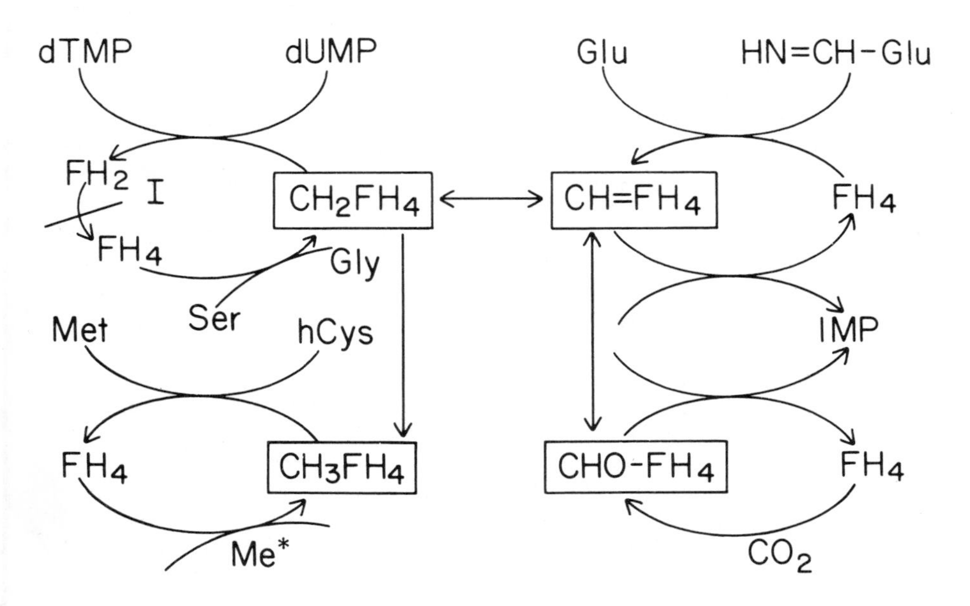

Figure 17.2 Diagrammatic representation of biosyntheses involving tetrahydrofolate (FAH_4). Combinations of FAH_4 with one carbon units produce the cofactors shown in boxes. Reading clockwise from the upper left, these are respectively, methylene-, methenyl-, formyl- and methyl-FAH_4. Abbreviations used: HN=CH-Glu, forminoglutamate; IMP, inosinate; hCys, homocysteine; Met, methionine; Gly, glycine. I stands for inhibitor. For other abbreviations, see legend to figure 17.1 (Figure reproduced from Hitchings (1978) with permission of Academic Press Inc.)

The exceptional reaction is catalysed by thymidylate synthetase, where the methyl group of the thymine moiety requires for its formation not only the methylene of methylene tetrahydrofolate, but also two hydrogen atoms from the pteridine nucleus (figure 17.2, upper left). A molecule of FH_2 is thus regenerated with each turnover of thymidylate synthetase. Thus DHFR is necessary for all cells to keep this cycle running. Microorganisms additionally synthesise FH_2 *de novo*. Since the reductase inhibitors are competitive with the substrate, the effectiveness of the inhibition is dependent on the size of FH_2 pool, and it is profitable to eliminate this contribution. The addition of a sulphonamide accomplishes this, and results in a strong potentiation, a broadening of the spectrum of activity, bactericidal versus bacteriostatic activity and a presumptive diminution in the incidence of resistant organisms. For those mathematically inclined, an elegant analysis of the conditions required for synergistic interaction of two inhibitors (including trimethoprim/sulphonamide) has recently appeared (Harvey, 1978). The cyclic, thymidylate synthetase, component is critically important to the synergism.

INHIBITOR ANALYSES OF REDUCTASES

In 1962, partial purification of reductases was just beginning. It was possible to show then that the K_Ds of five inhibitors with a mouse liver reductase preparation were consistent with the toxicities found in the whole animal (Hitchings, 1962). Shortly afterward, Burchall began the partial purification and inhibition analysis of a spectrum of microbial and other enzymes. These studies amply confirmed the interpretations put forward earlier (Hitchings *et al.*, 1952*a*), namely that 'Minor changes in their fine chemical structure may adapt them to close approximations. . . on protein surfaces of particular species . . . . An analogue more nearly related structurally to the metabolite would be less capable of species differentiation. . . '. Burchall and Hitchings (1965) published I_{50} values for six inhibitors acting on three bacterial and four mammalian enzymes. As predicted, the enzymes exhibited a rather uniform response to the structural analogue, amethopterin (methotrexate, MTX). Moreover, kinetic values differed but little from enzyme to enzyme. The most striking selectivities were exhibited by trimethoprim (several thousand times as active on the *Escherichia coli* as on the human enzyme) and a dimethyl dihydrotriazine (1000 times as active on the human enzyme as on that from *Escherichia coli*).

On close scrutiny, some very specific structure–activity correlations could be seen. Thus, with a 6-butylpyridopyrimidine, the addition of a 5-methyl group enhanced activity against *E. coli* and *Proteus vulgaris*, but diminished that against *Staphylococcus aureus*. Pyrimethamine did not consistently distinguish between bacterial and mammalian enzymes, and it was only later when the *Plasmodium berghei* reductase became available (Ferone *et al.*, 1969) that its striking specificity became apparent at the enzyme level. The gist of this work was to affirm the correlation between enzyme inhibition, whole cell inhibition and chemotherapeutic activity (Hitchings and Burchall, 1965).

An overview of the very broad differences to be found among reductases was provided by a taxonomic survey using the first two inhibitors mentioned above, trimethoprim and the dihydrotriazine. The ratio of the IC_{50} values of these gave

Table 17.1 Inhibition of various dihydrofolate reductases by two inhibitors

Source of enzyme	IC_{50}		
	TMP	DHT	Ratio TMP/DHT ($\times 10^8$)
Human	3×10^4	55	545
Guinea pig	2×10^3	4	500
Rat	2.6×10^4	14	1.86×10^3
Rabbit	3.7×10^4	16	2.3×10^3
Turtle	1.2×10^3	0.16	7.5×10^3
Frog	2×10^4	27	740
Carp	7×10^3	7	1×10^3
Tapeworm	560	1×10^4	5.6×10^{-2}
Lobster	3×10^3	10	300
P. berghei	17	0.8	146
S. cerevisiae	2.4×10^3	6.2×10^4	0.38
P. vulgaris	0.5	1×10^4	5×10^{-5}
S. aureus	1.5	5×10^4	3×10^{-5}
E. coli	0.5	6.5×10^4	7.6×10^{-6}

Abbreviations: DHT, 1-(*p*-butylphenyl)-1,2-dihydro-2,2-dimethyl-4,6-diamino-*s*-triazine; TMP, 2,4-diamino-5-(3′, 4′, 5′-trimethoxybenzyl)pyrimidine (trimethoprim).

a parameter in which the errors arising from incomplete purification would be expected to cancel one another, and provide a number characteristic of the enzyme studied. The total excursion of this parameter over the taxonomic range was of the order of $10^8 : 10^3$ for the human, $< 10^{-5}$ for *E. coli.* The values for mammals were grouped near the top of the range, those for bacteria near the bottom. Individual enzymes could be characterised by their responses to these two inhibitors (table 17.1). Data of this sort led to the expectation of varying sequences among DHFRs of different species, with however, conservation of the 'active centre' (Hitchings, 1969).

The data show clearly that a wide range of structures is consistent with ability to perform the catalytic function of the enzymes (Burchall, 1971). Just how wide has only become apparent as sequences of enzymes from various sources have accumulated.

PREPARATION OF PURE ENZYMES

The preparation of pure enzymes entailed initially the separation of very small amounts of specific proteins from large amounts of irrelevant but chemically similar molecules. The ultimate solution to the problem of isolating quantities of reductases sufficient for sequence and conformational studies involved improvement both in sources of starting materials and in the techniques of isolation.

By 1975, DHFR had been purified to homogeneity from *Lactobacillus casei* (Pastore *et al.*, 1974*a*; Gundersen *et al.*, 1972), *Streptococcus faecium* (Nixon and Blakley, 1968; D'Souza *et al.*, 1972) and T_4 bacteriophage (Erickson and

Mathews, 1973). Poe *et al.* (1972) and Burchall (1970) chose strains of *E. coli* which were enriched with enzyme in association with drug (MTX and trimethoprim, respectively) resistance. In both instances the possibility was addressed that the redundancy of enzyme had been accompanied by alterations, and in both instances it was concluded on the basis of kinetic properties that there were unlikely to have been any major differences from the native enzyme. Burchall, later, chose a strain of *E. coli* with a higher degree of resistance (to 500 $\mu g\ ml^{-1}$ of trimethoprim) and an extremely high activity of enzyme (300 times as much as the wild strain). The method of purification employed an ammonium sulphate cut, followed by affinity chromatography (using methotrexate coupled to AH-Sepharose 48).

The isolation procedures were not achieved without effort. Much of the early work and initial successes have been reviewed by Huennekens *et al.* (1976). Affinity chromatography was the key step, but the initial isolation of substantial quantities of bacterial enzymes was facilitated, as mentioned above, by the use of bacterial strains resistant to inhibitors because of the high concentrations of reductase they contain.

Several varieties of polymorphism have produced problems of interpretation, doubts about homogeneity and obstacles to crystallisation. Aggregates, the presence of bound ligands and the presence of minor amounts of slightly different species have all been observed, and not all of the multiplicities have been fully resolved. Nevertheless, as will be seen below, homogeneous preparations and crystalline structures suitable for X-ray conformation studies have now been achieved.

Perhaps an illustration of each of the complications will suffice. Huennekens and coworkers had encountered multiple forms of both chicken liver and *Lactobacillus casei* enzymes which eventually proved to be apoenzyme and enzyme-NADPH complexes (Mell *et al.*, 1968; Huennekens *et al.*, 1970, 1971, 1976; Dunlap *et al.*, 1971). This sort of complication derives from the kinetic properties of the enzyme. The reaction (Burchall, 1970) is random 'bi bi' with two dead-end complexes. In other words, two binary complexes give rise to a single ternary complex where the hydrogen exchange occurs, but the enzyme still has significant affinity for the products. Folic acid, also, binds fairly well to the enzyme whether or not it is a substrate. Another type of complication to the preparation of homogeneous enzymes is the tendency to form aggregates. This is largely prevented when dithiothreitol is included in the solutions during purification (Baccanari *et al.*, 1975). In addition, Pattishall *et al.* (1976) reported two forms of an *E. coli* enzyme with substantially different K_D values for binary complexes which, however, give the same ternary complex and are thereby converted to a single form. The explanation for these observations is still not entirely clear. Yet a further kind of multiplicity was described by Baccanari and Burchall (1977, 1978). Here electrophoretically similar forms differed substantially in binding to substrate and inhibitors. This effect stems from a single amino acid difference (A. Phillips, private communication). Bacterial and mammalian dihydrofolate reductases have been fairly homogeneous with regard to size (18 000-20 000) and, as far as one could tell, composition. With sequencing completed on several enzymes (v.i.), it is now apparent that mammalian enzymes possess 25-30 residues more than the bacterial, mostly by insertions. The reductases of protozoa, however, are in a class by themselves. Ferone *et al.* (1969) estimated the molecular weight of the *Plasmodium berghei* enzyme to be 180 000, and there was no evidence that it was polymeric in structure. Other protozoal reductases are also large (see for example Hitchings, 1978). Calf thymus, and T_4 phage of *E. coli* yield reductases in the 30 000 molecular weight range (see Huennekens *et al.*, 1971).

INHIBITOR RESISTANCE

A fascinating group of reductases have been identified as plasmid-R factors which convey trimethoprim resistance. They were first discovered by Fleming *et al.* (1972). One type has been characterised by Amyes and Smith (1974) and Skjöld and Widh (1974). Pattishall *et al.* (1977) described a second type and further characterised both, which have molecular weights in the 32 000–37 000 range. They have substrate and cofactor binding and kinetics not strikingly different from those of the chromosomal enzyme. Both bind folic acid and 2,4-diamino-pyrimidine. Type I is characterised by a high level of resistance to all standard diaminopyrimidine inhibitors, but less to pyrimethamine and *s*-triazine than to trimethoprim. Type II is almost completely unaffected by any of the standard inhibitors, including aminopterin and MTX.

SEQUENCES

Sequences and parts of sequences are available (counting both publications and private communications) on 13 pure enzymes from eight source tissues (Table 17.2). It may be of interest to consult additional references on isolation and sequence studies; for example Baccanari *et al.* (1975), Gauldie *et al.* (1973), Gupta *et al.* (1977), Kaufman and Kemmerer (1976, 1977). Reductases from two *E. coli* enzymes (trimethoprim and MTX resistant, respectively) have been sequenced in separate laboratories with virtually identical results, each showing 159 residues with two minor variations near the C-terminus. The resistance in each case appears to be the result of overproduction, rather than modification, of the enzyme. Two other bacterial enzymes, from *S. faecium* and *L. casei*, were found to have 167 and 162 residues, respectively, with molecular weights close to 18 000. The homology in sequences between these bacterial species is remarkably low. *E. coli* and *L. casei* reductases, for example, have only 30 per cent homology.

The differences between bacterial and mammalian species are even greater. The enzyme from leukemia L1210 contains 186 residues, giving a molecular weight of 21 458 daltons. The extra 30-odd amino acid residues are probably present as insertions along the chain for the most part, a conclusion reached after study of two reductases of known geometry (see below). The enzymes from homoiotherms (pig, ox, hen, mouse neoplasms) are much more uniform in composition and sequence than are the bacterial enzymes (for example Freisheim *et al.*, 1979), but even preparations from the same source (e.g. ox or hen liver) may not necessarily show complete identity. Some of the differences may be resolved as errors, but others appear to be valid. Among the 13 sequences and parts of sequences now available, only 16 identities have been found, even after adjusting the positions of the residues to give what appears to be the most reasonable fit

Table 17.2 Sequenced dihydrofolate reductases

Enzyme source	Number of residues	References
E. coli, TMP resistant	159	Stone *et al.*, 1977
E. coli, MTX resistant	159	Bennett, 1974; Bennett *et al.*, 1978
S. faecium Durans A, MTX resistant	167	Gleisner *et al.*, 1974; Peterson *et al.*, 1975
S. faecium, high catalytic coefficient	†	J. H. Freisheim, private communication
L. casei, MTX resistant	162	Bitar *et al.*, 1977; Batley and Morris, 1977*a,b*
L1210, mouse	186	Stone and Phillips, 1977*a*; D. Stone and A. W. Phillips, private communications
Sarcoma 180, mouse	†	Rodkey and Bennett, 1976
Hen liver 1	†	C. D. Bennett, private communication
Hen liver 2	186	J. H. Freisheim, private communication
Pig liver	186	Stone and Phillips, 1977*b*
Bovine liver 1	†	Peterson *et al.*, 1975,
Bovine liver 2	†	Baumann and Wilson, 1975
Bovine liver 3	185	R. L. Blakley, private communication; Lai *et al.*, 1979

†Partial sequence, from N-terminus.

Table 17.3 Fully conserved residues

E. coli	7	14	15	21	22	31
L1210	9	16	17	23	24	34
Residue	Ala	Ile	Gly	Pro	Trp	Phe
E. coli	43	46	54	57	62	95
L1210	53	56	67	70	75	116
Residue	Gly	Thr	Leu	Arg	Leu	Gly
E. coli	96	113	120	122		
L1210	117	136	143	145		
Residue	Gly	Thr	Glu	Asp		

Two numberings are given, each residue-by-residue from the amino-terminal end of the sequence of the respective enzyme. Gaps in the *E. coli* sequence are provided to maximise the number of homologies.

(Table 17.3). The differences in the number of residues from enzyme to enzyme make it awkward to identify any given residue. In this paper numbering as for the *E. coli* enzyme has been used throughout.

Figure 17.3 represents a short segment of the available sequences designed to convey a sense of the differences and correspondences. Bacterial and mammalian enzymes are shown one above the other. The numbers in the boxes represent the number of intragroup identities, for example (2,2,1) indicates that among five enzymes there are at this point two with one amino acid residue, two with another and a third with still another. Circled numbers mark identities between the indicated bacterial and mammalian enzymes. The purpose of this exercise is simply to provide a superficial view of resemblances and differences among the enzyme sequences so far in hand. For a more penetrating analysis, a knowledge of the three-dimensional structure is required.

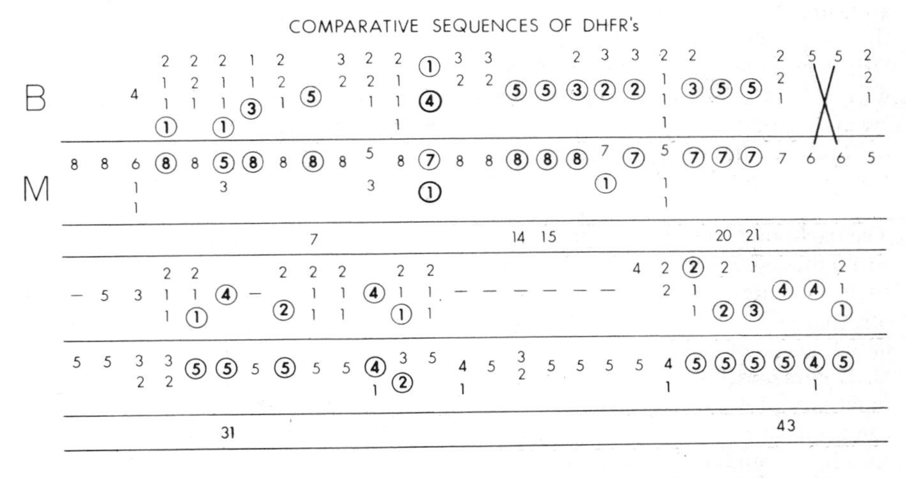

Figure 17.3 Representation of segment of amino acid sequences of dihydrofolate reductases from various sources. The numbers indicate the number of identities within the segment, and circled numbers above and below the horizontal line indicate identities between the two groups. Above the line are sequences from five bacterial sources; below the line from eight homoiotherm sources; see text for fuller explanation.

THE GEOMETRY OF DIHYDROFOLATE REDUCTASE

Announcement of the three-dimensional X-ray structure of DHFR from *E. coli* in binary combination with the inhibitor MTX (Matthews *et al.*, 1977) represented a quantum leap in knowledge concerning this enzyme. Over the years, various types of probe had been used to obtain information about the active site. Baker (1967), for example, prepared a plethora of compounds for inhibitor analysis and described in general terms a hydrophobic pocket. He made repeated attempts to reach functional residues on the protein from this site for covalent binding. Hansch *et al.* (1977) and Silipo and Hansch (1975) continued Baker's programme with QSAR studies of DHFR inhibitors. Nuclear magnetic resonance has been used by many workers in an effort to obtain information on the nature of substrate or inhibitor interactions with residues on the enzyme, and the geometry of the various species (Roberts *et al.*, 1974; Way *et al.*, 1975; Feeney *et al.*, 1975, 1977; Pastore *et al.*, 1974*a*; Kimber *et al.*, 1977). Similarly, circular dichroism and ultraviolet and fluorescence spectroscopy have given useful clues (for example,

Greenfield *et al.*, 1972; Freisheim and D'Souza, 1971; Reddy *et al.*, 1978; Poe *et al.*, 1974; Hood and Roberts, 1978; Otting and Huennekens, 1972). Chemical modification of certain amino acid residues has also provided information (for example, Freisheim and Huennekens, 1969; Liu and Dunlap, 1974; Freisheim *et al.*, 1977). By such means one could see through a glass darkly, but only with crystallisation of the enzyme was it possible to see face to face the nature of the protein and its active site. This feat took several years of trial and error in the accomplishment.

The crystallisation of an enzyme has always been something of a black art. A dash of this, a soupçon of that—a specific buffer, a trace metal, for example—may after long tedium produce a crystal which grows to a suitable shape and size for the X-ray beam. Matthews *et al.* (1977) accomplished this not only with *E. coli* DHFR in binary complex with MTX, but also with *L. casei* in ternary complex with MTX and NADPH (Matthews *et al.*, 1978). The scientific community eagerly awaits a report of the crystallisation of an apoenzyme. This would make possible diffusion studies of substrates as well as inhibitors, which could locate their positions in the enzyme cleft.

The backbone structure of DHFR and its relationship to other enzymes

From the standpoint of evolution, the relationship of DHFR to the NAD enzymes, such as the dehydrogenases, is of major interest. Matthews *et al.* (1977) pointed out that DHFR is the first NADPH-containing enzyme to have its structure elucidated, and they have therefore compared its conformations with those known for NAD enzymes. They find similarities in spatial arrangement but not in the conjunctivity of the $(\beta\alpha\beta)_2$ structure. They conclude that the two groups of enzymes cannot have evolved from a common ancestor, but that the nucleotide binding sites have come to resemble each other by convergent evolution.

Another possible relationship of DHFR is suggested by finding a high degree of homology in sequence between the carboxy terminal quarter of the β-galactosidase and the DHFR of *E. coli* (Hood *et al.*, 1978). This may or may not have evolutionary significance with respect to DHFR. One would feel that DHFR is the more primitive and more essential of the two enzymes.

The *L. casei* DHFR ternary complex (Matthews *et al.*, 1978) was found to have a remarkably similar backbone geometry to that of *E. coli* DHFR, despite large residue differences. In both cases the substrate and coenzyme binding functions are carried out by overlapping portions of the sequence, rather than by separate domains, in contrast with the dehydrogenases.

The cleft and binding sites of DHFR

E. coli DHFR is characterised by a deep cavity which cuts across one whole face of an otherwise nearly globular protein. MTX resides in this cavity. *L. casei* DHFR has a very similar cleft, with the cofactor NADPH neatly tucked under MTX in an extended conformation, and with the hydrogen on the *α-side* of the reduced nucleotide snuggled up to the pyrazine ring of MTX. The cleft conformation of the ternary complex differs from that of the *E. coli* DHFR in having the 'teen' loop (residues 12–21) and residues 125–128 moved as much as 3 Å to fold tightly around the cofactor by hydrogen bonding and van der Waal's interactions to exclude water from the pocket. This conformation change may be intimately in-

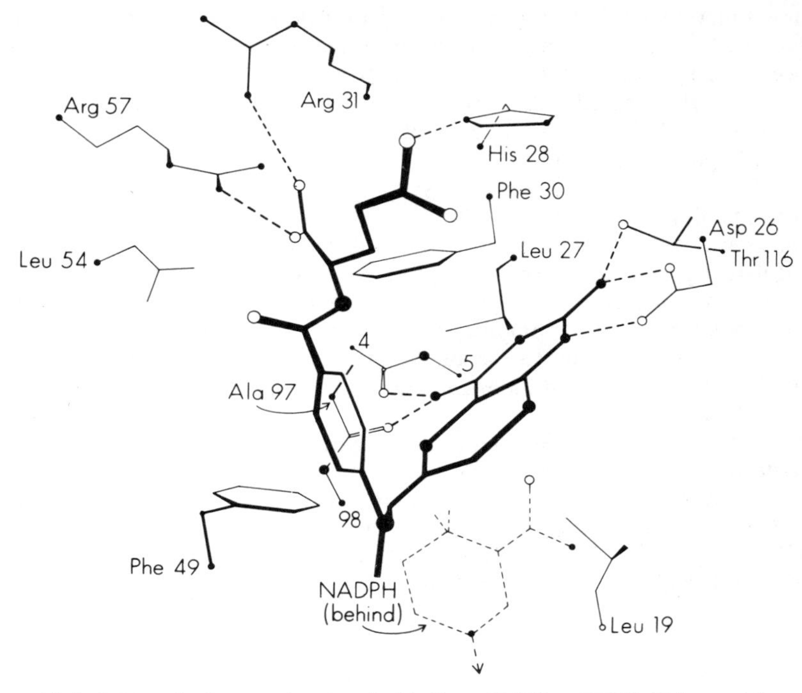

Figure 17.4 Schematic diagram showing the binding of MTX to DHFR of *L. casei* (ternary complex).

volved with the known cooperative effect of the coenzyme on binding to the MTX-enzyme complex.

Matthews *et al.* (1978) have identified 17 interactions between MTX and *L. casei*-NADPH complex (figure 17.4). Twelve of these are hydrophobic, and the rest are charge interactions and hydrogen bonds. Thirteen side-chain residues are involved, as well as the backbone and the reduced nicotinamide ring. In *E. coli*, nine of the 13 residues are identical and the remainder are similar (for example, Leu-4 versus Ile-5).

A comparison with the other known reductase sequences indicates that only five of the MTX interactions may possibly be identical in all cases. Arg-57, a conserved residue which has a charge interaction with MTX, may conceivably serve as the primary attractive force for the inhibitor. Asp-27, present in all sequenced bacterial species (but changed to Glu in those from mammals and birds) may then position the pteridine moiety in the cleft at the presumed active site. Conserved Phe-31 forms the roof of the cleft. Other conserved or nearly conserved hydrophobic side chains (mainly Leu and Ile) are positioned strategically around the aromatic rings of MTX to seal the pocket.

The residues listed in table 17.3 as conserved in all the enzymes studied to date are all found in or near the cleft. It is not clear in all cases why their conservation should be more essential (if indeed it is) than that of other residues involved in binding, function, or cleft geometry. The conservation of certain glycine residues is considered essential to cleft integrity, since side chains would cause steric interference for the substrate. NADPH surprisingly would appear to be bound to *L. casei* and *E. coli* along different strands of the β-sheet, according to the analysis of Matthews *et al.* (1978). Details of NADPH–enzyme interactions have not as yet appeared in published form.

Substrate binding

Matthews *et al.* (1977, 1978) have assessed the likelihood of the substrate, dihydrofolate, being bound in the cleft in the same manner as MTX. The geometry of the ternary complex strongly points to the conclusion that it closely resembles that of productive substrate binding, since the cofactor is positioned appropriately to deliver a hydrogen atom to C-6 of the pteridine ring. However, it is by no means clear that the orientation of the pteridine ring in MTX is the same as that in dihydrofolate.

In reducing the 5,6-double bond of 7,8-dihydrofolate to the tetrahydro derivative, a proton must be delivered to N-5, as well as a hydride to C-6. The only apparent proton sources in the active site are Asp-27 and Thr-113 (if one excludes water as a possibility) and Asp-27 would appear to be the more plausible of the two (Matthews *et al.*, 1977, 1978). It will be noted from figure 17.4 that MTX interacts with Asp-27 at its N-1 position, its most basic nitrogen. Dihydrofolate,

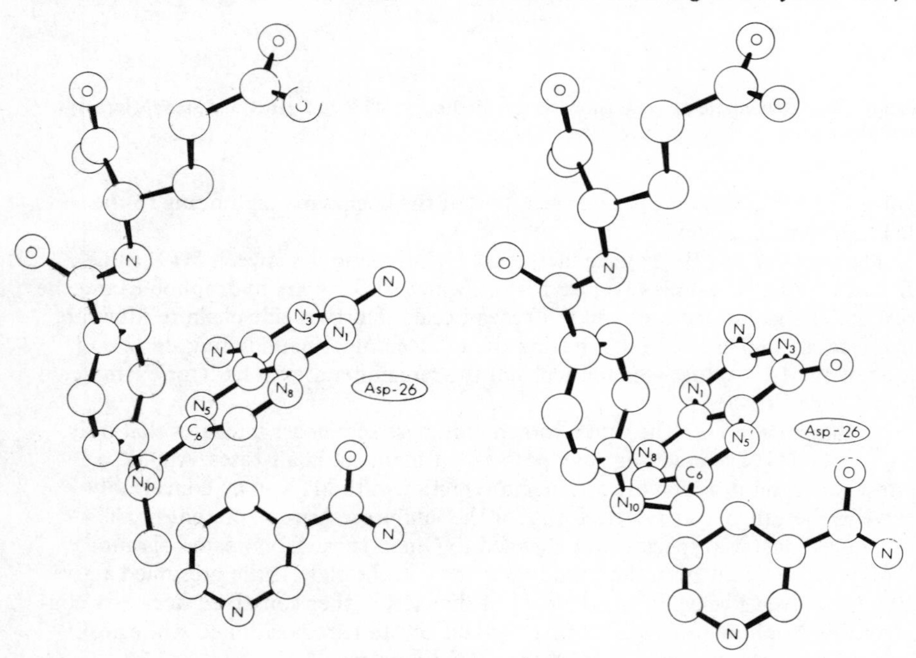

Figure 17.5 Diagram showing conformation of MTX in ternary complex of *L. casei* DHFR. (*Left*) Approximate conformation of substrate with rotation of pyrimidopyrazine by 180° around C_6–C_9 bond. (*Right*) Reduced ring of NADPH in probable position. N.B. Asp-26 (*L. casei*) is equivalent to Asp-27 (*E. coli*).

on the contrary, has its most basic nitrogen at N-5 (Poe, 1977; Kallen and Jencks, 1966) rather than at N-1, and this is the site which requires the proton, as stated above. If the pteridine ring is rotated 180° around its axis, leaving the rest of the molecule essentially intact, dihydrofolate would be in a position to receive the required proton from Asp-27 by direct exchange at N-5, as illustrated in figure 17.5.

The absolute configuration about C-6 of tetrahydrofolate is not known. If it were, the orientation of the dihydrofolate pteridine moiety would be self-evident, as can be seen by inspection of figures 17.4 and 17.5, which show the locus of delivery of the hydrogen to C-6, and hence the resultant stereochemistry about that atom. The problem has recently been addressed by Fontecilla-Camps *et al.* (1979) using the tetrahydrofolate derivative anhydroleucovorin. By X-ray crystallography they were able to assign its absolute configuration about C-6. The results suggest that the pteridine ring of dihydrofolate may indeed be rotated 180° about its axis from that of MTX in the cleft of DHFR.

Selectivity in inhibitor binding

It is necessary at this point to address the question of the nature of the selectivity observed in binding of small molecule inhibitors to reductases. We noted earlier that the antibacterial agent, trimethoprim, for example, is bound much more tightly to a wide variety of reductases from bacteria than those from mammals or birds. Since no three-dimensional structure is available for the latter type of DHFR enzyme, the nature of the specificity cannot be answered. Indeed, it may not be possible to draw a reliable conclusion when we do have this information. In the face of the tremendous sequence differences which we have described, one wonders indeed why a small molecule inhibitor can be so successful with a wide spectrum of bacteria.

One can search for clues by sequence study, in an effort to find loci where there is a common denominator for all bacteria which is different for the homoiotherms. There are several such sites, but most are at the surface, or not near the cleft. The one obvious difference which merits attention is the Asp/Glu change at the presumed active site in going from the known bacterial sequences to those of the homoiotherms. However, this does not alter MTX binding. A much greater difference, which may possibly lead to the answer, concerns the insertion of additional residues near the cleft in the mammalian and avian species. Extra loops near the edge of the cleft could alter the binding of some inhibitors markedly.

Practically all of the potent inhibitors of DHFR are diaminoheterocycles, and it has been generally assumed that these are bound at the active site, since binding has normally been found to be competitive, where tested. It does not necessarily follow that the analogues are bound with precisely the same geometry as that of the pyrimidine ring of MTX, however. One example which suggests that this is not necessarily the case is that of the 3-deaza analogue of trimethoprim (figure 17.6, **IIa**) (B. Rauckman and B. Roth, in preparation). This compound is about two orders of magnitude less active than trimethoprim (figure 17.6, **Ia**) against *E. coli* DHFR, despite its very close relationship. However, its analogue **IIb**, with its 4-methoxy group replaced by the smaller 4-hydroxy function, is four times more active than **IIa**; the trimethoprim congener **Ib** is only half as active as its parent. Thus **IIb** is only about *one* order of magnitude less active than **Ib**. The replacement of one

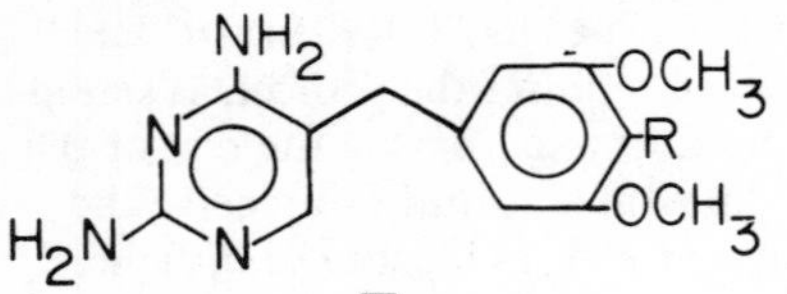

Figure 17.6 Structures of benzylpyrimidines and pyridines. **Ia** ($R=OCH_3$) trimethoprim. **Ib** (R=OH) the desmethyl analogue. **IIa** and **IIb** are the corresponding 3-deaza analogs of **Ia** and **Ib**.

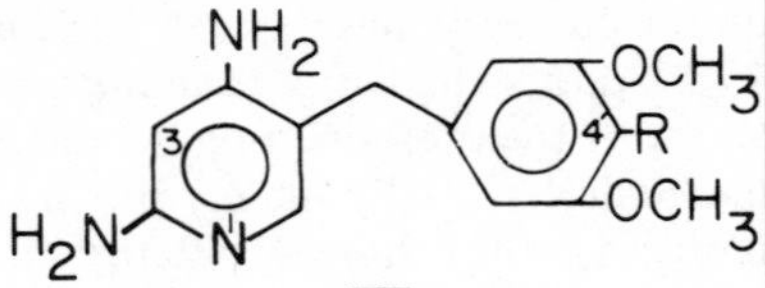

nitrogen by carbon probably alters the electron density pattern and thus its geometry in interaction with the enzyme, with the **II** orientation creating a poor fit for the aromatic moiety. The resultant increase in basicity would not be expected to be deleterious *per se* (Roth and Strelitz, 1969).

J. P. Albrand, B. Birdsall, J. Feeney, G. C. K. Roberts, E. A. Piper, A. S. V. Burgen and J. Cayley (private communications) have recently conducted some elegant nuclear magnetic resonance studies in an effort to elucidate trimethoprim geometry in interaction with the enzyme. They have concluded that there are two possible specifically described conformations for trimethoprim in complex, provided that trimethoprim is bound precisely as the pyrimidine ring of MTX is. We await the X-ray probe.

The new knowledge concerning the geometry of DHFR has anwered many questions, and raised many others. To date it has not told us how to tailor a perfect lock and key, selective to the target, and the answer may not be provided by additional X-ray models. However, it has created new dimensions in thinking in trying to create better inhibitors. There are many new hypotheses to test as we continue to probe the active site by the synthesis of new compounds.

CONCLUSIONS

Our programme of study of antimetabolites began to recognise very early that individual analogues produce a variety of responses in different species (Hitchings *et al.*, 1950*a,b*). Of the possible explanations of such divergences, one promised considerable practical as well as intellectual rewards, namely that in at least one type of antimetabolite, the 2,4-diaminopyrimidine family, the results indicated that cell receptors differed in structure from species to species (Hitchings *et al.*, 1952*a*). Further study revealed that these molecules could be shaped to inhibit specific targets, and a useful antimalarial, pyrimethamine, soon emerged (Hitchings *et al* 1952*b*).

There was a substantial measure of good fortune, not only in choosing a fruitful interpretation of the initial signals, but also in the characteristics of the biochemical pathway in which the target enzyme participates. To a major extent, the success of the diaminopyrimidines (as well as sulphonamides) depends on the inability of the pathogenic cell to utilise preformed folates, and on the fact that there are insufficient end-products of the folate-dependent systems to permit its growth in bypass fashion. These favourable conditions are the result of specific transport and incorporating systems which could not have been foreseen, but the practical results were observable and these encouraged the eventual experimental work required for documentation.

It has recently been proposed (Cohen, 1977) that a major multidisciplinary project be established to devise specific enzyme inhibitors for the chemotherapy of infectious disease. This proposal rests heavily on newly developed competence in isolation, sequencing and conformation determination of enzymes. It may at once underestimate the work of this sort that is in progress, and overestimate its possible contributions to the design of useful new agents. In the field described in this paper, it has not been easy to locate the positioning of inhibitors in the structures which have been determined, nor to understand with any certainty their selectivities. Such knowledge will surely come and improvements on the currently available inhibitors will probably result. This, however, is a far cry from the creation *de novo* of selective inhibitors from information on sequence and conformation. The productive chemotherapist eagerly makes use of whatever information will further his goals, and he can be depended upon to make good use of new tools. What the world does *not* need are more established projects.

As one surveys the literature for observed differences from species to species in isofunctional enzymes, it is evident that a myriad of such observations have been made (for example Hitchings, 1978). Many of the parasite-host differences are potentially exploitable. One hopes to find some selectivity as a starting point—either already discovered or initiated by probing with substrate, cofactor or transition-state analogues. The working out of such leads will necessarily be in part probing for hidden meanings, in part empirical testing. The art of innovation in the field of chemotherapy lies in the exercise of judgement as to which approaches will give the most meaningful answers with the least effort.

ACKNOWLEDGMENTS

A number of colleagues both intra- and extramural have been most helpful in providing access to unpublished data and interpretations. We should like to thank especially C. Bugg, J. J. Burchall, A. S. V. Burgen, P. Goodford, D. A. Matthews, F. Norrington and A. W. Phillips.

REFERENCES

Amyes, S. G. B. and Smith, J. T. (1974). *Biochem. Biophys. Res. Commun.*, **58**, 412
Anderson, J. R. (ed.) (1975). *Canad. Med. Assoc. J.*, **112**, 3S
Baccanari, D. P. and Burchall, J. (1978). In *Chemistry and Biology of Pteridines* (ed. R. L. Kisliuk and G. M. Brown), Elsevier/North Holland, New York, p. 361
Baccanari, D. P., Averett, D., Briggs, C. and Burchall, J. (1977). *Biochemistry*, **16**, 3566
Baccanari, D. P., Phillips, A., Smith, S., Sinski, D. and Burchall, J. (1975). *Biochemistry*, **14**, 5267
Baker, B. R. (1967). *Design of Active-Site-Directed Irreversible Inhibitors,* John Wiley, New York
Batley, K. E. and Morris, H. R. (1977*a*). *Biochem. Soc. Trans.*, **5**, 1097
Batley, K. E. and Morris, H. R. (1977*b*). *Biochem. Biophys. Res. Commun.*, **75**, 1010
Baumann, H. and Wilson, K. J. (1975). *Eur. J. Biochem.* **60**, 9
Bennett, C. D. (1974). *Nature, Lond.*, **248**, 67
Bennett, C. D., Rodkey, J. A., Sondey, J. M. and Hirschman, R. H. (1978). *Biochemistry*, **17**, 1328

Bernstein, L. S. and Salter, A. J. (eds) (1973). *Trimethoprim/Sulphamethoxazole in Bacterial Infections,* Churchill Livingstone, London
Bitar, K. G., Blankenship, D. T., Walsh, K. A., Dunlap, R. B., Reddy, A. V. and Freisheim, J. H. (1977). *Fedn Eur. biochem. Socs Lett.,* **80**, 119
Burchall, J. J. (1969). *Postgrad. med. J.,* **45**, Suppl. **29**, 29
Burchall, J. J. (1970). In *Chemistry and Biology of Pteridines* (ed. K. Iwai, M. Akino, M. Goto and Y. Iwanami), International Academic Printing Co., Tokyo, p. 351
Burchall, J. J. (1971). *Ann. N.Y. Acad. Sci.,* **186**, 143
Burchall, J. J. and Hitchings, G. H. (1965). *Molec. Pharmac.,* **1**, 126
Cohen, S. S. (1977). *Science, N.Y.,* **197**, 431
D'Souza, L., Warwick, P. E. and Freisheim, J. H. (1972). *Biochemistry,* **11**, 1528
Dunlap, R. B., Gunderson, L. E. and Huennekens, F. M. (1971). *Biochem. Biophys. Res. Commun.,* **42**, 772
Erickson, J. and Mathews, C. K. (1973). *Biochemistry,* **12**, 372
Feeney, J., Birdsall, B., Roberts, G. C. K. and Burgen, A. S. V. (1975). *Nature, Lond.,* **257**, 564
Feeney, J., Roberts, G. C. K., Birdsall, B., Griffiths, D. V., King, R. W., Scudder, P. and Burgen, A. (1977). *Proc. R. Soc.,* **196**, 267
Ferone, R., Burchall, J. J. and Hitchings, G. H. (1969). *Molec. Pharmac.,* **5**, 49
Ferone, R., Bushby, S. R. M., Burchall, J. J., Moore, W. D., and Smith, D. (1975). *Antimicrobial Agents Chemother.,* **7**, 91
Finland, M. and Kass, E. H. (eds) (1973). *Trimethoprim-Sulfamethoxazole*, University of Chicago Press, Chicago
Fleming, M. P., Datta, N. and Grüneberg, R. N. (1972). *Br. med. J.,* **i**, 726
Fling, M., Elwell, L. P. and Inamine, J. M. (1978). In *Genetic Engineering* (ed. H. W. Boyer and S. Nicosia), Elsevier/North Holland, New York, p. 173
Fontecilla-Camps, J. C., Bugg, C. E., Temple, C., Rose, J. D., Montgomery, J. A. and Kisliuk, R. L. (1979). In *Chemistry and Biology of Pteridines* (ed. R. L. Kisliuk and G. M. Brown), Elsevier/North Holland, New York, p. 235
Freisheim, J. H. and Huennekens, F. M. (1969). *Biochemistry,* **8**, 2271
Freisheim, J. H. and D'Souza, L. (1971). *Biochem. Biophys. Res. Commun.,* **45**, 803
Freisheim, J. H., Kumar, A. A., Blankenship, D. T. and Kaufman, B. T. (1979). In *Chemistry and Biology of Pteridines* (ed. R. L. Kisliuk and G. M. Brown), Elsevier/North Holland, New York, p. 419
Freisheim, J. H., Ericcson, L. H., Bitar, K. G., Dunlap, R. B. and Reddy, A. V. (1977). *Arch. Biochem. Biophys.,* **180**, 310
Garrod, L. P., James, D. G. and Lewis, A. A. G. (eds) (1969). *Postgrad. med. J.,* Suppl. 45
Garrod, L. P. (1971). *Drugs,* **1**, 3
Gauldie, J. and Hillcoat, R. L. (1972). *Biochim. biophys. Acta,* **268**, 35
Gauldie, J., Marshall, L. and Hillcoat, B. L. (1973). *Biochem. J.,* **133**, 349
Gleisner, J. M., Peterson, D. L. and Blakley, R. L. (1974). *Proc. natn. Acad. Sci. U.S.A.,* **71**, 3001
Greenfield, N. J., Williams, M. N., Poe, M. and Hoogsteen, K. (1972). *Biochemistry,* **11**, 4706
Gundersen, L. E., Dunlap, R. B., Harding, N. G. L., Freisheim, J. H., Otting, F. and Huennekens, F. M. (1972). *Biochemistry,* **11**, 1018
Gupta, R. S., Flintoff, W. F. and Siminovitch, L. (1977). *Can. J. Biochem.* **55**, 445
Hansch, C., Fukunaga, J. Y., Jow, P. Y. C. and Hynes, J. B. (1977). *J. med. Chem.,* **20**, 96
Harvey, R. J. (1978). *J. theor. Biol.,* **74**, 411
Hitchings, G. H. (1962). In *Drugs, Parasites and Hosts* (ed. L. G. Goodwin and R. H. Nimmo-Smith), Churchill, London, p. 196
Hitchings, G. H. (1969). *Cancer Res.,* **29**, 1895
Hitchings, G. H. (1974). *Adv. Enzyme Regul.,* **12**, 121
Hitchings, G. H. (1978). In *Tropical Medicine: From Romance to Reality* (ed. C. Wood), Academic Press, New York, p. 79
Hitchings, G. H. and Burchall, J. J. (1965). *Adv. Enzymol.,* **27**, 417
Hitchings, G. H., Rollo, I. M., Goodwin, L. G. and Coatney, R. G. (1952*b*). *Trans. R. Soc. trop. Med. Hyg.,* **46**, 467

Hitchings, G. H., Elion, G. B., Falco, E. A., Russell, P. B. and VanderWerff, H. (1950*b*). *Ann. N.Y. Acad. Sci.*, **52**, 1318
Hitchings, G. H., Falco, E. A., VanderWerff, H., Russell, P. B. and Elion, G. B. (1952*a*). *J. biol. Chem.*, **199**, 43.
Hitchings, G. H., Elion, G. B., Falco, E. A., Russell, P. B., Sherwood, M. B. and VandeWerff, H. (1950*a*). *J. biol. Chem.*, **183**, 1
Hood, K. and Roberts, G. C. K. (1978). *Biochem. J.*, **171**, 357
Hood, J. M., Fowler, A. V. and Zabin, I. (1978). *Proc. natn. Acad. Sci. U.S.A.*, **75**, 113
Huennekens, F. M., Vitols, K. S., Whiteley, J. M. and Neef, V. G. (1976). *Meth. Cancer. Res.*, **13**, 199
Huennekens, F. M., Mell, G. P., Harding, G. L., Gunderson, L. E. and Freisheim, J. H. (1970). In *Chemistry and Biology of Pteridines* (ed. K. Iwai, M. Akino, M. Goto and Y. Iwanami), International Academic Printing Co., Tokyo, p. 329
Huennekens, F. M., Dunlap, R. B., Freisheim, J. H., Gunderson, L. E., Harding, N. L. G., Levinson, S. A. and Mell, G. P. (1971). *Ann. N.Y. Acad. Sci.*, **186**, 85
Kallen, R. G. and Jencks, W. P. (1966). *J. biol. Chem.*, **241**, 5845
Kaufman, B. T. and Kemerer, V. F. (1976). *Arch. Biochem. Biophys.*, **172**, 289
Kaufman, B. T. and Kemerer, V. F. (1977). *Arch. Biochem. Biophys.*, **179**, 420
Kimber, B. J., Griffiths, D. V., Birdsall, B., King, R. W., Scudder, P., Feeney, J., Roberts, G. C. V. and Burgen, A. S. V. (1977). *Biochemistry*, **16**, 3492
Koetzle, T. F. and Williams, G. J. B. (1976). *J. Am. chem. Soc.*, **98**, 2074
Lai, P., Pan, Y., Gleisner, J. M., Peterson, D. L. and Blakley, R. L. (1979). In *Chemistry and Biology of Pteridines* (ed. R. L. Kisliuk and G. M. Brown), Elsevier/North Holland, New York, p. 437
Liu, J. K. and Dunlap, R. B. (1974). *Biochemistry*, **13**, 1807
Matthews, D. A., Alden, R. A., Bolin, J. T., Filman, D. J., Freer, S. T., Xuong, N. and Kraut, J. (1979). In *Chemistry and Biology of Pteridines* (ed. R. L. Kisliuk and G. M. Brown), Elsevier/North Holland, New York, p. 465
Matthews, D. A., Alden, R. A., Bolin, J. T., Freer, S. T., Hamlin, R., Xuong, N., Kraut, J., Poe, M., Williams, M. and Hoogsteen, K. (1977). *Science,N.Y.*, **197**, 452
Matthews, D. A., Alden, R. A., Bolin, J. T., Filman, D. J., Freer, S. T., Hamlin, R., Hol, W. G. J., Kisliuk, R. L., Pastore, E. J., Plante, L. T., Xuong, N. and Kraut, J. (1978). *J. biol. Chem.*, **253**, 6946
Mell, G. P., Martelli, M., Kirchner, J. and Huennekens, F. M. (1968). *Biochem. Biophys Res. Commun.*, **33**, 74
Nixon, P. F. and Blakley, R. L. (1968). *J. biol. Chem.*, **243**, 4722
Otting, F. and Huennekens, F. M. (1972). *Arch. Biochem. Biophys.*, **152**, 429
Pastore, E. J., Plante, L. T. and Kisliuk, R. L. (1974*a*). *Met. Enzymol.*, **24**, 281
Pastore, E. J., Kisliuk, R. L., Plante, L. T., Wright, J. M. and Kaplan, N. O. (1974*b*). *Proc. Natn. Acad. Sci. U.S.A.*, **71**, 3849
Pattishall, K. H., Burchall, J. J. and Harvey, R. J. (1976). *J. biol. Chem.*, **251**, 7011
Pattishall, K. H., Acar, J., Burchall, J. J., Goldstein, F. W. and Harvey, R. J. (1977). *J. biol. Chem.*, **252**, 2319
Peterson, D. L., Gleisner, J. M. and Blakley, R. L. (1975*a*). *J. biol. Chem.*, **250**, 4945
Peterson, D. L., Gleisner, J. M. and Blakley, R. L. (1975*b*). *Biochemistry*, **14**, 5261
Poe, M. (1977). *J. biol. Chem.*, **252**, 3724
Poe, M., Greenfield, N. J. and Williams, M. N. (1974). *J. biol. Chem.*, **249**, 2710
Poe, M., Greenfield, N. J., Hirshfield, J. M., Williams, M. N. and Hoogsteen, K. (1972). *Biochemistry*, **11**, 1023
Reddy, A. V., Behnke, W. D. and Freisheim, J. H. (1978). *Biochim. biophys. Acta*, **533**, 415
Roberts, G. C. K., Feeney, J., Burgen, A. S. V., Yuferov, V., Dann, J. G. and Bjur, R. (1974). *Biochemistry*, **13**, 5351
Rodkey, J. A. and Bennett, C. D. (1976). *Biochem. Biophys. Res. Commun.*, **72**, 1407
Roth, B. and Strelitz, J. Z. (1969). *J. org. Chem.*, **34**, 821
Silipo, C. and Hansch, C. (1975). *J. Am. chem. Soc.*, **97**, 6849
Sköld, O. and Widh, A. (1974). *J. biol. Chem.*, **249**, 4324
Smith, S. L., Stone, D., Novak, P., Baccanari, D. P. and Burchall, J. J. (1979).

J. biol. Chem., in press
Stone, D. and Phillips, A. W. (1977*a*). *Fedn Eur. biochem. Socs Lett.*, **74**, 85
Stone, D., Phillips, A. W. and Burchall, J. J. (1977*b*). *Eur. J. Biochem.*, **72**, 613
Way, J. L., Birdsall, B., Feeney, J., Roberts, G. C. K. and Burgen, A. S. V. (1975). *Biochemistry*, **14**, 3470

Index